Bioengineering and the Skin

Bioengineering and the Skin

Based on the Proceedings of the European Society for Dermatological Research Symposium, held at the Welsh National School of Medicine, Cardiff, 19–21 July 1979

Edited by

R. Marks

Department of Medicine, Welsh National School of Medicine, Cardiff

P. A. Payne

The University of Manchester Institute of Science and Technology, Manchester

MTP PRESS LIMITED
International Medical Publishers
LANCASTER · BOSTON · THE HAGUE

Published in the UK and Europe by
MTP Press Limited
Falcon House
Lancaster, England

Published in the USA by
MTP Press
A division of Kluwer Boston Inc
190 Old Derby Street
Hingham, MA 02043, USA

Library of Congress Cataloging in Publication Data
Main entry under title:
Bioengineering and the skin.
Includes index.
1. Skin—Congresses. 2. Biomedical Engineering—Congresses. I. Marks, R. (Ronald) II. Payne, P. (Peter A.) III. European Society for Dermatological Research. [DNLM: 1. Biophysics—Congresses. 2. Skin—Physiology—Congresses. WR 102 B6155 1979]
QP88.5.B555 612'.791 81–14288
ISBN 0–85200–314–5 AACR2

Typeset by Speedlith Photo Litho Limited, Manchester
and printed by Robert MacLehose Limited, Renfrew

Contents

MEASUREMENTS OF FUNCTION AND DIMENSION

THERMAL, ACOUSTIC, OPTICAL AND ELECTRICAL PROPERTIES

List of Contributors

T. ADAMS
Department of Physiology
Michigan State University
East Lansing
Michigan 48824, USA

H. ALEXANDER
Department of Orthopedics
College of Medicine and Dentistry of New Jersey
Newark
New Jersey 07103, USA

R. R. ANDERSON
Department of Dermatology
Harvard Medical School
Massachusetts General Hospital
Boston, Massachusetts 02114, USA

C. M. APT
Arthur D. Little, Inc.
25 Acorn Parkway
Cambridge
Massachusetts 02140, USA

E. J. ARTZ
Polymer and Material Science
Johnson & Johnson Products Inc.
New Brunswick
New Jersey 08903, USA

J. C. BARBENEL
Bioengineering Unit
University of Strathclyde
Wolfson Centre
106 Rottenrow
Glasgow G4 0NW, UK

D. J. BARKER
Department of Dermatology
Bradford Royal Infirmary
Bradford, UK

I. H. BLANK
Department of Dermatology
Massachusetts General Hospital
Boston
Massachusetts 02114, USA

A. J. M. BRAKKEE
Department of Dermatology and Laboratory of Medical Physics and Biophysics
University of Nijmegen
Nijmegen, The Netherlands

M. de la BRASSINNE
Department of Dermatology
University of Liège
B-4020 Liège
Belgium

G. S. BRODY
Division of Plastic and Reconstructive Surgery
Rancho Los Amigos Hospital
7601 E. Imperial Highway
Downey, California 90242, USA

Anne M. BROWN
Wound Care Research
Johnson & Johnson Products Inc.
New Brunswick
New Jersey, USA

T. E. BURLIN
Pro-Rector
The Polytechnic of Central London
35 Marylebone Road
London NW1 5LS, UK

M. A. CHIANTA
Aircraft and Crew Systems Technology Directorate
Naval Air Development Center
Warminster
Pennsylvania 18974, USA

T. H. COOK
Elizabeth Arden Research Center
Lilly Research Laboratories
Indianapolis
Indiana 46206, USA

J. A. COTTERILL
Department of Dermatology
Leeds General Infirmary
Leeds LS1 3EX, UK

W. J. CUNLIFFE
Department of Dermatology
General Infirmary at Leeds
Great George Street
Leeds, LS1 3EX, UK

T. CUNNINGHAM
Wound Care Research
Johnson & Johnson Products Inc.
New Brunswick
New Jersey 00893, USA

G. CURTIS
Department of Radiology
College of Medicine and Dentistry of New Jersey
Newark
New Jersey 07103, USA

D. DARR
Department of Biology
The Pennsylvania State University
University Park
Pennsylvania, USA

J. B. DAWSON
Department of Medical Physics
Leeds General Infirmary
Leeds, LS1 3EX, UK

S. DIKSTEIN
University of Cell Pharmacology
School of Pharmacy
Hebrew University, PO Box 12065
Jerusalem
Israel

A. G. EMSLIE
Arthur D. Little Inc.
25 Aγord Parkway
Cambridge
Massachusetts 02140, USA

J. W. FEATHER
Department of Medical Physics
Leeds General Infirmary
Leeds, LS1 3EX, UK

J. FERGUSON
Research Department
Beecham Products
Leatherhead
Surrey KT22 7RX, UK

A. Y. FINLAY
Department of Dermatology
University Hospital of Wales
Heath Park
Cardiff CF4 4XW, UK

C. FRANCHIMONT
Department of Dermatology
University of Liège
Belgium

P. FROST
Department of Dermatology
Mount Sinai Medical Center
Miami Beach
Florida, USA

T. GIBSON
Plastic Surgery Unit
Canniesburn Hospital
Bearsden
Glasgow, UK

M. GREAVES
Institute of Dermatology
Homerton Grove
London E9, UK

G. L. GROVE
Simon Greenberg Foundation
3401 Market Street
Philadelphia, Pa 19104, USA

E. GUIBARRA
Department of Medical Physics and Bioengineering
University Hospital of Wales
Heath Park
Cardiff CF4 4XW, UK

C. W. GUNNER
Division of Engineering
Polytechnic of Central London
115 New Cavendish Street
London W1M 8JS, UK

C. William HARGENS
The Laboratories
The Franklin Institute
Philadelphia, Pennsylvania, USA

LIST OF CONTRIBUTORS

A. HARTZSHTARK
Unit of Cell Pharmacology
School of Pharmacy
Hebrew University, PO Box 12065
Jerusalem, Israel

J. HU
Department of Dermatology
Harvard Medical School
Massachusetts General Hospital
Boston, Massachusetts 02114, USA

H. MG. HUSSEIN
Becton-Dickinson and Co
Rutherford,
New Jersey 07070, USA

W. C. HUTTON
Division of Engineering,The Polytechnic of Central London
115 New Cavendish Street
London W1M 8JS, UK

K. KAWASAKI
Shiseido Laboratories
1050 Nippa-cho
Kohoku-Ku
Yokohama-shi, Japan

J. N. KEARNEY
Department of Dermatology
General Infirmary at Leeds
Great George Street
Leeds LS1 3EX, UK

A. D. KEITH
Biophysics Program
The Pennsylvania State University
University Park
Pennsylvania, USA

R. M. KENEDI
Engineering Department
Hong Kong Polytechnic
Hung Hom
Kowtoon
Hong Kong

C. S. King
Department of Medicine
Welsh National School of Medicine
Heath Park
Cardiff CF4 4XW, UK

J. P. KUIPER
Department of Dermatology
University of Nijmegen
Nijmegen, The Netherlands

R. F. LANDEL
Jet Propulsion Laboratories
California Institute of Technology
4800 Oak Grove Drive
Pasadena, California 91109, USA

Y. LANIR
Department of Biomedical Engineering
Technion–Israel Institute of Technology
Haifa, Israel

Ch. M. LAPIERE
Department of Dermatology
University of Liège, B-4020 Liège
Belgium

J. C. van der LEUN
Institute of Dermatology
State University of Utrecht
Utrecht
The Netherlands

R. MARKS
Department of Medicine
Welsh National School of Medicine
Heath Park
Cardiff CF4 4XN, UK

D. L. MILLER
Polymer & Material Science
Johnson & Johnson Products Inc.
New Brunswick
New Jersey 08903, USA

A. J. MILLIGAN
Department of Radiation Oncology
Medical College of Ohio
Toledo, Ohio 43699, USA

R. MITCHELL
Department of Biology
The Pennsylvania State University
University Park, Pa, USA

S. NICHOLLS
Department of Medicine
Welsh National School of Medicine
Heath Park
Cardiff CF4 4XN, UK

S. OHTA
Shiseido Laboratories
1050 Nippa-cho,
Kohoku-ku
Yokohama-shi, Japan

J. A. PARRISH
Department of Dermatology
Harvard Medical School

Massachusetts General Hospital
Boston, Massachusetts 02114, USA

P. A. PAYNE
The University of Manchester Institute of Science and Technology
Manchester, UK

S. T. K. PENG
Jet Propulsion Laboratories
California Institute of Technology
4800 Oak Grove Drive
Pasadena, California 91109, USA

R. Q. PERRITT
Department of Electrical Engineering
Louisiana State University
Baton Rouge
Louisiana 70803, USA

G. E. PIERARD
Department of Dermatology
University of Liège
B-4020 Liège, Belgium

J. R. PIERGALLINI
Aircraft and Vrew Systems
Technology Directorate
Naval Air Development Center
Warminster
Pennsylvania 18974, USA

E. PINES
Wound Care Research
Johnson & Johnson Products Inc
New Brunswick, NJ 08903, USA

P. T. PUGLIESE
Xienta Research Inc
R.D. 1, Bernville
Pennsylvania 19506, USA

H. S. RANU
Wayne State University
Detroit
Michigan 48202, USA

E. ROBERTS
Department of Radiology
Welsh National School of Medicine
Heath Park
Cardiff CF4 4XW, UK

D. C. SALTER
Bernville Medical Center
Bernville
Pennsylvania 19506, USA

I. SIMON
Arthur D. Little Inc
25 Acorn Parkway
Cambridge
Massachusetts 02140, USA

N. B. SIMPSON
Department of Dermatology
General Infirmary at Leeds
Great George Street
Leeds LS1 3EX, UK

W. SNIPES
Biophysics Program
The Pennsylvania State University
University Park
Pennsylvania, USA

M. A. STEINMETZ
Department of Physiology
Michigan State University
East Lansing, MI 48824, USA

A. M. STOLL
Aircraft and Crew Systems
Technology Directorate
Naval Air Development Center
Warminster
Pennsylvania 18974, USA

M. TAKAHASHI
Shiseido Laboratories
1050 Nippa-cho, Kohoku-ku, Yokohama-shi
Japan

C. Y. TAN
Department of Medicine
Welsh National School of Medicine
Heath Park
Cardiff CF4 4XN, UK

M. TANAKA
Shiseido Laboratories
1050 Nippa-cho
Kohoku-ku
Yokohama-shi, Japan

D. E. THOMPSON
Department of Mechanical Engineering
Louisiana State University
Baton Rouge
Louisiana 70803, USA

A. M. TORGALKAR
Material Characterization Department
Loeser Ave, PO Box 858
Devro, Inc.
Somerville, NJ 08876, USA

LIST OF CONTRIBUTORS

Y. TSUDA
Department of Legal Medicine
Yokohama City University School of Medicine
Yokohama, Japan

A. J. H. VENDRIK
Laboratory of Medical Physics and Biophysics
University of Nijmegen
Nigmegen, The Netherlands

H. G. VOGEL
Hoechst Aktiengesellschaft
D-6230 Frankfurt/Main 80
W. Germany

T. G. WARSHAW
Department of Medicine (Dermatology)
College of Medicine and Dentistry of New Jersey
Newark, NJ, USA

P. F. F. WIJN
Department of Dermatology, and
Laboratory of Medical Physics and Biophysics
University of Nijmegen
Nigmegen, The Netherlands

E. Wyn WILLIAMS
Immunology Department
Glaxo Group Research Ltd
Greenford, Middx., UK

Introduction

R. MARKS

Biology has become a 'numbers game'. The advantages of being able to grade changes in tissue, submit results to statistical analysis and accurately record biological phenomena make measurement essential. This is as true for the various disciplines in applied biology as it is for the more esoteric aspects of the subject. Regrettably, skin biologists until recently had not seized the opportunities that the availability of their tissue of interest afforded and fell behind in the exploration of measurement techniques. Probably this resulted in part from the mistaken sentiment that 'to see is to know'. It also originated from the complexity of the skin which, as a closely interwoven mixture of tissue types, makes assessments technically difficult. However, we are optimistic about the future.

The International Society for Bioengineering and the Skin was formed in Cardiff in July 1979 in response to the wishes of the delegates who had attended the first International Symposium on the subject in Miami in 1976 and the second in Cardiff 3 years later. This volume is the proceedings of the Cardiff meeting. We believe that it demonstrates the brave efforts and variety of new ideas that characterise the studies of scientists who realise the importance of blending the physical sciences with skin biology.

The authors' individual contributions vary in their presentation. Some are formal scientific papers, some descriptions of apparatus or techniques and some contain reviews or appreciations of particular subjects. We decided to permit this variation and not edit heavily for the sake of uniformity, as we believed that the different styles were more appropriate to the various subjects. We trust that this will not irk and will make the book more readable.

The contributions have been grouped into four sections: (1) preliminary considerations; (2) mechanical properties; (3) measurements of function and dimension; and (4) thermal, acoustic, optical and electrical properties. Inevitably there is some overlap, but we thought it useful to impose some order for ease of referral.

Preliminary Considerations

1
Experiments and analyses: a retrospect

R. M. KENEDI and T. GIBSON

SUMMARY

Retrospective evaluation highlights the shortcomings of present-day efforts in analyses, particularly in clinical applicability. The load-deformation behaviour of the soft tissues is non-linear and time dependent, and it now seems clear that present attempts to describe the biomechanical behaviour of such living entities in terms of the mathematical concepts and techniques used for non-biological materials have failed.

A reorientation to alternative formulations requiring more effective collaboration between experimentalists and mathematicians than before is necessary to produce genuine progress. Hopefully this will go hand in hand with the re-establishment of the clinical problems of the human patient as both the start and the ultimate aim of all research.

PREFACE

The last two decades have seen wide acceptance of biomechanics as a pertinent and necessary part of any study of the diagnostic and therapeutic problems of the human patient.

The expansion of the relevant literature describing the experimental and analytical enquiries undertaken is reminiscent of the transformation (in the comic strip turned television programme) which produced the Incredible Hulk from its inoffensively insignificant alter ego. Correspondingly the records of the biomechanics of tissue, and in particular, that of skin, in their multifarious aspects have become intimidating in their extent[1,2]. Yet if one uses as a form of 'outcome measure' identifiable benefit to the human patient, the accomplishments of all this activity appear less than modest.

International meetings are regarded as opportunities for exchanging information and for presenting new material for the admiration (or criticism as the case may be) of one's peers: they should perhaps be used also as forums of retrospection, hopefully leading to redirection of research and development effort if that is adjudged to be desirable.

The writers, having been part of this explosive development of tissue biomechanics, particularly in its early phases, now propose to indulge in just such retrospection on two fronts: firstly in a *broad* sense to reconsider advances in biomechanical characterisation and modelling of tissue; and secondly in *specific* contrast to re-present some overlooked work of the Viennese anatomist Langer[3] of direct pertinence to the present discussion. The rediscovery, translation and publication in English of this 19th century masterpiece is the work of one of the writers of this paper (Gibson).

BIOMECHANICAL CHARACTERISATION AND MODELLING OF SKIN

There is now virtually no disagreement, certainly in a qualitative sense, on the characterisation of the structures and mechanical behaviour of skin as determined by *experiment*.

Subsurface human skin is regarded as a multicomponent microstructure consisting of intertwined networks of collagen, elastic and nerve fibres, small blood vessels and lymphatics, covered by a layer of partially keratinised epithelium and transfixed at intervals by hairs and the ducts of sweat glands. The networks are surrounded by interstitial fluid containing a varying amount of mucopolysaccharide ground substance, and all are mobile in this semi-fluid environment. As all tissues are living organs, biological effects can occur in addition to other influences of time dependence during tests modifying the properties being measured[2,4].

Regarding the *surface* of human skin, recent work[5] is effectively exploring the topography on the microscale with a clinically applicable replication technique.

The following are the main 'experimental' findings.

The load–deformation response on the macroscale shows a decrease in deformation with increase in load, the curve being concave to the load axis. The lateral (passive) deformations are comparable in magnitude with the direct (active) deformations.

Time dependence is apparent, creep, stress relaxation and other similar effects having been reliably established on the macroscale. Hysteresis (i.e. non-coincidence of the loading and unloading curves) is present and behaviour is thus non-elastic. Preconditioning, that is the establishment of a steady-state load–deformation response, after cyclic repetition of a given load or deformation programme appears to apply to all tissues, certainly *in vitro*. Its relevance to *in vivo* behaviour is less clear. Anisotropy is important and is particularly demonstrated by the variation in magnitude of the initial deformation at low loads, which is direction dependent. Skin, *in vivo*, is in a state of tension: the 'resting or residual' tension which varies both in direction and magnitude throughout the body.

All mechanical characterisations are environment dependent.

Generally, reliable experimental (and/or test) data are available under both uniaxial and biaxial, *in vivo* and *in vitro* conditions[1,6].

The use of the experimental data available to model tissue *analytically* has taken several forms.

The empirical curve fitting of results has mostly used exponential forms[7–10], and its latest version consists of two-term relationships to express the 'biphasic' nature of the experimental data. The second term in such relationships expresses the behaviour of the material (tissue) at high stress levels, the first term 'remedying' the situation at lower stress levels.

Since the bulk of the test data corresponds to uniaxial loading, with some restricted biaxial (no cases of triaxial) loading, numerous attempts have been made in the past two decades to generalize the concepts for the three dimensions of the real world. The aim of this is to[10,11] ... facilitate data collection and analysis: derive three-dimensional stress–strain history laws under finite deformation and unify the results of different types of experiments such as static, ... dynamic, ... stress relaxation, creep, hysteresis and cyclic fatigue[7].

Most of these attempts have been based on strain energy formulations and have on the whole been unsatisfactory. The reasons for this lack of success are many and lie essentially in the need either to simplify so as to obtain a *generalized* functional form of solution or to introduce numerics for *specific* cases and consequently lose any semblance of generality.

Thus elastic concepts are used to describe non-elastic behaviour, for example by restricting consideration only to the load increasing part of the data and excluding unloading paths; isotropy is assumed in analysing patently anisotropic results and incompressibility is postulated alongside results quoted to show the existence of compressibility[2].

In fact the present state of development of *analyses* may be explained by the fact that reality is sacrificed for mathematical expediency. An increasing number of analytical solutions are being produced under such constraining assumptions that they cannot be applied to matters of clinical importance (which presumably provided the motivation for their production). The somewhat depressing prospect that now faces biomechanical analysts is seeing their mathematically elegant, but in real terms over-simplified, solutions completed and *not* finding problems to which they may satisfactorily be applied.

To plunge, in Dickensian terms, further into the 'dismals', it might perhaps also be salutary to contrast the numerous investigations of the many researchers (including the writers') carried out in the period 1960 to date with the work of an individual, presented to the Viennese Imperial Academy of Science over 100 years ago.

KARL LANGER AND HIS LINES

Langer, born in Vienna in 1819, was Professor of Anatomy at Joseph's Academy in Vienna. His work was not only concerned with descriptive morphology but also, and perhaps even primarily, with function. 'Langer's Lines – known to every surgeon – partly resulted from his major study on the functional anatomy of skin. One of the writers of this paper (Gibson), in a search to unearth Langer's original paper, rediscovered not one but four papers, all presented to the Imperial Academy of Science in Vienna during April and November 1861 and forming the records of a complete study of the directional variations of the structural properties of human skin. The complete work translated, edited and annotated has now been published (Gibson, 1978)[3], and the following extracts (at times paraphrased) have been selected as of particular pertinence to the present discussion.

'*The cleavability of the cutis* – A round-bodied awl when thrust through the skin produces not a round stab wound but a linear cleft. If the clefts are placed as closely as possible they show a definite relationship to each other in the various parts of the body, defining by their sequential orientation lines which *are an expression of the fibre pattern of the skin.* Microscopic study showed tightly grouped fibres interwoven in bundles and arranged in the form of a three-dimensional lattice work. The fibres themselves are partly wavy and partly curly. The cleft orientation is dependent on the fibre arrangement; the stab clefts are widening the inter-fibre bundle spaces in the lattice, at times enlarged by actual tearing of the fibre bundles themselves.

'*Skin tension*'–When the skin is incised the wound gapes; this 'retractile' property means that the skin around the wound shrinks as the incision widens. Broadly, wound retraction is minimal when skin is incised along the lines and maximal when the incision is oriented at right angles to them. There are two major causes of skin (resting) 'tensions': *the contents and thus the degree to which the skin container is filled*-physiological and/or pathological, muscle and fat masses, water accumulations, pregnancy etc; and *the movement of joints* – a close relationship between the possible skin tensions and joint mechanics can be assumed.

'*The elasticity of the cutis*' – Langer cut a large number (100's) of cadaveric skin specimens (approximately 25 × 10 cm) along and transverse to the lines. The specimens were then tested in uniaxial tension under careful time control: 'In order to allow for the extension which continued after the weight was applied and obtain extension values which were as uniform as possible for the individual periods of loading, the loads were allowed to operate for identical time intervals, in some experiments for 2 minutes, in others 5. Further I allowed 2 minutes to elapse after each loading before a new weight was applied'. He then plotted his load–extension diagrams and commented that: 'skin does not stretch proportionally with applied load, indeed the amount of extension grows steadily smaller so that the course of progressive extension cannot be represented by a straight line but by a curve'. Langer also identified the existence of residual stretch after reloading but found 'virtually' total recovery (to about 1–3 mm in 25 mm) after leaving the unloaded specimens overnight in a moist environment. From his results he concluded that: 'the total amount by which a strip of skin lengthens (under tension) is the result of three factors: smoothing out of the curled fibres, rearrangement of the weave of fibres and extension of the fibres themselves. Obviously the uncurling of the fibres and at least part of the rearrangement of the network will take place during the first phase of extension'. He stipulated anisotropy: 'the difference in extensibility between the longitudinal and transverse strip is mainly an expression of the non-homogeneous nature of the tissue, i.e. the meshwork being stretched'. He repeatedly emphasised this non-homogeneity: 'the great extensibility of skin is not a function of the physical properties of its tissue elements but only of the arrangement of tissue'.

As can be seen, Langer demonstrated that:

(1) Human skin consists of an intertwined network of fibre bundles enclosed in a jell-like interstitial material. (2) The load–deformation response on the macroscale is curvilinear, concave to the load axis. (3) The load–deformation response is biphasic, the initial highly extensible phase at low load being primarily due to fibre and network shape changes. (4) The phenomenon is time dependent, creep is present with residual deformation manifested after loading. (5) Anisotropy is present, a *directional* indication of its character in the human body being given by the 'lines'; and (6) resting tensions are identifiable, varying throughout the body and broadly related to body bulk and joint movement.

The writers earlier described the present state of analysis as somewhat depressing from an applied point of view; so as not to allow the skin experimenters (which regrettably includes the writers of this paper) to escape entirely unscathed perhaps they might be permitted to express a modicum of reservation about their 'originality of effort' in the light of the hitherto overlooked work of Langer.

References

1. Kenedi, R. M., Gibson, T., Evans, J. H. and Barbenel, J. C. (1975). Tissue mechanics. *Phys. Med. Biol.*, **20**, 699.
2. Barbenel, J. C. (1979). Time dependent mechanical properties of skin. *PhD Thesis*, University of Strathclyde
3. Gibson, T. (1978). Translator, editor and annotator of 'On the Anatomy of Physiology of the Skin' by Professor K. Langer, 1861. The Imperial Academy of Science, Vienna. *Br. J. Plastic Surg.*, **31**, 93, 273
4. Evans, J. H. (1973). Structure and function – the influence on connective tissue mechanics of tissue contours and their micro-architecture. *PhD Thesis*, University of Strathclyde
5. Barbenel, J. C., Makki, S. and Agache, P. (1979). The variability of skin surface contours. (In preparation)
6. Lanir, Y. and Fung, Y. C. (1974). Two-dimensional mechanical properties of rabbit skin – I. Experimental system – II. Experiment results. *J. Biomechanics*, **7**, 29, 171
7. Morgan, F. R. (1960). The mechanical properties of collagen fibres: stress–strain curves. *J. Soc. Leather Trades Chemists*, **44**, 170
8. Kenedi, R. M., Gibson, T. and Daly, C. H. (1964). Bioengineering studies of the human skin. In Jackson, S. F., Harkness, J. M. and Tristram, G. R. (eds.) *Structure and Function of Connective and Skeletal Tissue*. (London, Butterworth)
9. Ridge, M. D. and Wright, V. (1964). The description of skin stiffness. *Biorheology*, **2**, 67
10. Fung, Y. C. B. (1967). Elasticity of soft tissues in simple elongation. *Am. J. Physiol.* **213**, 1532
11. Tong, P. and Fung, Y. C. (1976). The stress strain relationship for the skin. *J. Biomechanics*, **9**, 649

2
International standards for instruments, measurement methods and units for determination of physical properties and function of skin

P. A. PAYNE, R. MARKS and E. J. GUIBARRA

INTRODUCTION

There are many workers and scientific groups around the world who are concerned with the measurement of properties and/or functions of skin. Recent confirmation of this was provided by the delegate list for the Cardiff Symposium on Bioengineering and the Skin[1]. The biomechanical properties of skin have been studied and reported extensively.

Wilkes *et al.*[2], in a review written in 1973, listed 144 references to such work; and two years later, in a review of tissue mechanics Kenedi *et al.*[3] cited some 180 references. Much work has been carried out since these papers were written, but much remains to be accomplished before we can claim to have advanced towards the solution of any important clinical problem.

An understanding of the properties of skin and their relationship to its function is largely dependent on measurement of either (1) biochemical properties, or (2) mechanical, electrical, thermal or optical properties. It is this second group that this paper is concerned with, and emphasis will be on mechanical properties.

We assume that all measurements made are designed to produce data that can be used by other investigators in whatever field they may be working. This leads to the obvious conclusion that unless some agreed standards are established then, in addition to the possible language difficulties that are sometimes faced when work is discussed, we should face a further problem: we would have to convert measurements made in a wide variety of ways in an attempt to establish similar properties from one system of measurements to another. One approach in solving these problems is to agree on a set of standards.

STANDARDS FOR MEASURING PHYSICAL PROPERTIES OF SKIN

We may group most physical properties under one of the following headings: (1) mechanical; (2) electrical; (3) optical; and (4) thermal.

Of these four physical properties, a substantial amount of the published work relates to measurement of mechanical properties. A further division arises dependent on whether the measurement is performed *in vivo* or *in vitro*.

Standards for *in vitro* measurements

The methods used to collect sample materials and to subsequently store them before testing may have an important influence on the results obtained. There is a need to investigate these effects together with such variables as sample size, sample shape and method of sample fixation. Clearly, the temperature and humidity in which the measurement is made should also be standardised. If such environmental control is not possible then the ambient conditions must be recorded.

Standards for *in vivo* measurements

Listed below are those variables which are thought to be important in an *in vivo* method. We believe that control or measurement of all these is possible, but have asterisked those which we believe to be essential in recording data in the literature.

(1) *Type of test: (a) uniaxial (and direction); (b) biaxial; (c) torsional; and (d) other.
(2) *Sample area and shape.
(3) *Attitude and position of part during test.
(4) *Method of coupling of forces.
(5) *Rate of application and removal of forces.
(6) *Exact body site.
(7) *Existing stress.
(8) *Topical treatment history.
(9) Skin temperature.
(10) *Dermal thickness.
(11) Fat thickness.
(12) Blood flow conditions
(13) *History of exposure to sun or other source of radiation.
(14) Prior pressure or other loading history.
(15) *Environment: (a) temperature; and (b) relative humidity.
(16) *Age of subject.
(17) *Sex of subject.
(18) *State of health of subject.
(19) *Method of data reduction.

Any standard adopted for *in vivo* dermal property measurement must take into account the above variables. Similar lists may be drawn up for electrical, thermal or optical measurements.

MEASURING SKIN FUNCTION

At times the measurement of a function and a particular physical property may be identical. In most instances, however, function represents the sum of interactions between physical properties. A property is preferably tested *in vitro* and functions must of necessity be examined *in vivo*. By their nature, 'functions' will probably be more useful to clinicians and 'properties' more helpful to scientists. Regardless of the distinctions between the two, there is equal necessity for standardising the measurement techniques in both.

CONCLUSIONS

It has been generally agreed that international standards in this area are to be welcomed. Consequently, several complex programmes of investigation must be undertaken to determine how some of the above variables influence the measurements made. Co-ordination and monitoring will also be required to ensure that as techniques and devices are developed the agreed standards are updated to reflect these changes in technology.

References

1. (1979). *Bioengineering and the Skin Newsletter*, Vol. 2, No. 2, December
2. Wilkes, G. L., Brown, I. A. and Wildnauer, R. H. (1973). The biomechanical properties of skin. *CRC Critical Reviews in Bioengineering*, August, pp. 453
3. Kenedi, R. M., Gibson, T., Evans, J. H. and Barbenel, J. C. (1975). Tissue mechanics. *Phys. Med. Biol.*, **20**, 699

Mechanical Properties

3
Effects of fractionated doses of X-irradiation on the mechanical properties of skin: A long-term study

H. S. RANU

INTRODUCTION

The most common skin tumours are basal-cell epitheliomas. A cure rate of over 90% has been obtained by both surgery and radiotherapy. As these tumours occur most often on the face, the acceptability of treatment must depend not only on cure but also on the cosmetic effect obtained. Radiotherapy is often used to treat the more extensive lesions and those occurring in the older age group. Many irradiation fractionation regimens are in use. The factors which influence the choice of regimen are convenience to the patient, size and depth of the lesion, proximity to radiosensitive organs, and traditional and economic considerations. Little attempt has been made to compare the cosmetic effect of differing regimens, although this is clearly an important factor especially in younger patients. Besides cure, therefore, the most important considerations are the effect of the radiation on the normal tissue and the success with which the radiation injury is repaired. The cosmetic effect varies not only with dose, time and the number of fractions but also with the pigmentation of the patient's skin, the initial degree of atrophy and the site – face skin giving a relatively better cosmetic result than body skin. If depigmentation is pronounced the cosmetic result will appear worse in summer.

It was therefore thought desirable to establish a method of studying the cosmetic effects of radiation on normal skin, and a study was conducted on the cosmetic effect of clinically relevant doses on the skin of mice. This paper presents an approach which embraces the measurements of the mechanical properties of the skin. Since the elasticity and contraction consequent upon fibrosis influence cosmetic appearance of skin, a series of measurements of the mechanical properties of the irradiated skin were undertaken.

MATERIALS AND METHODS

Male mice reportedly enter their second telogen at 50 days, and their third anagen does not begin until day 92[1,2]. By selecting a mouse of suitable age, about 70 days old, all the cells in the follicle can reasonably be assumed to be in a prolonged resting phase (G_0 of the cell cycle). Plucking induces cell division in the resting population, providing an initially synchronous population[3].

Three groups of 15 male C_3H mice, each aged about 10 weeks and weighing about 20 g, had the flank hair plucked to ensure that the hair follicles were in the same state of the growth cycle on irradiation three days later. The mice were then placed in a cellophane harness in which windows had been cut (Figure 3.1), and two areas on each flank within the plucked region were irradiated without anaesthesia. The radiation was given at 50kV, 25mA with 1.0mm Al filter, 5 cm focus skin distance, and half-value thickness 0.77 mm Al. The calibration of the apparatus has been described[4]. Two areas were irradiated on each mouse on the anterior and posterior of the same flank using a cone of 1 cm diameter. The areas were located on the same flank so that an exit beam would not contribute to the dose.

The exposure rate was measured each time the mice were irradiated; it varied around 3800 R/min. The first group received five exposures of 750 rads with an interval of 1 day between each exposure (treatment A). The second

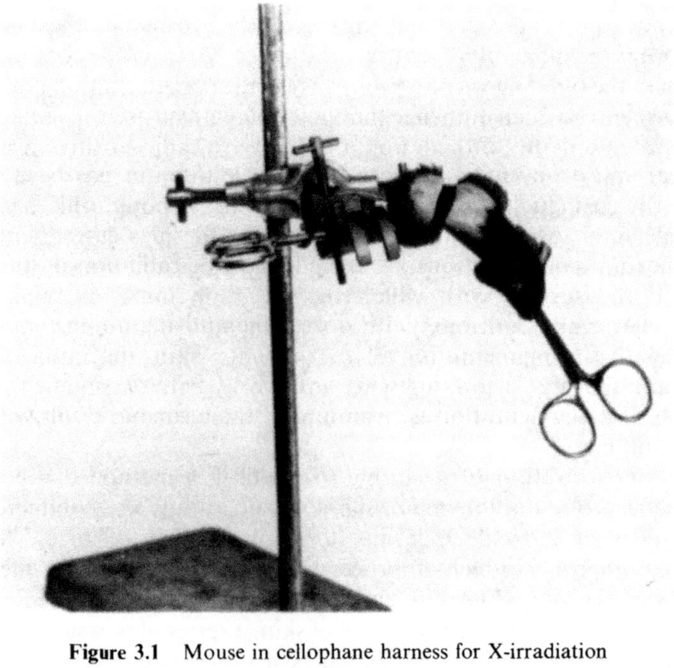

Figure 3.1 Mouse in cellophane harness for X-irradiation

group received nine exposures of 600 rads with an interval of 2 days between each exposure (treatment B). The third group received three exposures of 700 rads with an interval of 2 days between each exposure. For this third group a rest period of 5 days was permitted between the third and fourth exposures. After this four exposures of 600 rads were given with an interval of 7 days between each exposure (treatment C). A reference exposure of 2250 rads was received at a different site on the flank of every animal in addition to the fractionated dose on the flank.

These represent the range of accumulated doses relevant in radiotherapeutic treatments. The depth of the hair follicle below the skin surface three days after plucking is 230 μm. The depth–dose relationship shows that the dose at the follicle base is 98 % of the incident dose[5]. The back-scatter factor for a 1 cm diameter field is 1.00 at this quality[5], and the value of rads per röntgen is 0.90[6]. The calculated dose to the hair follicle base is therefore 0.88 rad per röntgen of exposure at the skin surface.

Since the investigation was designed to study the long-term if not permanent effects of radiation on the skin, observations were made several months after exposure. Previous work on the effects of X-rays[4] has indicated that, after about three months from completion of the irradiation, the response has reached a stable level and little if any further recovery occurs. Therefore, the mice (9 survived in each group) were killed 210 days after the first irradiation. Sagittal strips 2.5 cm long and 1 cm wide were cut from the skin of each animal by a specially designed cutter, so that the centre of the section was concentric with the 1 cm diameter circle of irradiated skin. There was complete alopecia in the irradiated area at the time the specimen was cut. This hairless area was 1 cm in diameter, hence it was concluded that no contraction of the skin had occurred because of irradiation. A similar section was cut from an unirradiated portion of the mouse flank, care being taken to ensure the skin specimen was not cut from the area covered by the transmitted X-ray beam, then all specimens were placed in normal saline before subsequent mechanical testing, which occurred within an hour of excision.

The apparatus used for measuring the tensile mechanical properties of skin is shown in Figure 3.2. A schematic arrangement of the apparatus has been shown previously[7]. The tests were carried out on a Mayes hydraulic servo-controlled testing machine (W. H. Mayes and Son (Windsor) Ltd, Windsor, England). The skin specimen was mounted in two specially designed jaws which had matching blunt serrations to facilitate gripping but prevent cutting (Figure 3.3). The lower jaw and a Perspex immersion tank were attached to the main hydraulic actuator. The tank enabled the tissue to be tested in saline, since the mechanical properties of skin are dependent on the degree of hydration.

The thickness of the skin sample was measured with a Mercer dial gauge. The gauge lightly compressed the skin on measurement, causing slight inaccuracy in the measured value of absolute thickness; this measurement error was estimated as ± 5 %. On the testing machine the gap between the jaws was set at 1 cm. This gauge length of 1 cm was chosen to provide a test area which had received the most uniform radiation dose. The skin specimen was clamped in the jaws. The Perspex immersion tank was filled with saline

Figure 3.2 General view of the Mayes hydraulic servo-controlled universal testing machine and *in vitro* apparatus attached to it: (1) movable clamp; (2) jaws and skin specimen; (3) fixed clamp; (4) Perspex immersion tank: (5) heater and saline reservoir; (6) peristaltic pump; and (7) X–Y recorder

Figure 3.3 Jaws used for gripping skin specimen (matching serrations can be seen)

solution kept at 37 ± 0.5 °C and the specimen was allowed to equilibrate in the fluid. A tensile load was applied to the skin sample at an extension rate of 0.25 mm/s. This extension rate was selected to correspond to the rate at which skin recovers when stretched to blanching *in vivo*. The skin was stretched to rupture and the load–extension characteristics were recorded on an X–Y plotter.

For each specimen the following mechanical properties were measured:

(1) skin thickness; (2) load at rupture; (3) extension at rupture; (4) mechanical stress at rupture, which was calculated from the load at rupture divided by the cross-sectional area of the specimen; and (5) stiffness.

The final part of the load-extension curve is primarily associated with the stretching of the aligned collagen fibres[8]. This property will be referred to as the stiffness of the skin and is defined as slope of the last quarter of the load–extension curve, $\Delta F/\Delta E$. To give the property units (N/mm^2) and meaning consistent with other values of stiffness, the values of original length (L), and cross-sectional area (A), are brought into each calculation with the equation:

$$\text{Stiffness} = \left[\frac{\Delta F}{\Delta E}\right]\frac{L}{A}$$

where F is the load and E the extension.

RESULTS AND CONCLUSIONS

The results for the measurements of the mechanical properties of skin are presented in Table 3.1. The mean (± SE) extension, load and stress at rupture together with mean (± SE) skin thickness and stiffness for the nine animals in each group are tabulated for unirradiated skin and for the skin which received a single dose of 2250 rads in the three treatment schedules.

One objective was to determine which treatment schedule caused the least cosmetic damage, since all schedules are highly effective in curing skin tumours. The criterion selected for judging the optimum treatment for minimum cosmetic damage was the treatment which resulted in a particular variable having a value after radiation closest to that in unirradiated skin. In order to facilitate comparisons the measurements of the mechanical properties of skin contained in Table 3.1 were normalised in Table 3.2 on the value for unirradiated skin for each animal. These have been expressed as a percentage, and therefore the treatment giving the value closest to 100 % would be judged best.

It will be seen from Table 3.2 that treatment B was the optimum treatment with regard to the mechanical properties of the skin. When the scarring of skin was examined treatment B again apparently gave the best cosmetic result[7]. Treatment C seemed to be the next most favourable with regard to the general cosmetic appearance of the skin. Thus the clear implication for nearly all practical cases of radiotherapeutic treatments of skin malignancy is that

Table 3.1 Results (as means ±SE) of variables for different treatment schedules

Variable studied	*Unirradiated skin (n = 27)*	*Skin receiving 2250 rads (n = 27)*	*Skin receiving Treatment A (n = 9)*	*B (n = 9)*	*C (n = 9)*
Extension at rupture (mm)	9.2 ± 0.4	6.0 ± 0.5	3.6 ± 0.3	7.1 ± 0.5	5.2 ± 0.6
Load at rupture (N)	26.2 ± 1.4	20 ± 1.2	24.8 ± 2.9	29.0 ± 1.7	13.6 ± 1.6
Skin thickness (mm)	0.7 ± 0.01	0.7 ± 0.02	0.6 ± 0.04	0.8 ± 0.02	0.6 ± 0.02
Stress at rupture (N/mm^2)	3.7 ± 0.2	3.0 ± 0.2	4.8 ± 0.8	3.7 ± 0.3	2.4 ± 0.2
Stiffness (N/mm^2)	8.1 ± 0.5	13 ± 1.2	27 ± 5.0	10.4 ± 1.3	9.8 ± 1.8

treatment B is more likely to preserve the cosmetic appearance and functional activity of skin than the other two treatments. In treatment B a fractionation scheme was used which permitted two days between fractions. Thus a possible explanation is that this interval permitted more recovery and repair of normal tissues than malignant tissue, the net result being cure of the cancer but appreciably less damage to the normal tissue.

This observation is verified by *in vivo* clinical survey[9–11] of female patients receiving radiotherapy of the breast, thus confirming that radiotherapy given three times a week causes least damage to the cosmetic and elastic properties of skin because this number of fractions a week permits a preferential recovery of the skin.

It is suggested that a seven-month period in the mouse is comparable to a much longer period in man. The method described has an advantage in that

Table 3.2 Normalized results (%) of variables studied for different treatment schedules

Normalised variable	*Skin receiving 2250 rads (n = 27)*	*Skin receiving treatment*		
		A (n = 9)	*B (n = 9)*	*C (n = 9)*
Extension at rupture	69.5	47.6	71.7	70.1
Load at rupture	86.5	138.8	82.4	65.1
Skin thickness	98.7	85.2	107.2	79.8
Stress at rupture	96.6	173	90.6	81.7
Stiffness	153	215	141	145

low-energy X-rays can be used conveniently. It is concluded that the method is a suitable model for studying the cosmetic effects of irradiation and the relative merits of different fractionation schemes used in treating malignant disease of the skin.

References

1. Silver, A. F., Chase, H. B. and Arsenault, C. T. (1969). Montagna, W. (ed.) *Advances in Biology of Skin, Hair Growth*, p. 9.265. (New York: Pergamon Press)
2. Silver, A. F. and Chase, H. B. (1970). DNA synthesis in the adult hair germ during dormancy (telogen) and activation (early anagen). *Dev. Biol.*, **21**, 440
3. Potten, C. S. (1971). Tritiated thymidine incorporation into hair follicle matrix and epidermal basal cells after stimulation. *J. Invest. Dermatol.*, **56**, 311
4. Ranu, H. S., Burlin, T. E. and Hutton, W. C. (1975). Effects of X-irradiation on the mechanical properties of skin. *Phys. Med. Biol.*, **20**, 96
5. British Journal of Radiology (1972). Central axis depth dose for use in radiotherapy. *Br. J. Radio. Suppl.*, **11**
6. ICRU (1964). *Physical Aspects of Irradiation.* ICRU, Report 10b or NBS Handbook 85. (Washington, D.C.: ICRU Publications)
7. Ranu, H. S. (1975). Effects of ionizing radiation on the mechanical properties of skin. *PhD Thesis*, CNAA, London
8. Ranu, H. S. (1979). The mechanical and structural response of skin to irradiation. *J. Biomech.* **12**, 601
9. Ranu, H. S., Hutton, W. C., Burlin, T. E. and Ranu, H. K. (1976). Radiation and its effect on the elastic properties of the human skin. In *Digest 11th International Conference on Medical and Biological Engineering, Ottawa, Canada*, p. 52
10. Ranu, H. S., Hutton, W. C. and Burlin, T. E. (1976). Erythema and the elastic properties of human skin during radiotherapy. In *Digest 4th International Conference on Medical Physics, Ottawa, Canada*, p. 35.7
11. Ranu, H. S. and Ranu, H. K. (1977). Effects of ionizing radiation on the physical properties of skin. In *Digest 1st Mediterranean Conference on Medical and Biological Engineering, Sorrento, Italy*, p. 19.5

4
Flexural properties of human fingernails

A. Y. FINLAY, P. FROST, A. D. KEITH and W. SNIPES

INTRODUCTION

Nail is a specialised epidermal structure which has evolved with physical properties differing widely from normal stratum corneum. Studying the physical properties of nail may contribute to the understanding of nail disease, and the techniques may be used to assess the value of topical nail treatments. With comparative studies, further understanding of epidermal physical properties may also be gained. The study of the physical properties of nail has commanded relatively little attention compared with current interest in physical properties of epidermis: this is surprising in view of the cosmetic interest in nails and the widely advertised claims of cosmetic manufacturers that their products alter nail properties as well as appearance.

The physical properties of nails are commonly subjectively assessed and described in such ill-defined terms as hardness, toughness or brittleness. However, for objective assessment the following properties of nail have been described.

(1) Water diffusion through nail plate[1,2].
(2) Water binding capacity of nail plate[1].
(3) Modulus of elasticity:
  (a) by direct measurement of nail deflection under different loads[1,3]; and
  (b) by measurement of the velocity of sound in nail plate[1].
(4) X-ray diffraction patterns of nail[1].
(5) Tensile strength[3].
(6) Ability to withstand tearing[3].
(7) Ability to absorb impact[3].
(8) Hardness[4,5]
(9) Rate of nail growth[6,7].

Our studies were concerned with assessing flexural properties of nails: we have designed an instrument (nail flexometer) to measure the number of flexions required for nail sections to fracture. This method allowed us to study the role of different structural components in maintaining normal flexibility.

MATERIALS AND METHODS

Longitudinal paired sections of human cadaver fingernails were used in all experiments, with one section acting as control. In most experiments nails from only one specific finger position were used. When being tested in the nail flexometer each nail section was held proximally and the tested end of the nail was bent repeatedly through 90° in each direction by being swept past a plastic wheel (Figures 4.1 and 4.2). The number of flexions up to the time of nail breakage was recorded as the fracture number (FN).

Nail sections were treated with water, mineral oil or a phospholipid preparation (PLW) containing 22% phosphatides. The effect of water, mineral oil and PLW immersion on nail flexibility was measured by subsequently immersing nail sections in water for up to 21 hours and immediately studying them on the nail flexometer. Paired sections served as controls. Ten further nail sections were weighed and reweighed at intervals during 24 hours of water immersion after initial treatment.

The effects of various topical preparations in prolonging flexibility of hydrated nails were assessed by first immersing sections of test and paired control nails in water for 2 hours. The test specimens were then immediately painted with mineral oil or PLW and the controls left in air unpainted. The nails were then tested at various time intervals after subsequent reimmersion in water.

Phospholipids were extracted from paired nail sections by immersion in a solution of 90% acetone, 9% water and 1% acetic acid: this procedure was

Figure 4.1 Nail flexometer, general view: (A) nail; (B) steel shaft; (C) rotating cam; (D) meter showing number of nail flexions; and (E) spring

Figure 4.2 Nail flexometer, close-up view: (A) nail, held between bevelled brass plates; (B) steel shaft; (F) plastic wheel; and (G) light beam intermittently interrupted by metal flange (H)

designed to remove lipids without structural modification of nail proteins. These lipid-free nails were left in air for 18 hours at 20 °C and then one section from each pair was immersed for 1 hour in water, mineral oil or PLW before being tested.

Other lipid-free paired nail sections were placed in water or PLW for 1 hour and then left in air for 3 hours before testing.

RESULTS AND DISCUSSION

The nail flexometer measures the ability of a nail section to withstand repeated flexions. The long axis of the α-keratin proteins in nail plate is perpendicular to the direction of growth, so when nail is tested in the nail flexometer the fracture is parallel to the long axis of the α-keratin proteins[1].

Effect of water

After immersion in water for 1 hour, nail weight is increased by 21% (Figure 4.3): nail flexibility is also considerably increased (Figure 4.4). The degree of nail hydration is therefore a major factor in determining nail flexibility, perhaps because water molecules hydrate the α-helical protein and allow them to move more freely with respect to each other. Stratum corneum also takes up

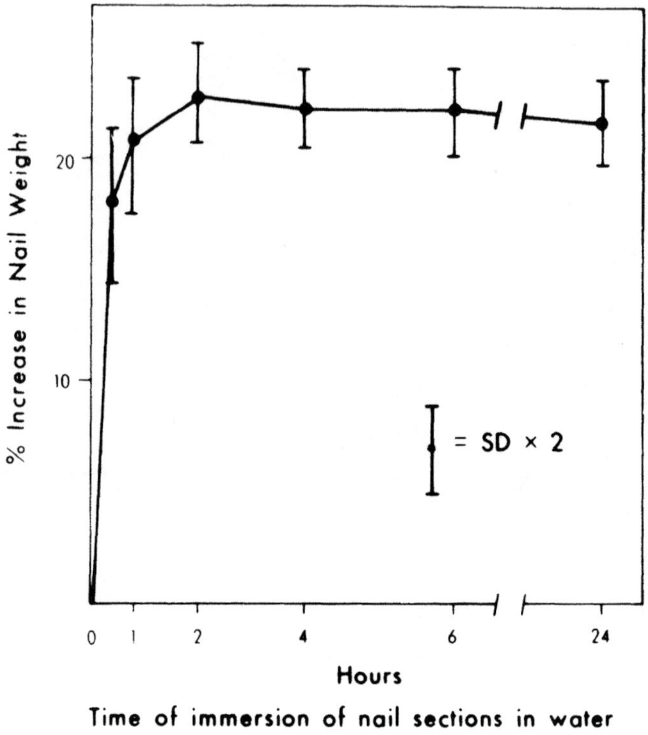

Figure 4.3 Effect of water immersion on nail weight

water rapidly, and the elastic modulus of stratum corneum is a function of water content[8].

Effect of mineral oil

Mineral oil treated nails show no increase in flexibility (Figure 4.5), but when mineral oil is applied to previously hydrated very flexible nails, their flexibility is prolonged (Figure 4.6). The exact explanation for this is not known, but this could result from an occlusive effect of mineral oil which would delay water evaporation.

Effect of phospholipids

Membrane phospholipids strongly bind water[9] and may therefore contribute to nail flexibility. Immersion of nails in PLW greatly increases nail flexibility (Figure 4.5), but this effect may be due almost entirely to the effect of the water

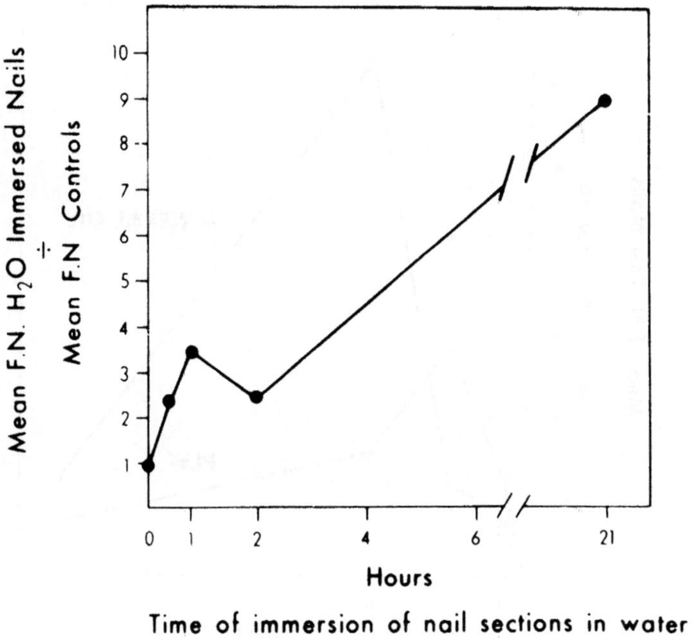

Figure 4.4 Effect of water immersion on nail flexibility

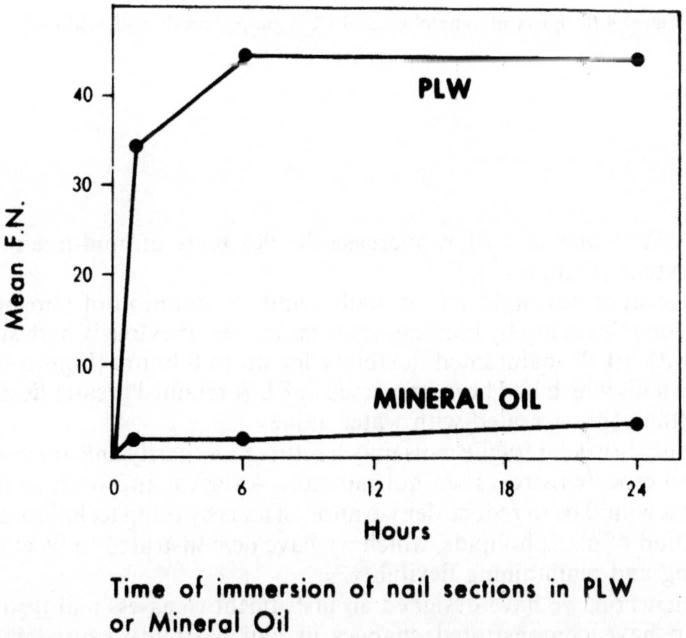

Figure 4.5 Direct comparison of effect of mineral oil with effect of PLW

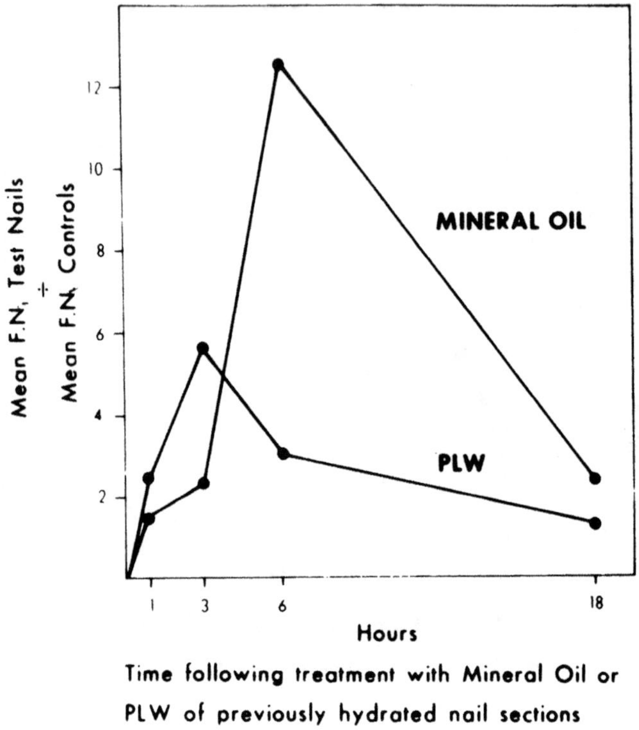

Figure 4.6 Effect of mineral oil and PLW on previously hydrated nails

in the PLW: water and PLW increase the flexibility of lipid-free nails to a similar extent (Figure 4.7).

Water evaporates rapidly from nails[1], and the addition of phospholipids may prolong flexibility by binding water molecules. Previously hydrated nails coated with PLW maintained flexibility for up to 6 hours (Figure 4.6), and lipid-free nails which had been immersed in PLW retained greater flexibility at 3 hours than those treated with water alone.

Clinically brittle or fragile nails may fracture more easily and are less able to withstand repeated stress than normal nails. A logical approach to reducing brittleness would be to reduce dehydration of nails by using techniques such as the addition of phospholipids, which we have demonstrated to be effective in increasing and maintaining flexibility.

In conclusion, we have designed an instrument to assess nail flexibility *in vitro* and have demonstrated changes in nail flexibility caused by water, mineral oil and a phospholipid preparation.

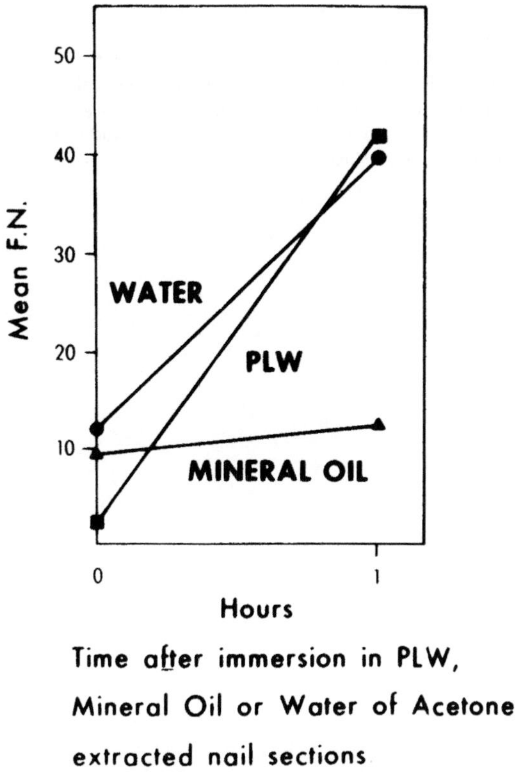

Figure 4.7 Effect of different compounds on lipid-free nails

ACKNOWLEDGEMENT

Figures 4.1, 4.2, 4.3, 4.4, 4.6 and 4.7 are reproduced by permission of the editor of the *British Journal of Dermatology*.

References

1. Baden, H. P. (1970). The physical properties of nail. *J. Invest. Dermatol.*, **55**, 115
2. Burch, G. E. and Winsor, T. (1946). Diffusion of water through dead plantar, palmar and torsal human skin and through toe nails. *Arch. Dermatol. Syph.*, **53**, 39
3. Maloney, M. J., Paquette, E. G. and Shansky, A. (1977). The physical properties of fingernails. 1. Apparatus for physical measurements. *J. Soc. Cosmetic Chem.*, **28**, 415
4. Michaelson, J. B. and Huntsman, D. J. (1963). New aspects of the effects of gelatin on fingernails. *J. Soc. Cosmetic Chem.*, **14**, 443
5. Newman, S. B. and Young, R. W. (1967). Indentation hardness of the fingernail. *J. Invest. Dermatol.*, **49**, 103

6. Hillman, R. W. (1955). Fingernail growth in the human subject. Rates and variations in 300 individuals. *Hum. Biol.*, **27**, 274
7. Bean, W. B. (1963). Nail Growth. *Arch. Intern. Med.*, **111**, 476
8. Van Duzee, B. F. (1978). The influence of water content, chemical treatment and temperature on the rheological properties of stratum corneum. *J. Invest. Dermatol.*, **71**, 140
9. Keith, A. D., Snipes, W., Chapman, D. (1977). Spin-label studies on the aqueous region of phospholipid multilayers. *Biochemistry*, **16**, 634

5
Effects of treatment with prednisolone and PUVA on the mechanical properties of human skin *in vivo*

C. W. GUNNER, E. W. WILLIAMS, M. GREAVES, W. C. HUTTON and T. E. BURLIN

SUMMARY

Brief descriptions are given of apparatuses which are being used to determine changes in the mechanical properties of human skin *in vivo*. Results are presented from a study of five patients who received large doses of prednisolone for the treatment of alopecia. Preliminary results in three patients treated with PUVA (photochemotherapy with long-wavelength ultraviolet radiation) for psoriasis are also included. The results of treatment with prednisolone showed that no significant change occurred in any of the mechanical properties. The preliminary results of PUVA treatment indicated a tendency to increased 'stiffness' (decreased extensibility) and an increase, or no change, in residual deformation.

INTRODUCTION

Burton and Shuster[1] reported an increase in skin extensibility (decrease in nominal elasticity modulus) in patients treated with large doses of prednisolone, and they suggested that other soft tissues may be similarly affected. We are studying this possibility with experiments on the rat aorta, and we have also tested patients receiving this treatment in an attempt to verify their results.

PUVA treatment uses long-term dosing with ultraviolet radiation, which may cause structural changes in the skin[2]. We are investigating the long-term effects of PUVA treatment on the mechanical properties of the skin of patients to identify these possible structural changes, and we give preliminary results in three patients.

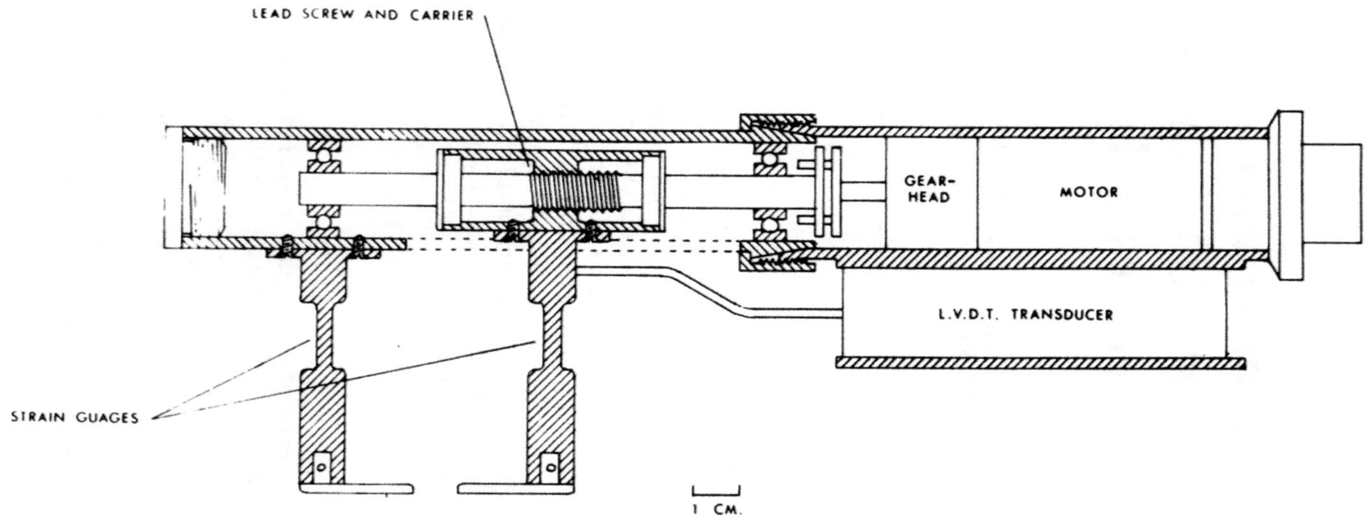

Figure 5.1 Sectional view of *in vivo* extensometer

IN VIVO APPARATUS

Extensometer

A hand-held extensometer[3,4] has been built for *in vivo* studies of humans. The device is shown in section in Figure 5.1. The tabs at the ends of the two strain-gauged arms are stuck to the skin surface with double-sided adhesive tape. A motor and gear-head combination (housed in the apparatus) drives one arm by means of a lead screw and carrier away from the other at a constant rate, so stretching the skin between the tabs. The separation of the tabs is measured by a linear variable differential transformer (LVDT) transducer, and the force developed in the skin is measured by strain-gauges attached to the reduced sections of the arms. Figure 5.2 compares the force–extension characteristics

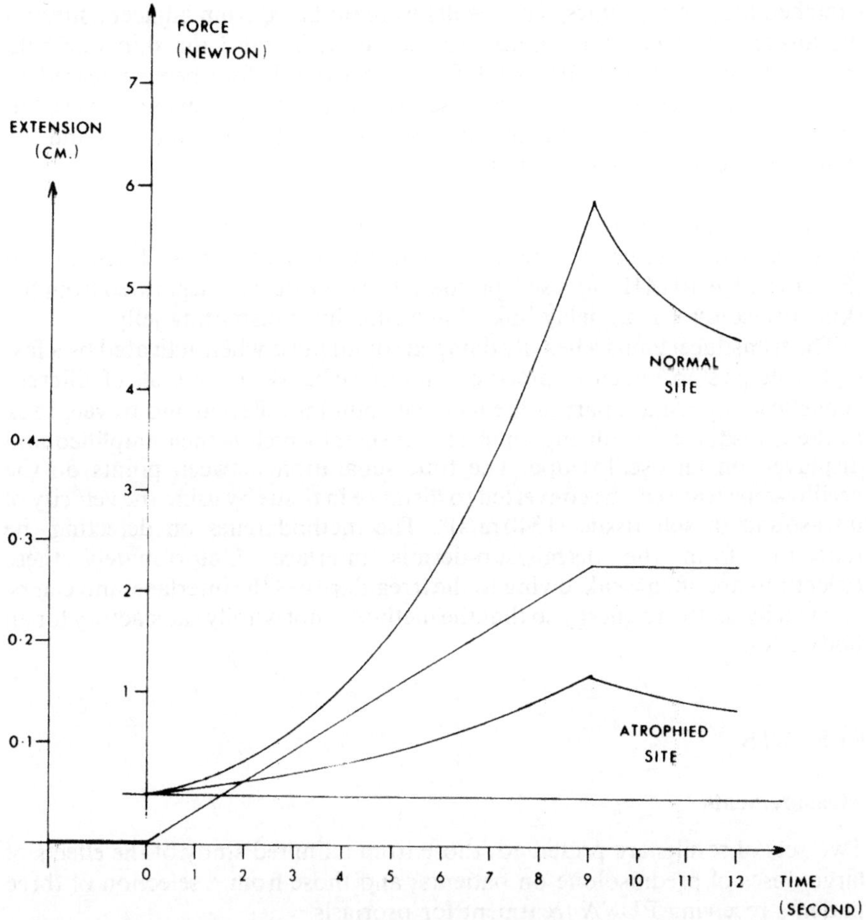

Figure 5.2 Comparison of extensometer outputs obtained from skin severely affected by steroid atrophy and adjacent normal skin near the axilla

obtained from skin severely affected by steroid atrophy with adjacent normal skin near the axilla.

Recoil apparatus

A recoil apparatus[4,5] has been constructed to measure the extension–time characteristic of skin *in vivo* on removal of a deforming force. The device (Figure 5.3) has two tabs, one fixed, the other attached to a carrier which is free to slide in a slot in the body of the apparatus. The separation of the tabs is gauged by a miniature LVDT transducer. Output from this is fed to an ultraviolet recorder. The tabs are stuck to the skin surface with double-sided adhesive tape; the skin is then stretched by driving the tabs apart either manually or by means of the extensometer described above. The tabs are then released to obtain the recoil characteristic.

Figure 5.4 shows the effect on the recoil characteristic of holding the skin stretched for varying times. The results were obtained from adjacent sites on the forearm of a normal subject. For curves 1, 2 and 3 the skin was held stretched for times of 1, 10 and 100 s respectively before being released to obtain the recoil characteristic. In these cases strain-gauge cement was used to stick the tabs to the skin to eliminate viscous creep between the tabs and the skin over the longer time intervals.

Ultrasonic apparatus for determining skin thickness in vivo

An ultrasonic pulse-echo system was built to measure skin thickness *in vivo* (Figure 5.5). A 10 MHz, focused, piezoelectric transducer is separated from the skin surface by a 1 cm delay-link of acoustically transmitting jelly.

The transducer emits a heavily damped sound pulse when activated by a fast high-voltage step. On encountering an interface between materials of different acoustical impedance, part of the incident sound is reflected and travels back to the transducer, producing an electrical signal which is then amplified and displayed on an oscilloscope. The time separation between points on the oscilloscope trace can be converted to distance in tissue by using the velocity of ultrasound in soft tissue (1540 m/s)[6]. The method relies on detecting the reflection from the dermis/sub-dermis interface. Unfortunately these reflections are often weak, owing to the irregularity of the interface, and can be masked by scattered energy so that the method is not wholly satisfactory for all body sites.

RESULTS

Measurements

Two sets of results are presented: those from a limited study of the effects of large doses of prednisolone on patients; and those from a selection of three patients receiving PUVA treatment for psoriasis.

Our standard method was to select three body sites. For the prednisolone study the volar surface of both forearms and the middle of the back were tested.

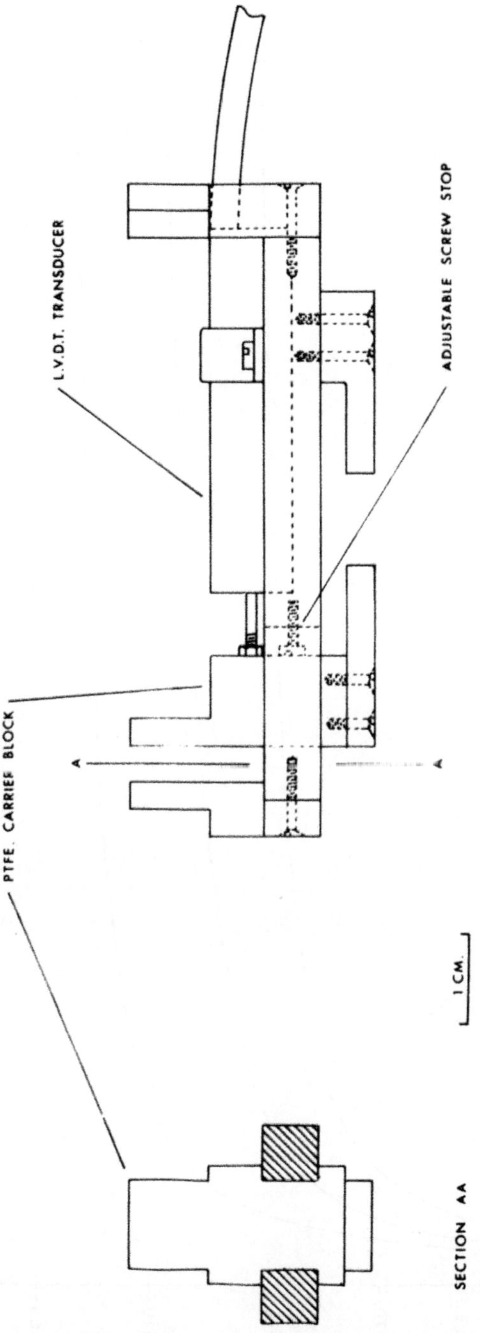

Figure 5.3 *In vivo* recoil apparatus

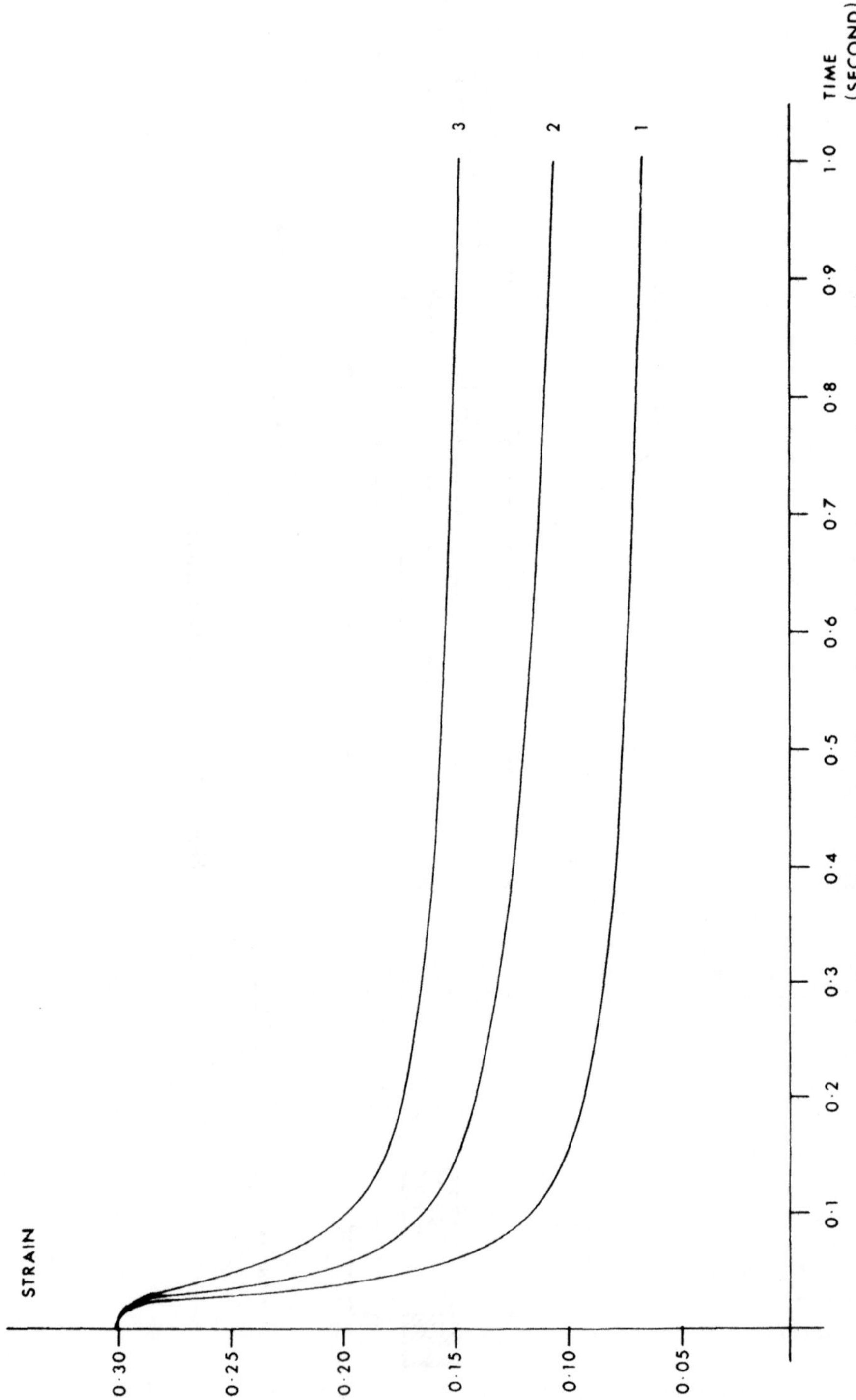

Figure 5.4 Recoil characteristics obtained from adjacent sites on forearm of a normal subject

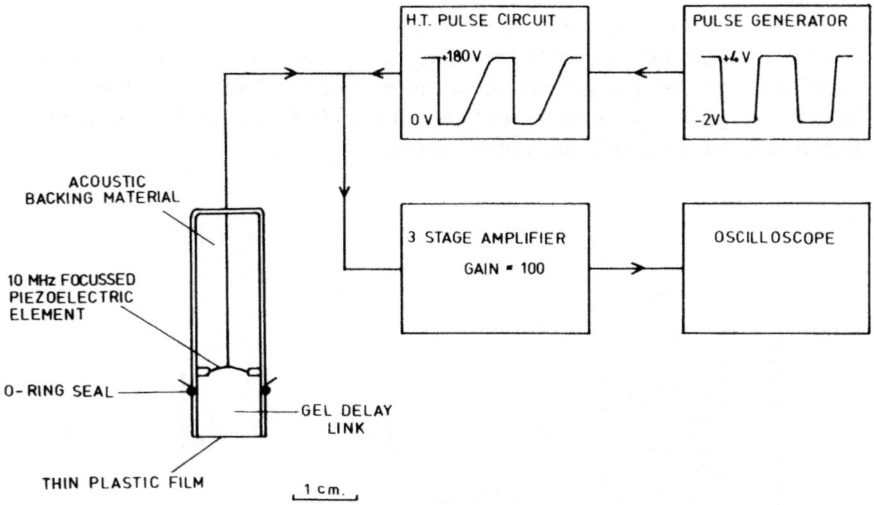

Figure 5.5 The ultrasonic pulse-echo system

For the PUVA tests the mid-back, buttocks and the back of the thighs were used as these should have been fairly uniformly exposed to the radiation. At each site extensometer and recoil tests were made both perpendicular and parallel to the limb–body axis, and one ultrasonic skin-thickness measurement was taken.

The gradient, or nominal elasticity modulus, of the force–extension characteristic was measured at a particular extension (as large as possible) for each body site. This value was then expressed as a percentage of the pretreatment value for that site, and the mean and standard error of all six values was calculated.

Two parameters were calculated from the recoil characteristics: the maximum velocity, equivalent to the maximum gradient; and the residual deformation, measured as the average of the deformation at 0.5 and 1.0 s. These parameters were also expressed as percentages of the pretreatment values, and the means and standard errors were calculated.

Effects of large doses of prednisolone

Burton and Shuster[1] reported an increase in the skin extensibility of patients treated with large doses (2 g i.v.) of prednisolone. Force–extension characteristics were obtained from the forearm, using a uniaxial testing machine, before dosing and at various times afterwards. The maximum gradient of the characteristic showed a decrease as soon as 2 hours after dosing, and the effect remained for at least seven days. This led to speculation that other tissues might be similarly affected. We are exploring this possibility by studying the effects of large systemic doses of steroids on the rat aorta. We have also carried out a limited study in five patients receiving prednisolone treatment for alopecia (two doses of 1 g separated by 24 hours and given orally) in an attempt to verify the changes reported by Burton and Shuster[1]. Results in the five patients are shown in Figure 5.6. Each patient was tested before dosing, four patients were tested once after dosing, and one patient was tested at three different times after dosing.

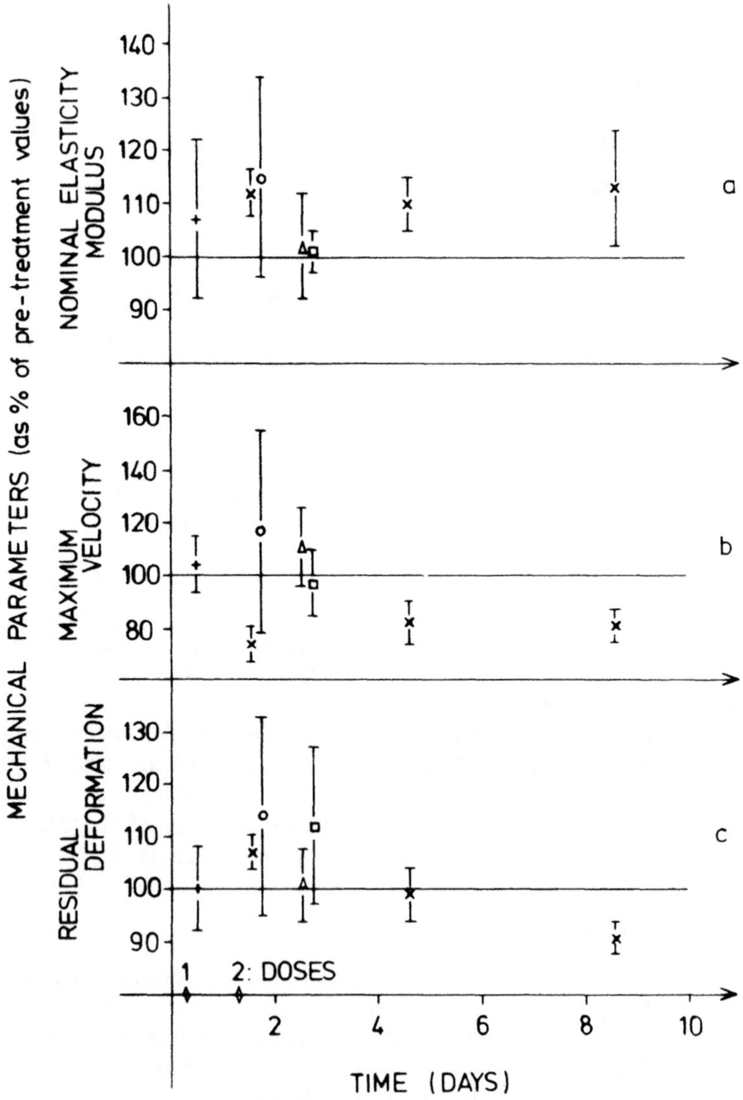

Figure 5.6 Results (as means ±SE) in five patients treated with large doses of prednisolone

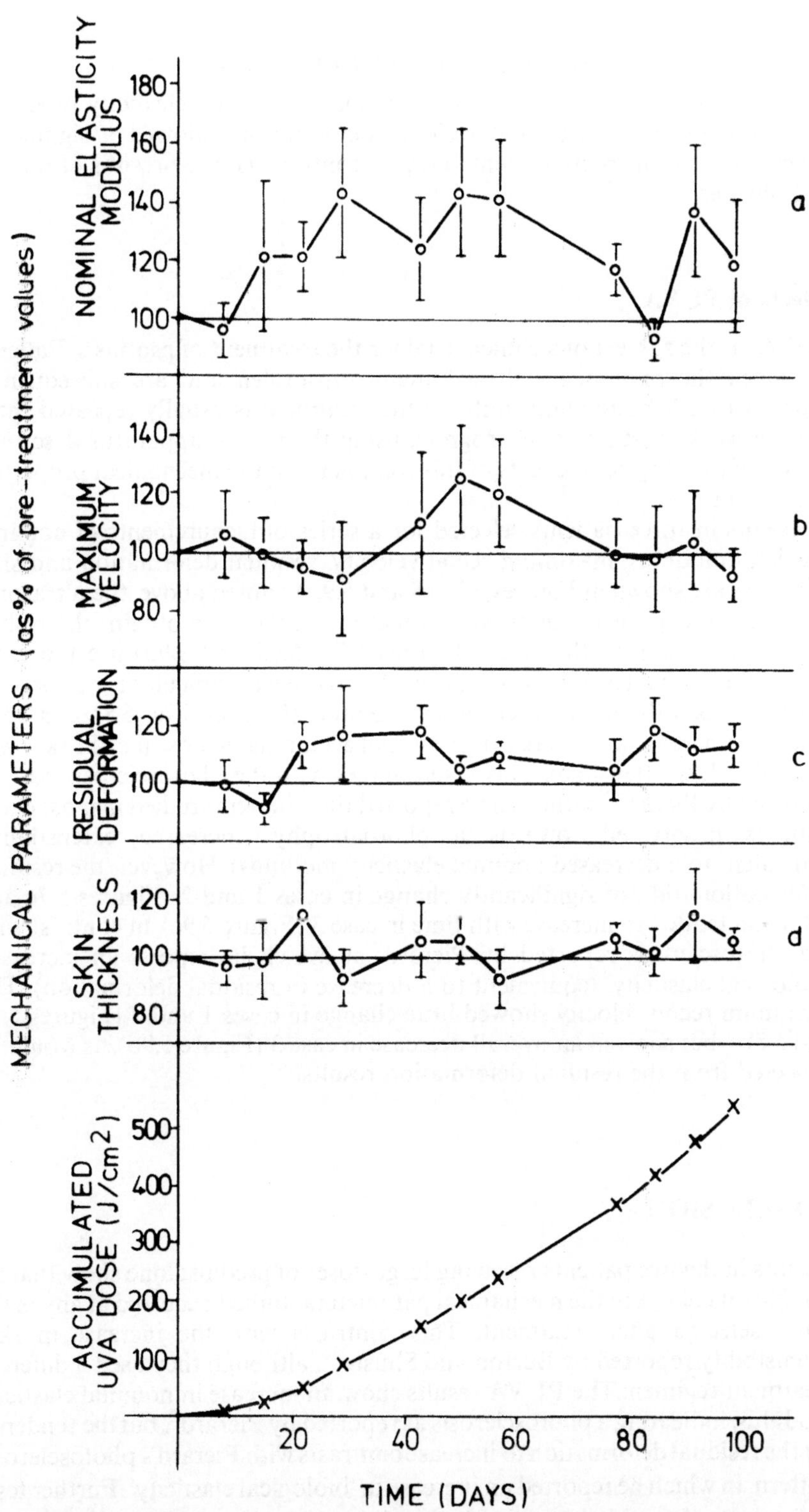

Figure 5.7 Results (as means ±SE) in case no. 1 receiving PUVA treatment

None of the three variables after treatment (nominal elasticity modulus, maximum recoil velocity and residual deformation) showed a significant difference from the pretreatment value, in contrast with the results of Burton and Shuster[1].

Effects of PUVA

PUVA[7] is the subject of a clinical trial for the treatment of psoriasis. Patients are given the photoactive drug 8-methoxypsoralen and are subsequently exposed to UVA radiation; initially the treatment is usually repeated three times a week. A study is in progress, using the *in vivo* apparatus described above, of the long-term effects of this treatment on the mechanical properties of the skin.

Results in three patients selected for a series of meaurements of nominal elasticity modulus, maximum recoil velocity, residual deformation and skin thickness are shown in Figures 5.7, 5.8 and 5.9. As noted above, the ultrasonic skin thickness measurements were sometimes difficult to obtain; the results shown here for skin thickness (Figures 5.7d, 5.8d and 5.9d) are the most reasonable estimates, but in some cases the trace was difficult to interpret. In all three cases the nominal elasticity modulus (Figures 5.7a, 5.8a and 5.9a) tended to increase with increasing time and accumulated dose (i.e. the skin was becoming less extensible). This result agrees with the photosclerosis pattern reported by Pierard[2], although he reported that the skin studied in most of the patients progressed towards a photoatrophy (increased extensibility, equivalent to a decreased nominal elasticity modulus). However, the residual deformation did not significantly change in cases 1 and 2 (Figures 5.7c and 5.8c) and tended to increase with time in case 3 (Figure 5.9c), in contrast with the photosclerosis reported by Pierard[2] in which he reports an increased 'biological elasticity' (equivalent to a decrease in residual deformation). The maximum recoil velocity showed little change in cases 1 and 2 (Figures 5.7b and 5.8b), but showed an overall decrease in case 3 (Figure 5.9b), as would be expected from the residual deformation results.

CONCLUSION

Results in the five patients receiving large doses of prednisolone show that no significant change in the mechanical parameters studied occurred at any of the times selected after treatment. This contrasts with the increase in skin extensibility reported by Burton and Shuster[1], although they used a different treatment regimen. The PUVA results show an increase in nominal elasticity modulus, indicating a photosclerosis as reported by Pierard[2], but the tendency for the residual deformation to increase contrasts with Pierard's photosclerosis pattern, in which he reported an increase in 'biological elasticity'. Further tests on patients should help to clarify this discrepancy.

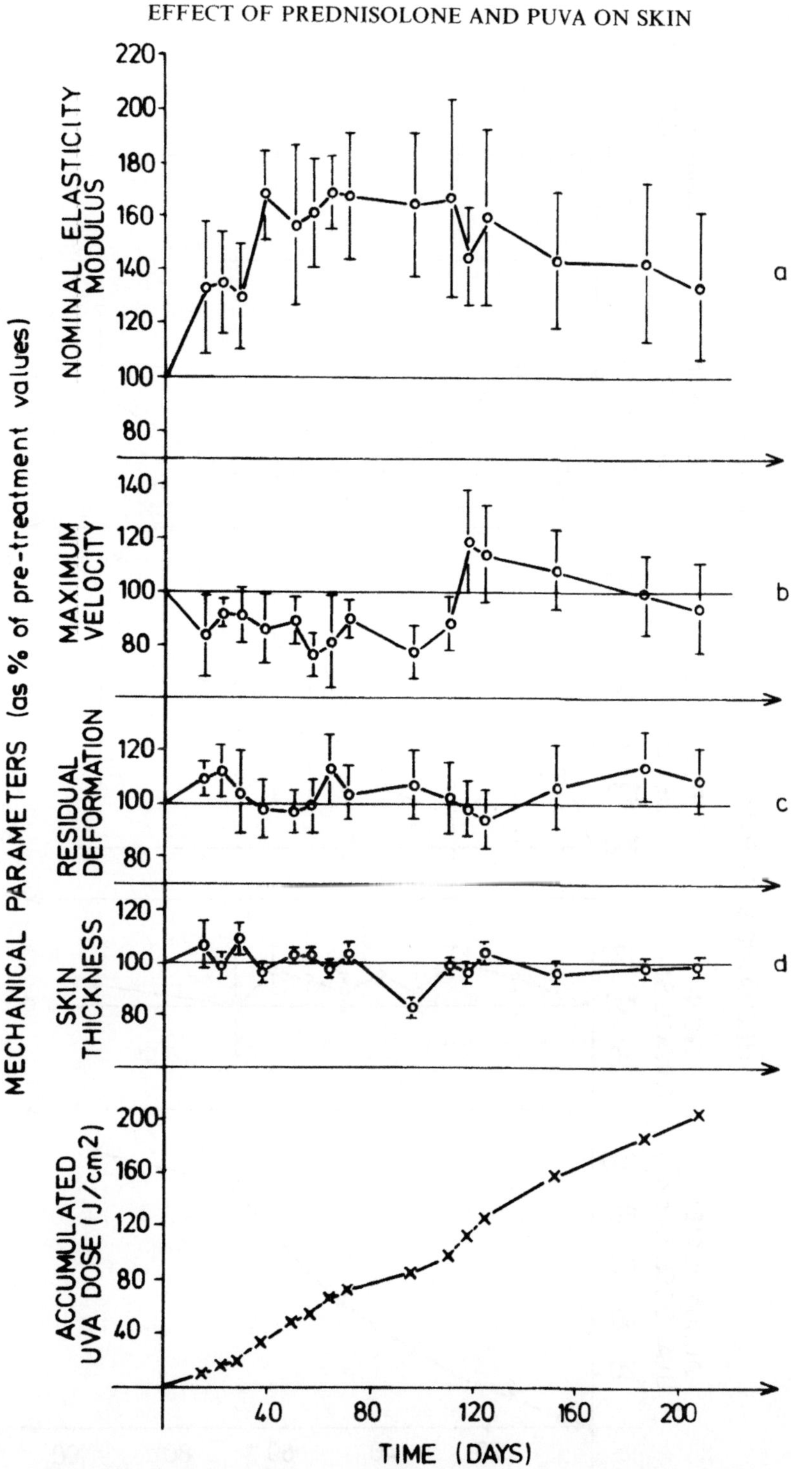

Figure 5.8 Results (as means ±SE) in case no. 2 receiving PUVA treatment

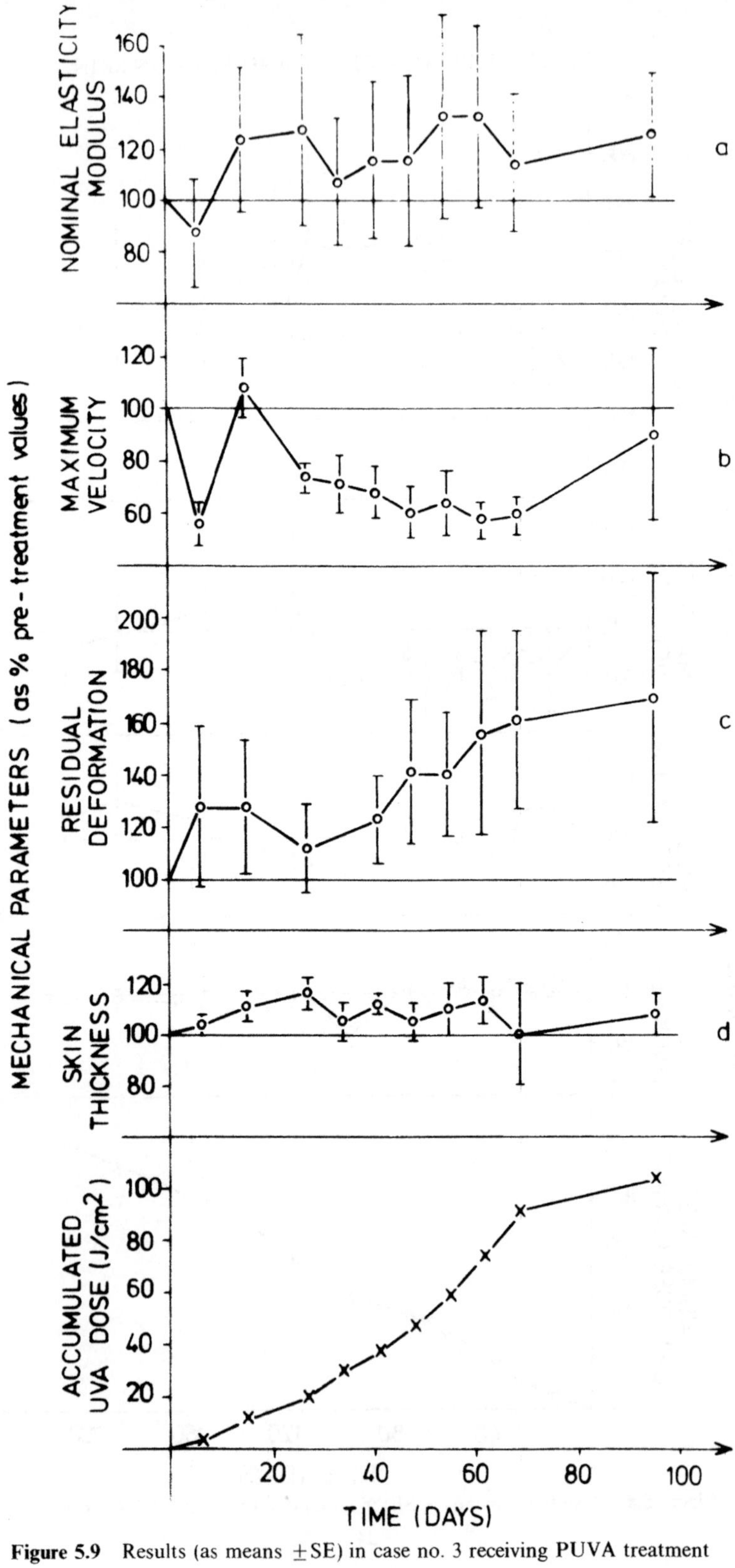

Figure 5.9 Results (as means $\pm$SE) in case no. 3 receiving PUVA treatment

References

1. Burton, J. L. and Shuster, S. (1973). A rapid increase in skin extensibility due to prednisolone. *Br. J. Dermatol.*, **89**, 491
2. Pierard, G. E. (1978). Iatrogenic alterations of the biochemical properties of human skin. *Br. J. Dermatol.*, **98**, 113
3. Gunner, C. W., Hutton, W. C. and Burlin, T. E. (1979). The mechanical properties of skin *in vivo* – a portable hand-held extensometer. *Br. J. Dermatol.*, **100**, 161
4. Gunner, C. W., Hutton, W. C., Burlin, T. E. and Wyn Williams, E. (1979). The effects of some clinical treatments on the mechanics of skin. *Eng. Med.*, **8**, 58
5. Gunner, C. W., Hutton, W. C. and Burlin, T. E. (1979). An apparatus for measuring the recoil characteristics of human skin *in vivo*. *Med. Biol. Eng. Comput.*, **17**, 142
6. Wells, P. N. T. (1969). *Physical Principles of Ultrasonic Diagnosis.* (London: Academic Press)
7. Wolff, K., Fitzpatrick, T. B., Parrish, J. A., Gshnait, F., Gilchrest, R., Honigsmann, H., Pathak, M. A. and Tannenbaum, L. (1976). Photochemotherapy for psoriasis with orally administered methoxsalen. *Arch. Dermatol.*, **112**, 943

6

In vivo measurement of some elastic properties of human skin

S. DIKSTEIN and A. HARTZSHTARK

SUMMARY

We developed an *in vivo* indentometer exerting an initial pressure of 1 g/cm^2 and final pressure of 10 g/cm^2. The indentation so measured (by indentometry) on the forehead skin was 0.04–0.07 cm. When slightly modified (using a perpendicular pull of 10 g/cm^2) the same equipment measured skin slackness. We called this technique levarometry.

The effect of over 20 cosmetic raw materials from widely different sources was evaluated. Only lactic acid at pH 4.0, vegetable oil and lanolin showed any effects lasting for more than 90 minutes.

The instrument may be used to develop better, longer lasting medico-cosmetic agents.

INTRODUCTION

The elastic properties of the human skin are known to change with age. One of the major aims of 'therapeutical cosmetics' is to restore the youthful appearance – increased elasticity and humidity – of the skin[1]. To evaluate the effects of cosmetic preparations on the human skin we used a modified Schade's instrument[2,3] having increased sensitivity of at least ten-fold. This was achieved by using modern electronic techniques rather than the mechanical means of the original instruments. This increased sensitivity permitted us to analyse the skin *in vivo* in the low section of the stress–strain curve, which is more interesting physiologically[4]. It also permitted semi-quantitative evaluation of the effect of some cosmetic raw materials.

DESCRIPTION OF THE INSTRUMENT AND MEASUREMENT OF INDENTATION

The measuring area is a circular piece of plastic material (made of Teflon) having a surface area of 0.2 cm^2. It is connected to a light metal measuring rod counterbalanced so that the net pressure of the system is less than 1 g/cm^2. The measuring rod can be loaded with specially constructed weights, thereby increasing the pressure from the starting pressure to any desired value. The measuring rod is connected to a linear variable differential transformer (LVDT). The output of the LVDT was graphically recorded. A paper velocity of 6 cm/min gave sufficiently clear recordings.

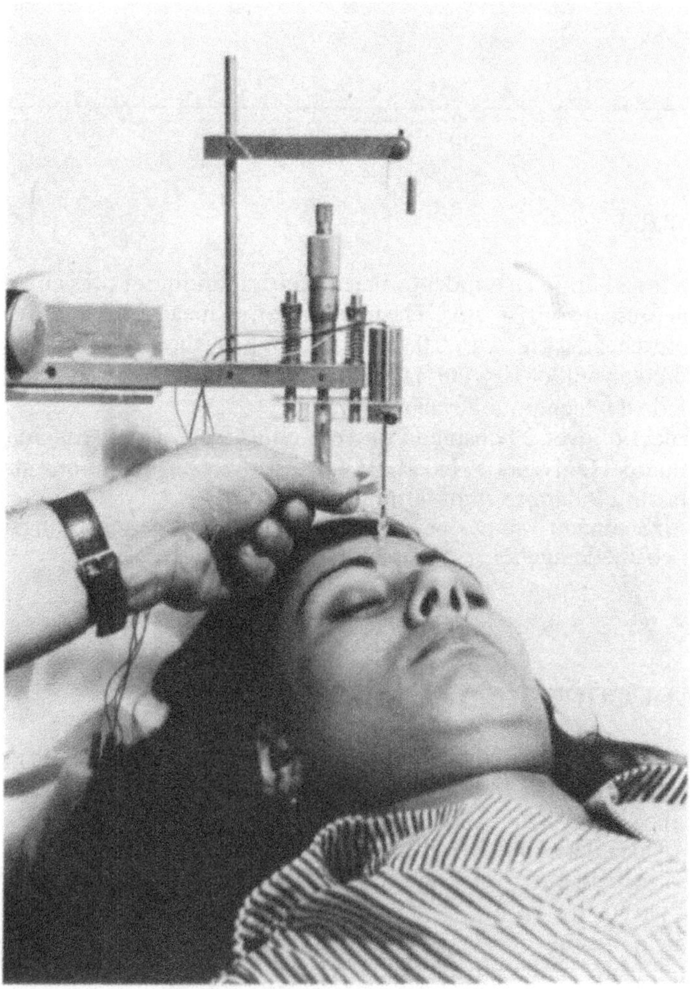

Figure 6.1 Indentometry

For the sake of convenience we chose to measure the skin in the middle of the forehead. The volunteer lay on his or her back with eyes closed. The measuring rod was adjusted to touch the skin, the electronic system was zeroed and the recorder started (Figure 6.1). After the baseline was stabilised (in about 15 seconds) a standard weight was suddenly applied. Indentation was recorded for 10 seconds. The weight was then removed and the rebound recorded for a further 10 seconds. For preliminary comparison we measured: (1) the total indentation recorded in 10 seconds (I_{max}); (2) the percentage rebound after 6 seconds, as:

$$\frac{\text{distance the measuring rod moved back}}{I_{max}}$$

Figure 6.2 shows a typical curve.

To measure the effectiveness of some cosmetic raw materials the compound was applied to the forehead, usually as a 10% solution (depending on the solubility or dispersibility either in water or in ethyl alcohol), using a piece of cotton wool. The indentation curve was recorded at half-hour intervals after

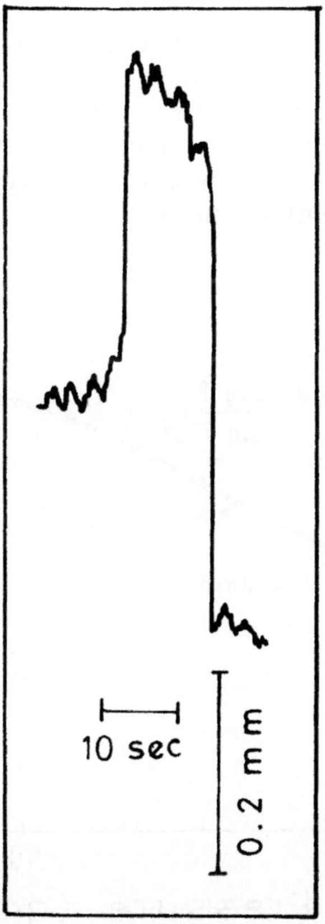

Figure 6.2 A typical indentation curve

treatment. In preliminary experiments we found that any effect of plain tap water or alcohol disappeared after 30 minutes, although the effect of water was obvious in shorter periods. Each measurement was repeated at different points, and if results differed more than 20% a third measurement was made. Care was taken not to choose an area which covered the line of a wrinkle.

The mean value for a person on the same day or on several consecutive days agreed within 10%. It was arbitrarily decided that any treatment which still showed an effect 90 minutes after application would be called 'effective'. Usually 6–10 people were needed to supply significant data.

Since it was impossible to find enough female volunteers who had not used some cosmetic treatment before the experiment, we decided to use male volunteers from the laboratory and student staff aged between 30 and 50 years to test each agent. Between measurements the volunteers continued their normal activity. The control values for all the volunteers in the 'cosmetic treatment' experiments, using 10 g/cm^2 standard pressure, were: I_{max}, 0.5–0.7 mm; rebound, 55–70%.

RESULTS OF INDENTOMETRY

To ascertain the pressure–indentation relationship of the instrument we measured indentation using pressures of 10, 20 and 40 g/cm^2. Indentation is approximately linearly related to the square root of the pressure, reminiscent of Hooke's law[5] modified for small indentations[6] (Figure 6.3). We chose a pressure of 10 g/cm^2, since we wished to analyse the low section of the stress–strain curve (see introduction).

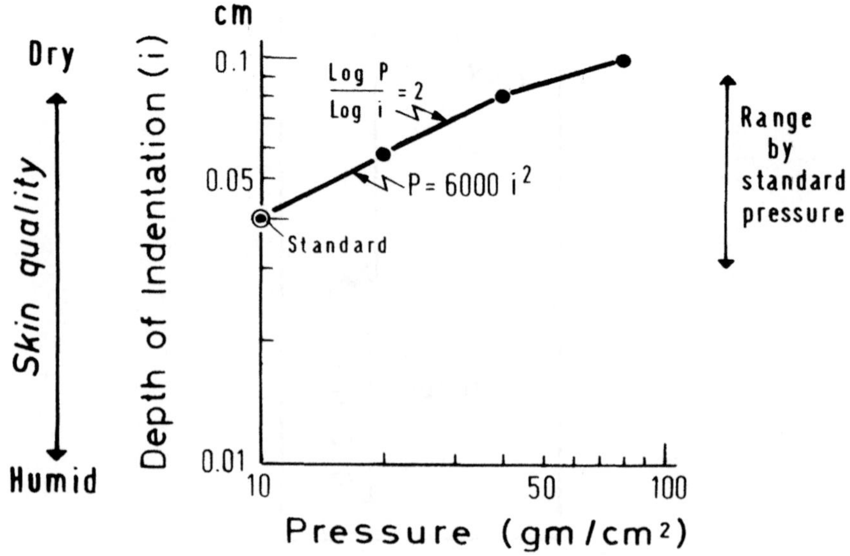

Figure 6.3 A typical pressure–indentation relationship in indentometry

Table 6.1 Effects of typical 'effective' treatments

Agent	*Change after 90 min in:*		*Change after 120 min* in:*	
	I_{max}(%)	*Rebound* (%)	I_{max}(%)	*Rebound* (%)
Olive oil	−20	0	−20	0
10% Sodium lactate pH 4.0 in water	−20	+10	−10	+5
Lanolin, hydrous	−30	+15	−20	0
Glycerol 25% in water	−30	0	0	0

* No agent was effective after 180 minutes.

Some basic cosmetic ingredients were then tested. We found that the following agents were not effective: 10% lecithin in water, 10% cetylalcohol in ethyl alcohol, 10% stearic acid in ethyl alcohol, 10% Tween 20 in water, 20% urea in water, 10% sodium L-pyroglutamate pH 4.0 in water, 10% sodium acetate pH 8.6 in water, 10% sodium acetate pH 4.5 in water, 10% honey in water, light mineral oil, Vaseline, butter, 10% calcium chloride in water, 10% sodium bicarbonate in water, 10% sodium caprylate pH 4.0 in water, silicon oil MS200-20 cp, 10% cholesterol in alcohol, isopropyl palmitate, diisopropyladipate, perhydrosqualene, and M-199 tissue-culture fluid.

Table 6.1 shows those treatments which were found to be effective. No agent was effective 3 hours after application.

DISCUSSION OF INDENTOMETRY

It is quite clear that the *in vivo* measurements contrast with *in vitro* results. For example, in the *in vitro* study of skin moisturizers measuring elastic modulus and stress relaxation modulus, 4% sodium pyroglutamate (2-pyrolidone-5-carboxylate) was found to be highly effective[7]. Although these results cannot strictly be compared with ours, in our case 10% sodium pyroglutamate was certainly among the less effective agents. On the other hand, in our tests lactate pH 4.0 was highly effective, in accordance with the *in vitro* and *in vivo* studies by Middleton[8]. Although a recent *in vivo* study by Middleton[9] seemed to show a beneficial effect of long-term treatment with 5% sodium pyroglutamate, we wish to point out that even in his study it was less effective than lactate[8].

The following factors could explain the discrepancy between *in vitro* and *in vivo* results.

(1) Penetration. Animal fats and vegetable oils penetrate more than mineral oils[10] although mineral oils are excellent occlusives *in vitro*[7]. It seems that penetration *in vitro* and *in vivo* differ considerably.

(2) Self-regulation. *In vivo* many factors – e.g. blood supply, hormones, and neurohormones – might regulate the water and fat supplies to the stratum corneum. Any external influence might be counterbalanced by these factors. *In vitro* no counter-regulation exists.

(3) *In vivo* the skin is stretched[11].

It is now generally accepted that the so-called 'moisturization' of the skin *in vivo* is determined mainly by the amount of water in the stratum corneum. In normal conditions the water-soluble 'natural moisturizing factor' (NMF), which is about a fourth part of the stratum corneum, has been claimed to be partially responsible for water retention[12,13]. Our results seem to suggest that in *in vivo* conditions (at least in healthy males) lactic acid is the main efficient natural constituent of the NMF which can be applied externally. It is more effective than an acid alone, since acetate at the same pH is much less effective.

It is unlikely, however, that our indentometry measures any property connected with the stratum corneum: the degree of indentation ($\simeq 0.05$ cm) suggests a property connected with at least the whole epidermis and possibly with the dermis as well.

The likeliest explanation of our results is that indentometry compresses the skin, and the resulting indentation is strongly dependent on the fluid interchange with the surrounding tissue[14]. Hence the decrease of compression by increased moisture content of the forehead skin (water is much less compressible than protein). On the other hand, water or other agents may somehow increase the stretch of the forehead skin causing decreased indentation and slackness.

MEASURING SKIN SLACKNESS BY PERPENDICULAR PULL (LEVAROMETRY)

We have described above an apparatus capable of measuring parameters of skin compressibility. These parameters were found to be strongly influenced by the skin's state of humidity and possibly also elasticity. No rigid connection, however, exists between skin quality and number of wrinkles[15].

It was therefore desirable to try to quantify wrinkles. We assumed that a stretched skin would wrinkle less than a less stretched skin. We have built an instrument which applies a perpendicular pull to the skin. The height of the pull is measured by a standard weight. This measure quantifies skin slackness which in turn is assumed to represent the tendency of the skin to wrinkle.

DESCRIPTION OF THE INSTRUMENT USED IN LEVAROMETRY

The instrument we used is a modified form of the one described above. The following modifications were introduced.

(1) The measuring probe head was made hollow, permitting the application of vacuum to 'glue' the skin to the probe.

(2) The measuring weight was applied to the counterbalance and not to the measuring rod. (Figure 6.4)

Other experimental details were exactly as described above.

Figure 6.5 shows a volunteer's skin response to different pulls. It is clear that with a pull of over 20 g/cm^2 the linear relationship of Hooke's law no longer applies. We therefore chose 10 g/cm^2 for our standard measurement. Here, too, we are interested in the physical variables under minimal deformation.

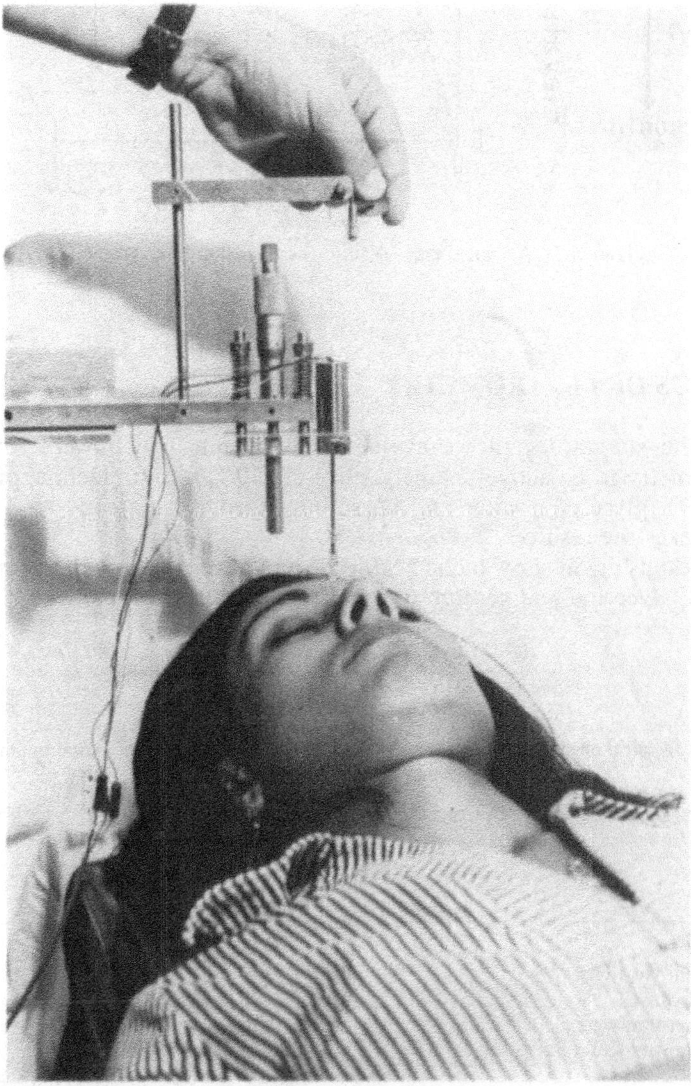

Figure 6.4 Levarometry

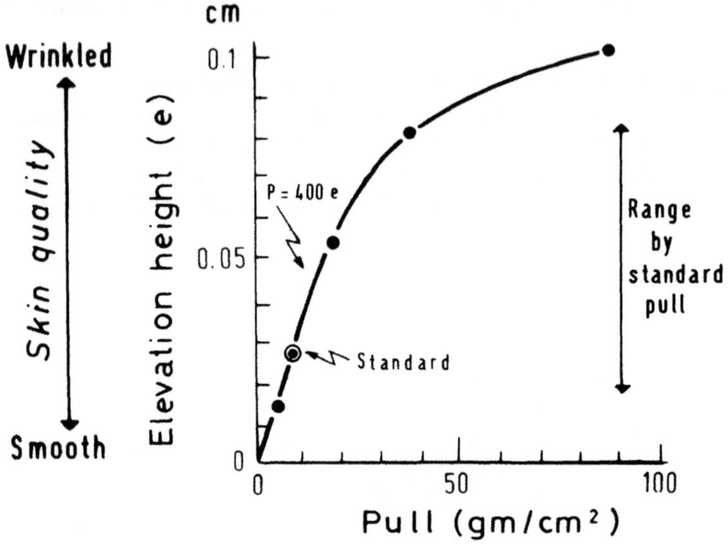

Figure 6.5 A typical pull–elevation relationship in levarometry

RESULTS OF LEVAROMETRY

Using the standard pull we evaluated all agents which were found by indentometry to be active – namely, olive oil, 10% sodium lactate, pH 4.0 in water, 25% glycerol in water, and 5% sodium pyroglutamate pH 4.5. Table 6.2 summarizes the results.

The results again show lactic acid to be an active compound. Surprisingly, olive oil, glycerine and sodium pyroglutamate each had no effect.

Table 6.2 Typical effect of various agents on skin slackness. Results given as maximum elevation by standard pull*

Agent	*Time after application (min)*				
	0	60	90	120	150
Olive oil	25	25	25	25	—
4% Sodium lactate pH 4.5	20	10	14	16	18
Lanolin, hydrous	30	28	26	24	20
25% Glycerol in water	25	26	28	27	27
5% Sodium pyroglutamate pH 4.5	25	26	28	27	27

* 50 units = 0.1 cm.

DISCUSSION

Does our approach satisfy all the scientific criteria for measuring the effect of medicocosmetic agents on the skin *in vivo*? Clearly, the answer is 'not yet'; a lot remains to be done. Scientifically, we have yet to show the exact form of correlation between water content *in vivo* and our measurements. Statistically, we have yet to find out how many categories of skin can be observed with our technique. We must also ascertain whether we can match a proper cosmetic treatment to each type of skin, since over-humidification and over-fatting are also undesirable[11].

Two criticisms of our selection of active agents may be levelled.

(1) It does not take into account the possibility that the agents when emulsified might have a different effect, nor does it examine the possibility of interaction between the components.
(2) It deals only with acute effects, whereas therapeutic cosmetic treatments, by their nature, are long-term.

In conclusion, we believe that cosmetic treatments should be validated on volunteers on the same area of skin for which they are intended, since both human *in vitro* experiments and animal models are always open to serious criticism. 'Therapeutic cosmetics' unlike drug development, can in most cases, go straight to human models.

References

1. Weirich, E. G. (1972). Die Moderne Kosmetik als Arbeitspartnerin der Medizin. *Z. Kosmetik Wiss. Praxis*, **5**, 1
2. Schade, H. (1912). Die Elasticitätsfunction des Bindegewebes und die intravitale Messung ihrer Störungen. *Z. Exp. Pathol. Therapie*, **11**, 369
3. Kirk, E. and Koorning, S. A. (1969). Quantitative measurements of the elastic properties of the skin and subcutaneous tissue in young and old individuals. *J. Gerontol.*, **4**, 273
4. Vogel, H. G. and Hilgner, W. (1977). Analysis of the low part of stress–strain curves in rat skin. Influence of age and dermotropic drugs. *Arch. Derm. Res.*, **258**, 141
5. CRC. (1976). *Handbook of Chemistry and Physics*. 47th Edn, p. F-107. (Cleveland, Ohio: CRC)
6. Fink-Jensen, P. (1965). Hardness testing of organic coatings. *Pure Appl. Chem.*, **10**, 259
7. Rieger, M. M. and Deem, D. E. (1974). Skin Moisturizers. II. The effects of cosmetic ingredients on human stratum corneum. *J. Soc. Cosmetic Chem.*, **25**, 253
8. Middleton, J. D. (1974). Development of a skin cream designed to reduce dry and flaky skin. *J. Soc. Cosmetic Chem.*, **25**, 519
9. Middleton, J. D. and Roberts, M. E. (1978). Effect of a skin cream containing the sodium salt of pyrollidone carboxylic acid on dry and flaky skin. *J. Soc. Cosmetic Chem.*, **29**, 201
10. Barnett, G. (1957). Emollient creams and lotion. In Sagarin, E. (ed.) *Cosmetics: Science and Technology*, p. 11. (New York: Interscience Publications Inc.)
11. Danielson, D. A. (1973). Human skin as an elastic membrane. *J. Biomechanics*, **6**, 539
12. Strianse, S. J. (1974). The search for the ideal moisturizer. *Cosmetics and Perfumery*, **89**, 57
13. Padberg, G. (1972). Einfluss der Bindung von Kohlenhydraten an die Skleroproteine auf die Wasserbindung der Hornschicht. *J. Soc. Cosmetic Chem.*, **23**, 271
14. Daly, H. C. and Odland, G. F. (1979). Age related changes in the mechanical properties of human skin. *J. Invest. Dermatol.*, **73**, 84
15. Wright, E. T. and Shellow, W. D. R. (1973). The histopathology of wrinkles. *J. Soc. Cosmetic Chem.*, **24**, 81

7
A resonance frequency technique to determine the energy absorbed in stratum corneum *in vivo*

A. M. TORGALKAR

SUMMARY

A cantilever-mode transducer was designed to determine the energy absorbed in stratum corneum. The measurements were made in a frequency range of 700–800 Hz. The softness of stratum corneum was measured in terms of loss tangent (tan δ), a dimensionless number. An acoustic spectrometer was used to determine the resonance frequency, and the results obtained were recorded on an X–Y recorder. The vertical displacement of the arm of the transducer was controlled by attaching a ring weight to the horizontal portion of the arm. This permits the measurement of stratum corneum softness whether it is close to a bone or to fatty tissue.

The method was used to determine the energy absorbed in normal and dry skin and was also useful for evaluating the effect of various treatments on skin softness. In a controlled environment, the data was reproducible.

INTRODUCTION

A method based on measuring mechanical resonance frequency to determine the elastic modulus originated about 40 years ago[1]. Since then, numerous investigators have used the technique successfully to determine the dynamic mechanical properties of materials. These investigators have improved and refined the basic method so that today resonance frequency can be determined to a high degree of accuracy[2]. An instrument has been developed to determine elastic modulus, loss tangent and other mechanical properties, and is called the acoustic spectrometer[3–5].

This paper describes a technique for exciting, detecting, and measuring the resonance frequency of the material. It also describes a transducer designed to

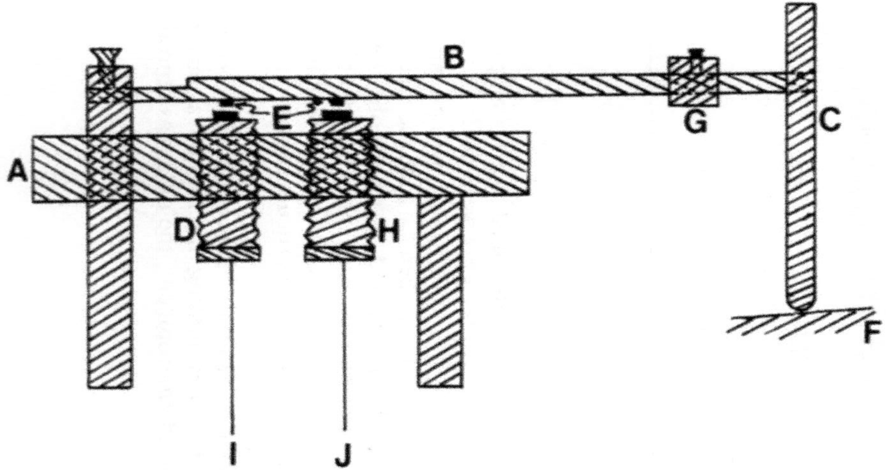

Figure 7.1 Diagrammatic sketch of a cantilever-mode transducer. A = Platform. B = Horizontal arm. C = Vertical arm. D, H = Electromagnetic coils. E = Ferromagnetic material discs. F = Skin. G = Adjustable weight. I, J = Cables to an acoustic spectrometer

pass sound waves of a predetermined resonance frequency through skin and a method for calculating the energy absorbed by it.

METHOD

The instrument consists of two parts: (1) a cantilever mode transducer; and (2) an acoustic spectrometer, which measures and records resonance frequency.

Transducer

The cantilever-mode transducer used to transfer frequency through skin is shown in Figure 7.1.

Two electromagnetic coils (D, H) are screwed onto a platform (A). This arrangement permits them to be adjusted to the desired level. One of these coils serves as a vibrator and the other as a receiver. The vertical and horizontal arms (C, B) of the transducer are made of aluminum. One end of the horizontal arm is fixed to a vertical support of the platform in such a way that the ferromagnetic metal discs (E) are positioned just above the electromagnetic coils. The vertical arm (C) is fixed to the other end of the horizontal arm, which is free to resonate. The frequency passing through the vertical arm is measured by the frequency-counter unit of an acoustic spectrometer. The resonance frequency of the arm can be adjusted to any desired value by moving a ring weight (G), attached to the horizontal arm. The lower tip of the vertical arm is rounded and is placed in contact with the skin to pass sound waves of resonance frequency through it.

Acoustic spectrometer

This is used to determine the resonance frequency and the internal friction of the testing material. A diagrammatic sketch of an acoustic spectrometer is shown in Figure 7.2. An audio-oscillator, audio-amplifier, pick-up amplifier, cathode ray oscilloscope, frequency counter and $X-Y$ recorder are the major components of the acoustic spectrometer.

The output of the audio oscillator is amplified and fed to the driver, the mechanical energy of which in turn is transmitted to the arm. At the same time that the driver is energized, the oscillator output also goes to the horizontal plates of the oscilloscope. When the oscillator is not close to the resonance frequency of the arm, a horizontal line is seen on the oscilloscope (this is the electron beam moving back and forth at the frequency of the oscillator). The oscillatory frequency is scanned and eventually reaches one of the mechanical resonance frequencies of the arm.

The predominant characteristic of the arm is a large increase in the amplitude of its vibrations when resonating. The increased amplitude detected by the pick-up closely adjacent to the arm is then amplified and fed to the vertical plates of the oscilloscope. If the same frequency is received by horizontal and vertical plates of the oscilloscope at the same time a Lissajou

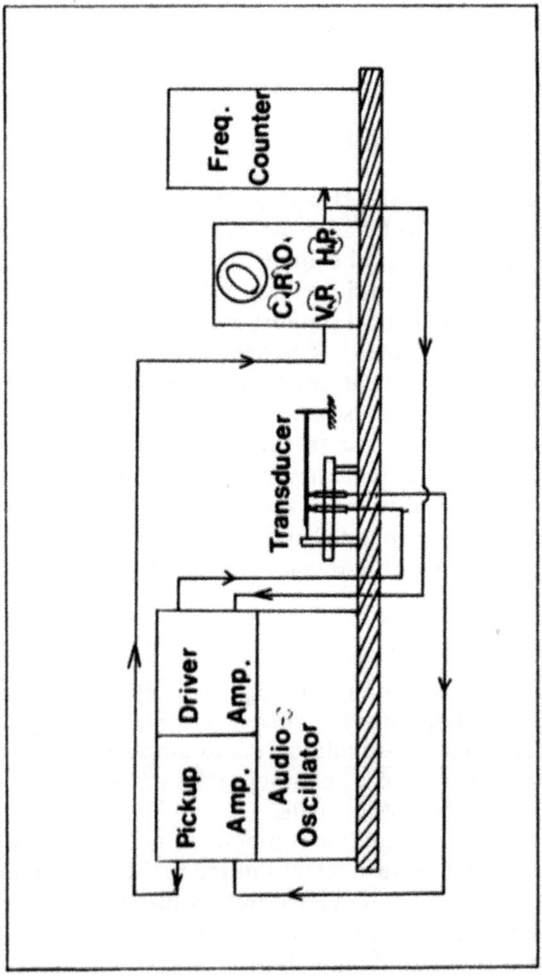

Figure 7.2 Diagrammatic sketch of an acoustic spectrometer. CRO = Cathode ray oscilloscope. VP = Vertical plate. HP = Horizontal plate

pattern, as shown in Figure 7.2, occurs. The oscillator frequency setting at which this pattern has its maximum vertical amplitude is one of the resonance frequencies of the arm. This frequency is counted accurately by the frequency counter connected to the oscillator.

To determine the skin softness, a cantilever mode transducer was connected to an acoustic spectrometer (Figure 7.2). One of the electromagnetic coils was used as a driver while the other was used as a pick-up, so as to excite and detect the resonance. The ring weight on the horizontal arm of the transducer was fixed at a desired position, and the resonance frequency of the transducer arm was determined in air.

Once the resonance frequency of the transducer arm was determined, adjustments were made on the acoustic spectrometer to sweep the resonance frequency automatically.

The resonance frequency was scanned through a ± 50 Hz range for a high degree of accuracy. The scanning of the frequency was recorded on the X–Y recorder to determine the energy loss. A typical energy-loss curve is shown in Figure 7.3.

The rounded tip of the transducer arm was then brought in contact with skin, and the energy loss in skin was recorded. The increase in absorbed energy

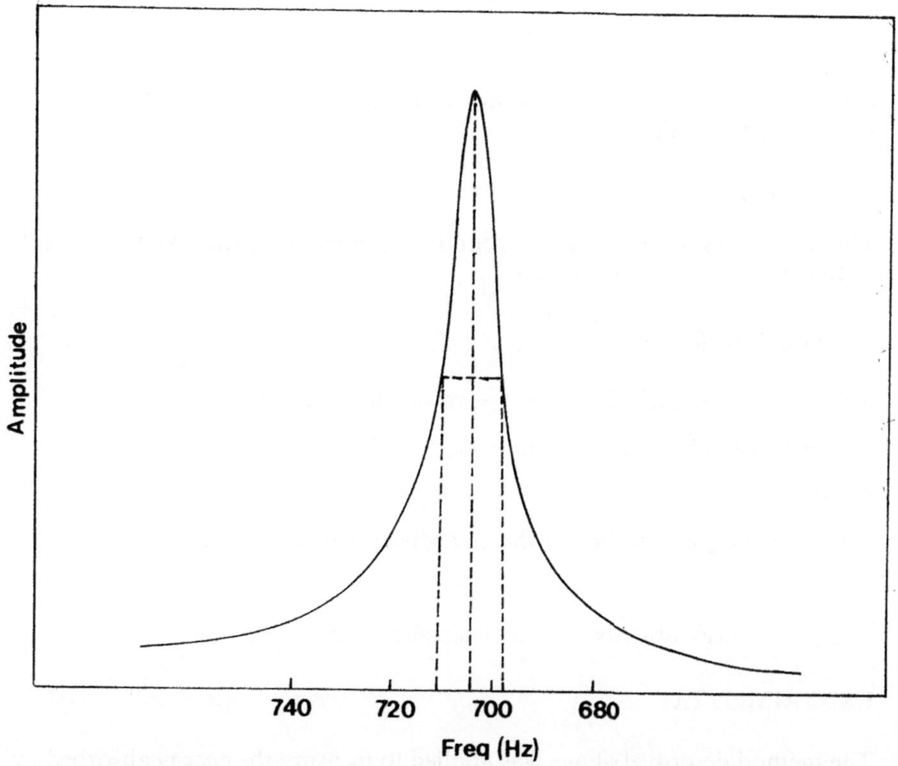

Figure 7.3 Typical energy-loss curve recorded by an acoustic spectrometer

Figure 7.4 Typical experimental arrangement for testing forearm skin, showing transducer-arm placement on skin surface

is a measure of the energy transmitted through skin. The testing of forearm skin is shown in Figure 7.4.

Calculations

The energy loss was determined from the resonance frequency and half-width of the resonance frequency curve[2]:

$$\text{energy loss, } Q^{-1} = \frac{(f_2 - f_1)}{3^{1/2} f}$$

where $f_2 - f_1$ = half width and f = resonance frequency.

The energy absorbed in skin = $(Q_S^{-1} - Q_A^{-1})$,

where

Q_S^{-1} = energy absorbed in the transducer arm and skin,

and

Q_A^{-1} = energy absorbed in the transducer arm in air.

EXPERIMENTAL

The method described above was applied to measure the energy absorbed by skin under various controlled conditions.

The effect of moisture build-up on the skin

To study the effect of moisture build-up on the skin, one-inch diameter areas on the forearms of a 30-year-old man were selected. At least five measurements of energy absorbed by the skin in the selected area were recorded. The average of these five readings was noted as the average energy absorbed by the skin in normal conditions. These skin areas were then wrapped with Saran-Wrap (Dow Chemical Company). The subject was then allowed to enter the room, which was at 49 °C, for 15 minutes. Testing was performed 2 hours after the arms were wrapped with Saran-Wrap. A stop-watch was started, and 3 minutes làter air at room temperature was blown directly onto each skin area and measurements taken as a function of time.

The effect of various treatments of the skin

The forearm skin of two males and one female was tested. The energy absorbed by the forearm skin in normal conditions was recorded. The skin was then treated with cream. Excess cream was removed and energy absorbed by the skin recorded. The remaining cream was then removed by washing with soap, and energy absorbed by the skin was recorded again. The same procedure was then repeated, except that the cream was removed by using alcohol.

Energy absorbed by the skin at different ages

About 30 females, aged from 35 to 60 years, were selected for this test. Energy absorption measurements were made on the right and left sides of the face. At least five measurements were taken on each side of a 1 square inch skin area around the cheek. The faces were washed with soap 15 minutes before measurements. The subjects were divided into age groups of 5-year spans, and the energy absorbed by the skin was determined in each group. Calculations were made for the energy absorbed in each group for the left side of the face, the right side of the face, and the whole face.

Similar measurements were also carried out for male (aged 35–65 years) and female subjects on the hands.

RESULTS AND DISCUSSION

The results obtained in the forearms of a 30-year-old male subject are given in Table 7.1. The energy absorbed by skin as a function of time is shown in Figure 7.5 for the left arm and Figure 7.6 for the right arm. The energy absorbed by the skin due to moisture build-up was 127 % more on the left arm than the energy absorbed by normal skin. The energy absorbed by the right arm skin, after moisture build-up, was 145 % more than the energy absorbed by normal right forearm skin. Results from the study of energy absorbed by the skin with timc showed that as the moisture was evaporated from the skin by air blower energy absorption was reduced drastically. The skin-softening effect due to moisture build-up was decreased by 90 % in 11 minutes when air was blown on the left

Table 7.1 Energy absorbed by air-blown skin of arms of male subject

*Time (min)**	*Energy absorbed* × 10^4
	Left arm
3	4.65
4	4.72
5	3.40
6	3.10
8	2.35
11	2.22
Normal skin	2.05
	Right arm
3	5.15
4	4.82
6	3.80
7	3.25
8	2.75
13	2.35
Normal skin	2.10

* After start of experiment: air first blown at 3 minutes.

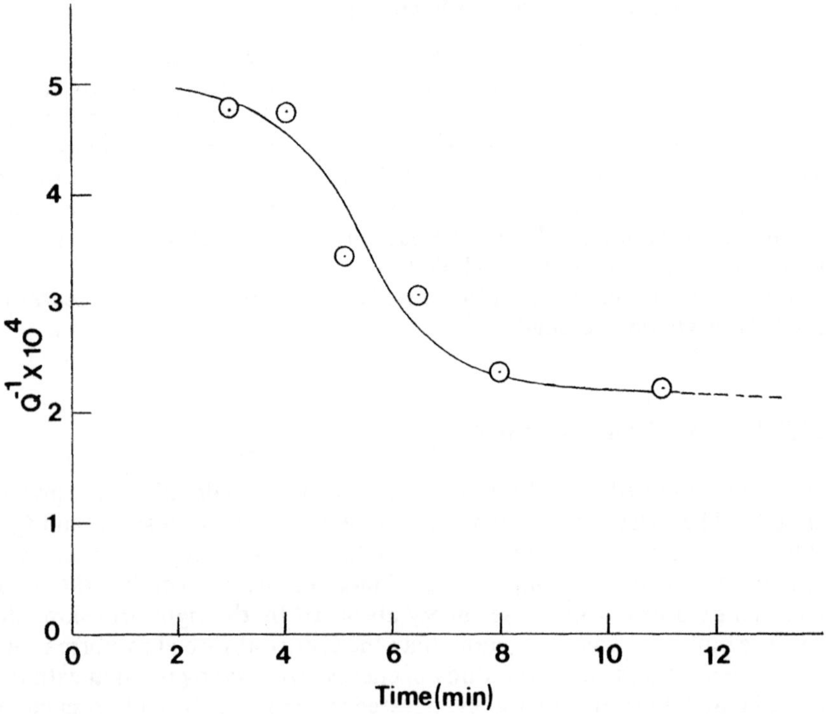

Figure 7.5 The effect of moisture build-up on energy absorbed (Q^{-1}) by left arm

Table 7.2 Effect of cream, soap and alcohol on energy absorbed ($\times 10^4$) by skin. Numbers in parentheses give percentage change from value in normal skin

Skin	*Subject*		
	Male	*Male*	*Female*
Normal	3.98	3.75	3.92
With lotion	4.88 (+22)	4.24 (+13)	4.68 (+19)
With lotion removed by:			
Soap	3.94 (−1)	3.70 (−1)	3.94 (0)
Alcohol	3.86 (−3)	3.55 (−5)	3.75 (−4)

forearm skin. A similar effect was observed on the right forearm in 13 minutes.

The effect of treating skin with cream, soap, and alcohol on the energy absorption in two male and one female subject is shown in Table 7.2. Two male and one female subject showed 18, 12 and 19% increases respectively in the energy absorption after skin treatment with cream. It was observed that after washing skin with soap, energy absorption was reduced to a normal level. A decrease of about 3–5% below normal values of energy absorption was observed when alcohol was used to remove the cream instead of soap.

The energy absorbed on the right side of the face, left side of the face and on the total face of female subjects was calculated. The average energy absorption on the face in various age groups was determined. These energy values are shown in Table 7.3. The energy absorbed by the facial skin was found to decrease with increase in age. The relationship was, however, non-linear.

Similar studies were also performed on the hands of both male and female subjects. The results obtained are shown in Figure 7.7. It was found that energy absorbed by the hands of male subjects decreased more rapidly with increase in age compared with those of female subjects.

Table 7.3 Average energy absorption ($\times 10^4$) by faces of female subjects according to age group

Age group (yr)	*Face Left side*	*Face Right side*	*Total face*
35–40	1.91	2.01	1.96
41–45	1.72	1.65	1.68
46–50	1.60	1.50	1.55
51–55	1.71	1.38	1.55
56–60	1.37	1.28	1.33

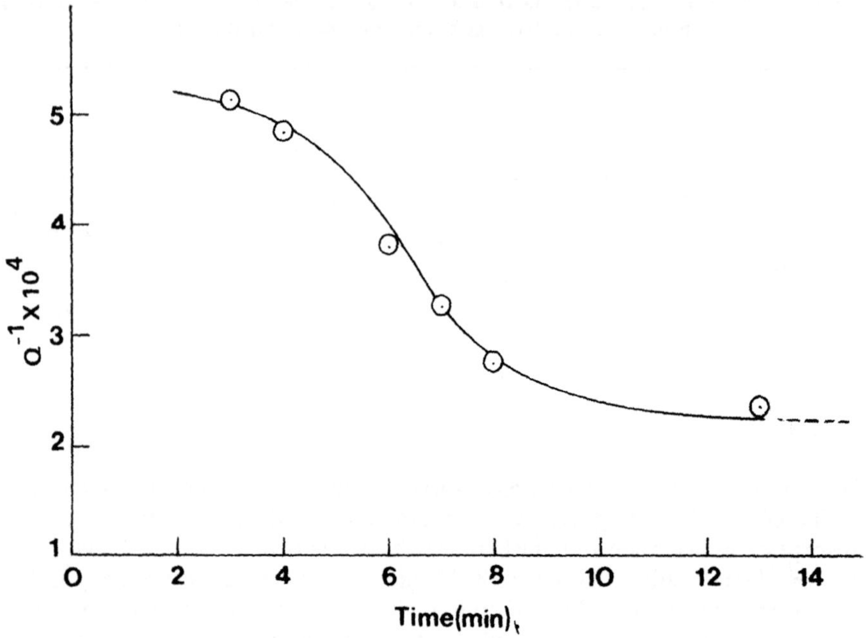

Figure 7.6 The effect of moisture build-up on energy absorbed (Q^{-1}) by right arm of one volunteer

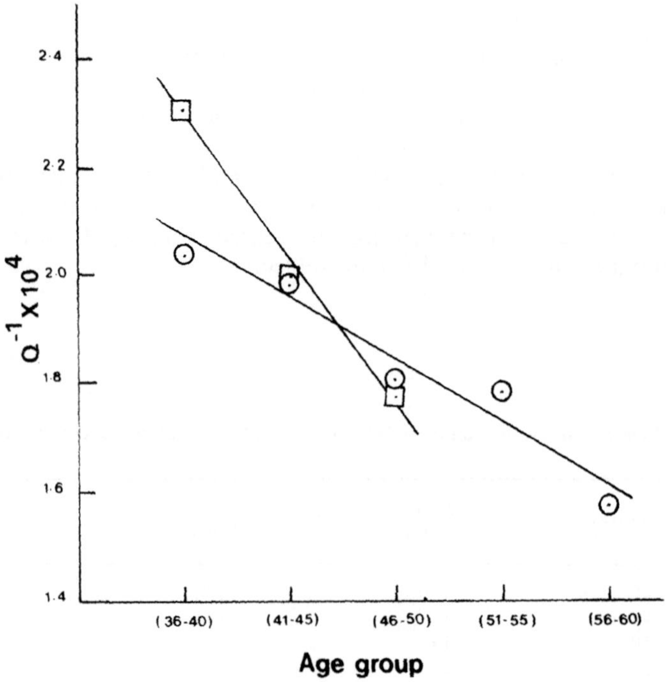

Figure 7.7 Effect of age on the energy absorbed (Q^{-1}) by the hand skin of male and female subjects

CONCLUSIONS

A cantilever beam resonance frequency technique can be applied to the measurement of the softness of the skin *in vivo* by determining the mechanical energy absorbed by the skin tissue.

The softness of the skin as determined by this technique was dramatically increased by hydration from sweating under occlusion and increased somewhat less by an emollient cream. The outer stratum corneum of the skin is thus the predominant material examined with this technique.

A study of subjects in different age groups indicated decreased softness with increasing age. Whether this was a result of softer 'material' or thinner stratum corneum is not clear.

References

1. Ide, J. M. (1935). Some dynamic methods of determination of Young's modulus. *Rev. Sci. Instruments*, **6**, 10
2. Spinner, S. and Tefft, W. E. (1961). A method for determining mechanical resonance frequencies and for calculating elastic moduli from these frequencies. *Proc. ASTM*, **61**, 1221
3. Muller, R. H. (1963). Acoustical spectrometry. *Analyt. Chem.*, **35**, 115A
4. Fitzgerald, J. V. (1963). Acoustic spectrometry of adhesives. *Adhesive Age*, November, 36
5. Hargens, C. W. (1964). Sonic testing. *Ind. Res.*, February, 67

8
The mechanism of stratum corneum plasticization with water

M. TAKAHASHI, K. KAWASAKI, M. TANAKA, S. OHTA and Y. TSUDA

SUMMARY

Water sorption and dielectric properties of stratum corneum (human and guinea-pig) were measured *in vitro* in relation to relative humidity (RH). The dynamic viscoelastic properties of stratum corneum were also measured to investigate its plasticization mechanism.

The pliability of human stratum corneum increased rapidly above 60 % RH. The water sorbed in human stratum corneum was found to be 'bound water' in the range 0–60 % RH but to include 'free water' above 60 % RH. In guinea-pig stratum corneum the same change occurred at 20 % RH. Stratum corneum was plasticized with not only water but also alcohols and hydrogen-bond breakers such as lithium bromide and urea.

It is concluded that the plasticization of stratum corneum by water is caused by 'free water' which breaks hydrogen bonds in keratin.

INTRODUCTION

The importance of water in plasticizing stratum corneum is well known, and it is also known that the extensibility of stratum corneum depends on its content of water and hygroscopic substances[1–6].

The mechanical properties of stratum corneum and their relationship to water content have been extensively investigated, mainly because of their possible relationship to skin condition. Dry skin is thought to lack the ability to retain water, thereby reducing the pliability of stratum corneum.

However, the mechanism of the plasticization has been studied only rarely, and the way that water causes plasticization is not known.

The purpose of the present work was to study the state of water molecules sorbed in the stratum corneum and to clarify the plasticization mechanism.

MATERIALS AND METHODS

Stratum corneum was removed from skin obtained from human abdomen and guinea pig back by the heat-trypsinization method described by Kligman and Christophers[7].

The thickness of the stratum corneum was examined in transverse section with light microscopy. The stratum corneum was embedded in paraffin wax through dehydration and fixation and 5 μm sections were cut, deparaffinized and stained with methylene blue. Photographs were taken at 400× magnification.

The mechanical properties of stratum corneum were measured by a dynamic method using a dynamic piezoelectrical analyser (Toyo Seiki Ltd., Japan). The strips of stratum corneum (20 × 5 mm) were excised with a razor. Because of its orderly appearance, the stratum corneum was expected to be mechanically anisotropic. However, the mechanical parameters were largely unaffected by the direction of cutting.

The sorption isotherm of water vapour on the stratum corneum was examined with a Cahn Electrobalance. The dry weight of a sample of 10 mm diameter was measured in an evacuated system, then water vapour was introduced into the system as the vapour pressure approached the predetermined value. After equilibrium was reached, the moistened weight was measured.

The dependence of dielectric properties of stratum corneum on relative humidity (RH) was determined with a dynamic piezoelectrical analyser. An area of silver measuring 5 × 5 mm was evaporated on to the centre of both surfaces of a sample measuring 20 × 10 mm to serve as electrodes. The measurement of the capacitance between the electrodes gave the dielectric constants and losses at several frequencies.

RESULTS

Figure 8.1a shows the viscoelastic moduli in human stratum corneum in relation to RH. At low RH range, the dynamic elastic modulus (E') decreased gradually with increasing RH, and at about 60% RH it started to decrease abruptly. On the other hand, the loss modulus (E'') had a maximum at a particular RH. The result in the case of guinea pig stratum corneum is shown in Figure 8.1b. These moduli behaved in an almost similar manner to those in human stratum corneum except that the gradient of the curves drastically changed at 20% RH.

The RH dependence of dielectric properties in human stratum corneum is shown in Figure 8.2. Both dielectric constant (ε') and loss (ε'') in human stratum corneum scarcely changed with increasing RH up to 50% RH then rose exponentially. The results in guinea pig stratum corneum (Figure 8.3) were similar to those obtained in human samples except that these variables increased abruptly from about 20% RH with increasing RH.

These results suggest that the water molecules sorbed below 50% RH in human stratum corneum (20% in guinea pig samples) are strongly bound at

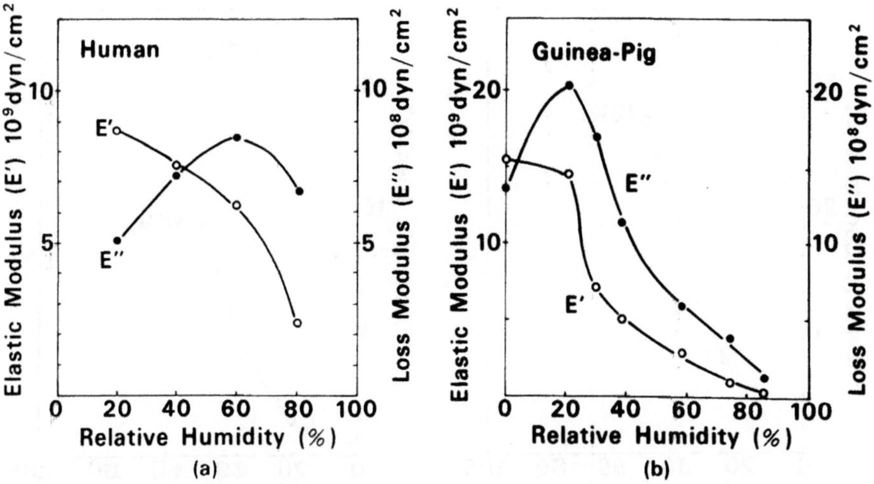

Figure 8.1 Dynamic elastic modulus (E') and loss modulus (E'') for human (a) and guinea pig (b) stratum corneum versus relative humidity at 25 °C

the polar sites in the stratum corneum and have little freedom of movement, since there is not much change in the dielectric properties in these ranges of RH. On the other hand, at RH above these values the water molecules seem to be loosely bound.

Both water vapour sorption isotherms for human and guinea pig stratum corneum were sigmoidal, but guinea pig stratum corneum took up more water than human at higher RH. The clustering theory proposed by Zimm[8,9] was

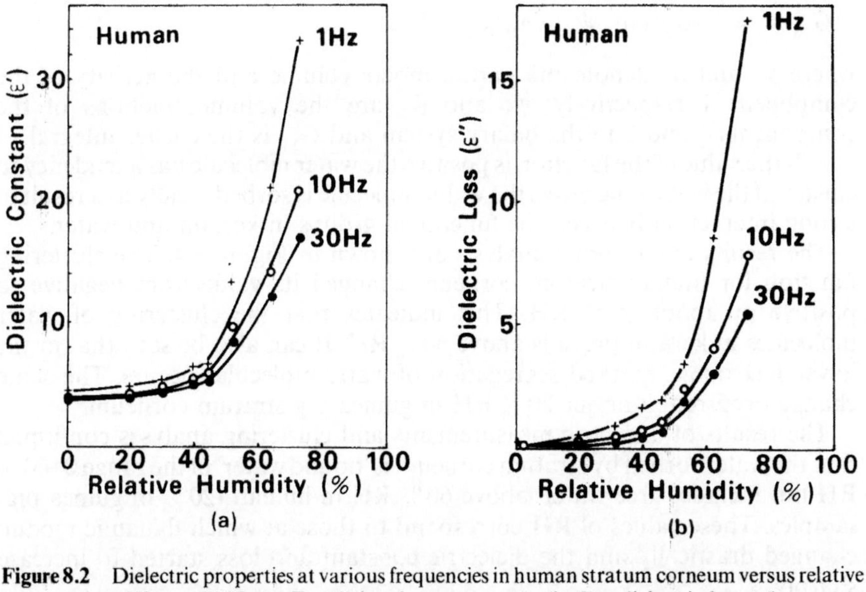

Figure 8.2 Dielectric properties at various frequencies in human stratum corneum versus relative humidity at 25 °C. (a) For dielectric constant (ε'). (b) For dielectric loss (ε'')

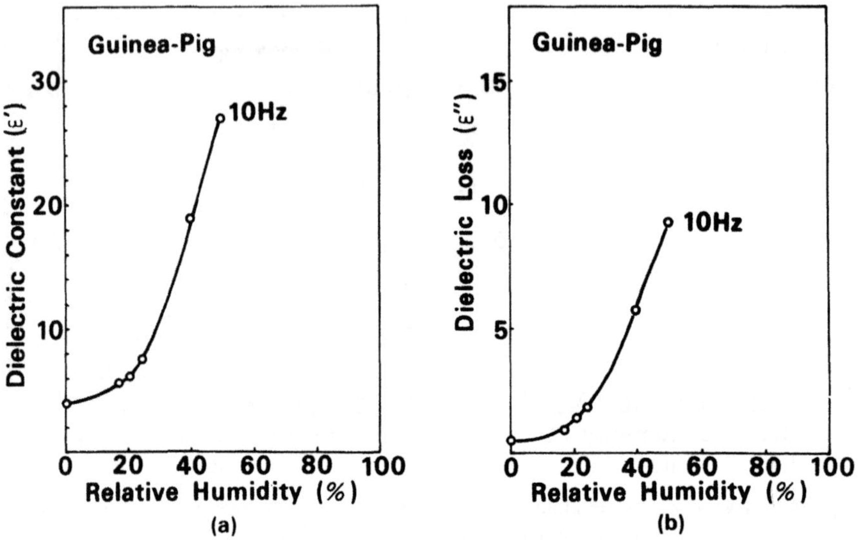

Figure 8.3 Dielectric properties at 10 Hz in guinea pig stratum corneum versus relative humidity at 25 °C. (a) For dielectric constant (ε'). (b) For dielectric loss (ε'')

applied to these results to examine the state of water sorbed in stratum corneum. This method was developed to determine the extent of clustering in any binary system at sorption equilibrium. The clustering function can be calculated according to equation:

$$G_{11}/V_1 = -\psi_2[\partial(a_1/\psi_1)/\partial a_1]_{P,T} - 1$$

where V_1 and a_1 denote the partial molar volume and the activity of the component 1 respectively, ψ_1 and ψ_2 are the volume fractions of the components 1 and 2 in the binary system and G_{11} is the cluster integral.

If the value of the function is positive the water molecule has a tendency to cluster; if the value is negative the water molecule is sorbed locally as a result of strong interaction between the functional groups in keratin and water.

The results of clustering analysis are shown in Figure 8.4. The clustering function for human stratum corneum changed its value from negative to positive at about 60% RH. This indicates that the clustering of water molecules in keratin prevails above 60% RH. It can also be seen that in the lower RH range, marked segregation of water molecules occurs. The same change occurred at about 20% RH in guinea pig stratum corneum.

The results of dielectric measurements and clustering analyses confirmed that the water sorbed by stratum corneum is 'bound water' in the range 0–60% RH but includes 'free water' above 60% RH in human (20% in guinea pig) samples. These values of RH correspond to those at which dynamic moduli changed drastically and the dielectric constant and loss started to increase sharply.

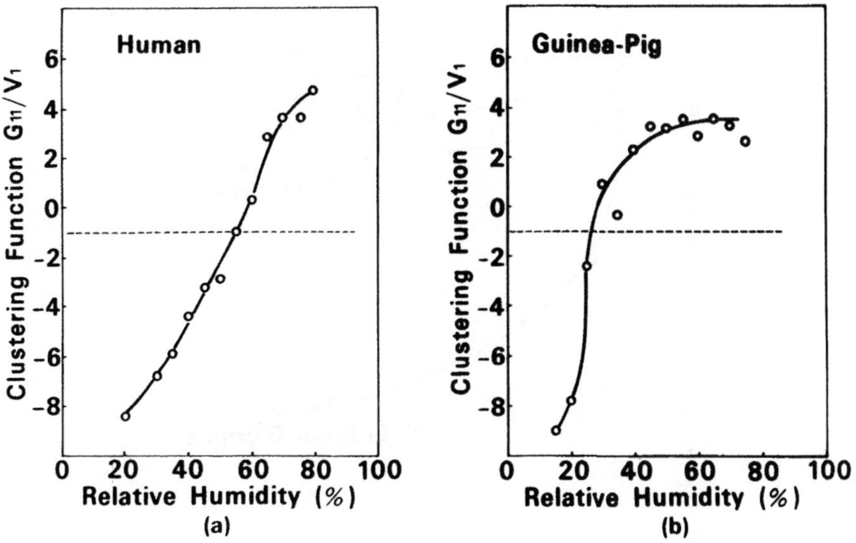

Figure 8.4 Clustering function for water vapour in human (a) and guinea pig (b) corneum at 25 °C

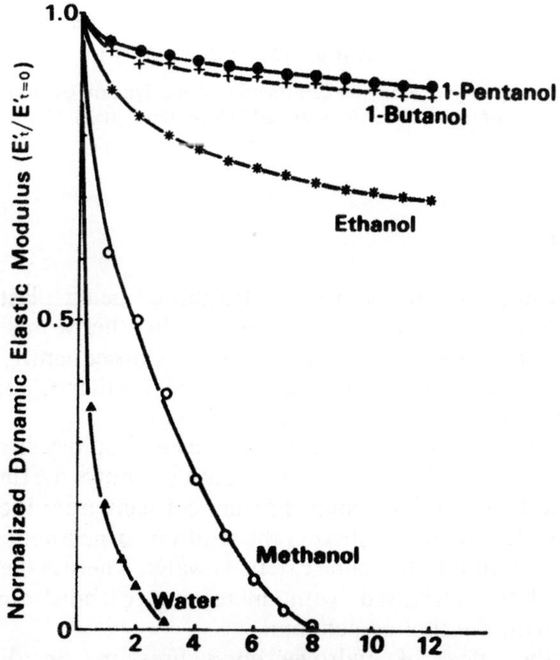

Figure 8.5 Time dependence of normalized dynamic elastic modulus in human stratum corneum after immersion in water and low-molecular-weight alcohols at 25 °C

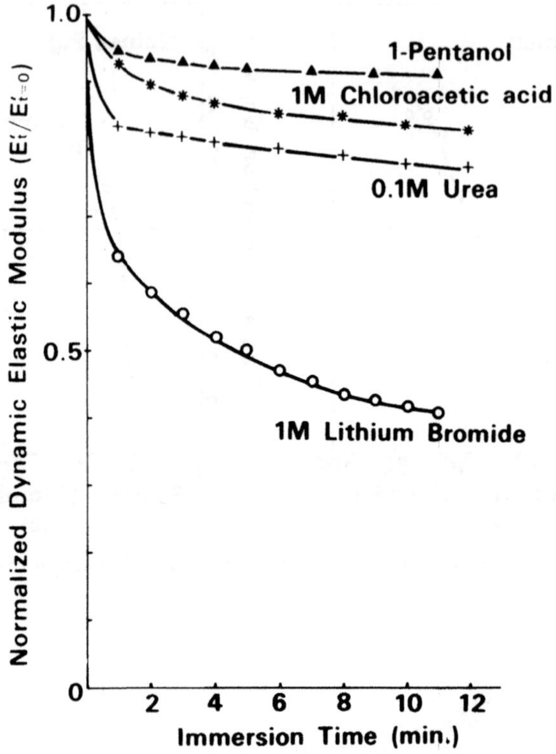

Figure 8.6 Effect of hydrogen-bond breakers on normalized dynamic elastic modulus in human stratum corneum immersed in l-pentanol at 25 °C

DISCUSSION

From these results, the mechanism of stratum corneum plasticization with water can be considered as follows. Free water – in other words loosely bound water – breaks hydrogen bonds in keratin. Consequently, the stratum corneum is plasticized. Therefore we investigated whether or not the hydrogen bonds affected the mechanical properties of keratin.

Firstly, the mechanical properties were measured on dry stratum corneum immersed in water and several alcohols. Figure 8.5 shows the time dependence of dynamic modulus for dry human stratum corneum after the immersion in alcohols and water. Water plasticized the stratum corneum most quickly and methanol plasticized it to the same extent as water, whereas the plasticization effect of alcohols decreased with increasing carbon number, almost disappearing with butanol or pentanol.

Secondly, the effect of hydrogen-bond breakers on the mechanical properties of stratum corneum was studied. Figure 8.6 shows the effect of chloroacetic acid[10], urea[11] and lithium bromide[12] (known hydrogen-bond breakers) on the dynamic modulus of human stratum corneum. Pentanol solutions of these hydrogen-bond breakers plasticized the stratum corneum remarkably quickly. This also suggests that the alteration of hydrogen bonds is closely related to the change of mechanical properties in stratum corneum.

At lower RH the keratin chains are immobilized with hydrogen bonds, hence the stratum corneum is highly elastic. However, when RH is raised to about 60 % or above for human samples, or to 20 % or above for guinea pig samples, the water in the stratum corneum starts to cluster and breaks hydrogen bonds in keratin. Consequently the interaction among keratin chains becomes weak, and slippage among them occurs when a deforming force is applied. Thus the loss modulus increases at this stage.

At higher RH most of the hydrogen bonds have already been broken, therefore both dynamic and loss modulus decrease as RH increases.

It is concluded that the plasticization of stratum corneum with water is caused by 'free water' which breaks hydrogen bonds in keratin.

References

1. Blank, I. H. (1952). Factors which influence the water content of the stratum corneum. *J. Invest. Dermatol.*, **18**, 433
2. Blank, I. H. (1953). Further observations on factors which influence the water content of the stratum corneum. *J. Invest. Dermatol.*, **21**, 259
3. Middleton, J. D. (1968). The mechanism of water binding in stratum corneum. *Br. J. Dermatol.*, **80**, 437
4. Park, A. C. and Baddiel, C. B. (1972). Rheology of stratum corneum. I. A molecular interpretation of the stress–strain curve. *J. Soc. Cosmetic Chem.*, **23**, 3
5. Park, A. C. and Baddiel, C. B. (1972). Rheology of stratum corneum. II. A physicochemical investigation of factors influencing the water content of stratum corneum. *J. Soc. Cosmetic Chem.*, **23**, 13
6. Van Duzee, B. F. (1978). The influence of water content, chemical treatment and temperature on the rheological properties of stratum corneum. *J. Invest. Dermatol.*, **71**, 140
7. Kligman, A. M. and Christophers, E. (1963). Preparation of isolated sheets of stratum corneum. *Arch. Dermatol.*, **88**, 702
8. Zimm, B. H. (1953). Simplified relation between thermodynamics and molecular distribution functions for a mixture. *J. Chem. Phys.*, **21**, 934
9. Zimm, B. H. and Lundberg, J. L. (1956). Sorption of vapors by high polymers. *J. Phys. Chem.*, **60**, 425
10. Read, R. E. (1949). Permanent waving of human hair: The cold process. *J. Soc. Cosmetic Chem.*, **1**, 109
11. Levy, M. and Magoulas, J. P. (1962). Effect of urea on hydrogen bonding in some dicarboxylic acids. *J. Am. Chem. Soc.*, **84**, 1345
12. Alexander, P. (1951). Changes in the physical properties of wool fibers produced by breaking hydrogen bonds with lithium bromide solutions. *Ann. N.Y. Acad. Sci.*, **53**, 653

9
Photosclerosis induced by long wave length ultraviolet light and psoralens

G. E. PIERARD, C. FRANCHIMONT, M. de la BRASSINNE and Ch. M. LAPIERE

SUMMARY

In highly photosensitive patients, PUVA therapy can injure the superficial microvasculature, which results in oedema, fibrosis of the dermis and coating of capillary venules by an amorphous material.

The functional importance of this pathological process has been evaluated by measuring *in situ* physical properties of the skin. We have measured the extensibility and the biological elasticity of the skin by exerting a vertical traction of 1 N on 11.8 cm^2 of skin surface. The dermoepidermal adhesion and vascular resistance were estimated from 'minimal blistering time' and 'minimal purpura time' respectively by applying a suction of 66.7×10^3 Pa.

Photosclerosis developed in 11 patients treated for 3 years with PUVA. During the phase of oedema, extensibility and biological elasticity were decreased, corresponding to a high viscosity. Blistering time and purpura time were not significantly modified. When sclerosis developed, extensibility and blistering time were decreased while biological elasticity and purpura time were increased.

The presence of abnormal deposits in the fibrotic dermis explains the increased elastic stiffness. The increased resistance to bleeding results from the reduced mobility of the papillary dermis and from the coating sheaths around vessels. The increased susceptibility to blistering results from altered basal epidermal cells and fibrosis of the dermis.

INTRODUCTION

Oral photochemotherapy with PUVA[1,2] modifies the structure and function of the superficial dermis[3-12]. These effects, which are not identical with those of classical actinic damages, can induce photoatrophy or photosclerosis[6,7,9,12].

The aim of this study was to conduct a long-term evaluation of the development of photosclerosis induced by treatment with PUVA.

PATIENTS AND METHODS

Among 320 patients treated with PUVA[13] according to the schedule of the European Cooperative Clinical Trial[14], 68 underwent rheological tests of the skin. Biomechanical properties were evaluated by the ability of skin of the flexor forearm to withstand vertical force of extension[15,16]. A traction of 1 N was exerted for 20 s on 11.8 cm^2 of skin surface. Loading deformation (*LD*) was then recorded as well as the residual deformation (*RD*) remaining 20 s after releasing the mechanical stress. The skinfold thickness (*S*) was measured at the same site using Harpenden skinfold calipers. Extensibility of the skin (*SE*) is the ratio between *LD* and log *S*. Biological elasticity (*BE*), expressed as a percentage, is the time-dependent recoil to the initial position of the skin after deformation; it represents $100\,(LD - RD)LD^{-1}$. Extensibility of the skin and biological elasticity were expressed in rectangular co-ordinates or in polar co-ordinates, giving rise by the modulus R and the angle θ to other evaluations of rheological properties (Fig. 9.1).

Using an angiosterrometre, we estimated the dermoepidermal adhesion and the vascular resistance by the minimal blistering time (*BT*) and by the minimal purpura time (*PT*) respectively, induced by a suction of 66.7×10^3 Pa. Normal values were previously determined[17].

Skin biopsy samples were examined by optical microscopy after staining with haematoxylin–eosin–safran, PAS–haematoxylin–safran, orcein–Giemsa and Foot silver impregnation.

In each patient, several determinations of the rheological properties were made during a 3 year follow-up. The results were considered in relationship to many parameters: age and sex of the patient, hydroxyproline and proline urinary excretion, duration of the PUVA treatment, number of exposures to

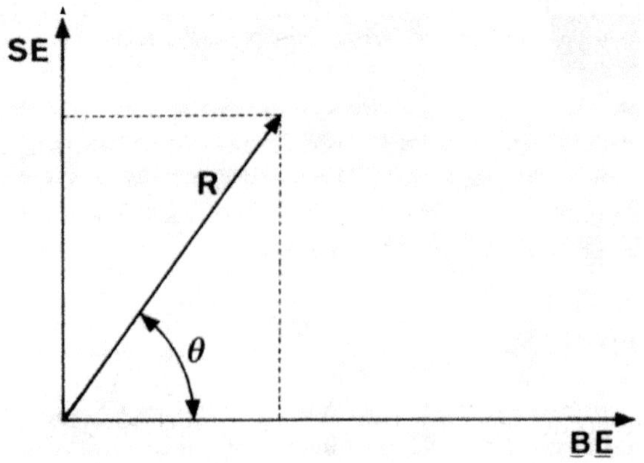

Figure 9.1 Principle of the graphic representation of the rheological properties of skin. For key see text

UVA, total dose of UVA, mean dose of UVA received in one exposure and in one week, number of exposures to UVA and total dose delivered during the past month, mean dose received in one exposure during the past month, minimal erythematous dose at 72 and 120 hours and the photosensitivity–pigmentation index as used in the European Cooperative Clinical Trial[14].

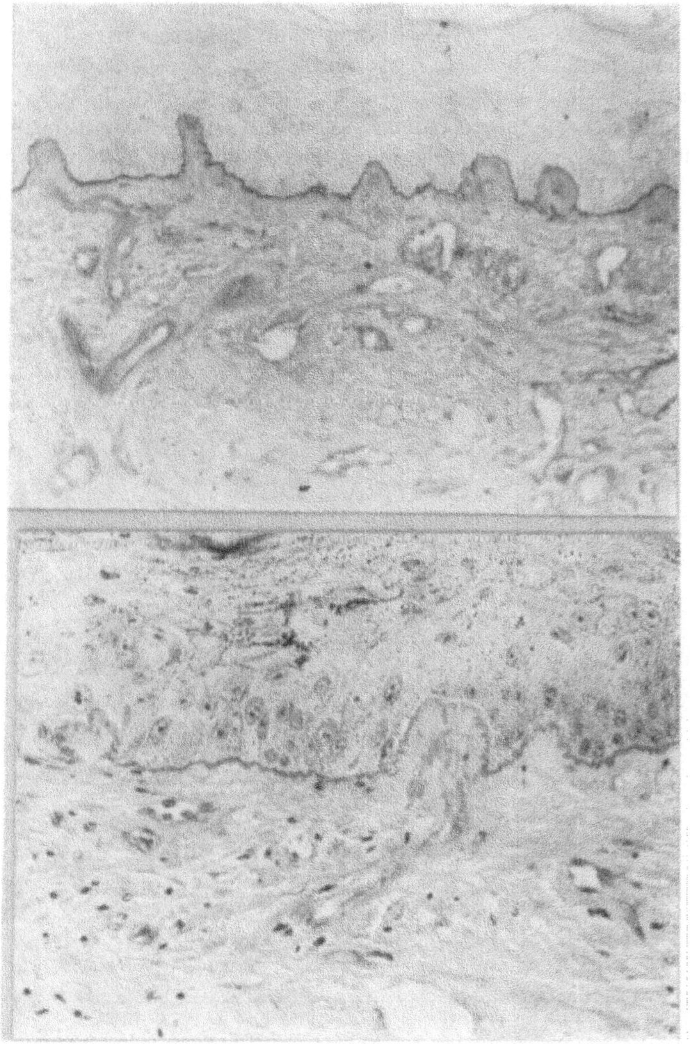

Figure 9.2 Fibrosis of the superficial dermis and coating sheaths around vessels. Dermal papillae are abnormal, dome-shaped and scalloped, containing an amorphous eosinophilic material (stained with PAS with diastase)

RESULTS

As observed by optical microscopy, PUVA therapy can injure the superficial microvasculature and induce oedema; fibrosis of the dermis may then also occur, with a coating of capillary venules with an amorphous material (Figure 9.2).

The functional importance of this pathological process can be substantiated by the rheological measurements. Photosclerosis was recognized in 11 patients. Photosensitive young adults seem to be more susceptible to developing photosclerosis. During the phase of oedema, extensibility and elasticity were decreased, corresponding to an increased viscosity. Blistering time and purpura time were not significantly modified. When sclerosis developed, extensibility and blistering time were decreased while elasticity and purpura time were increased (Figures 9.3 and 9.4). Blistering times were less than 1 hour in 70% of normal subjects but in all patients receiving PUVA

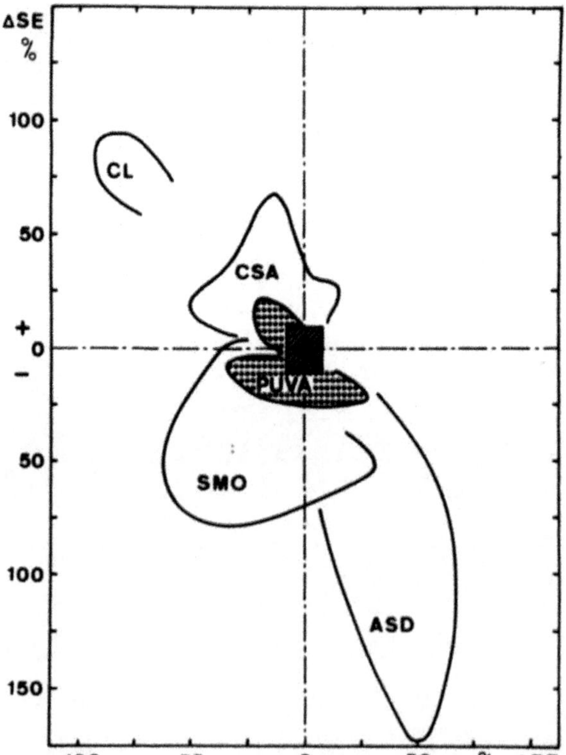

Figure 9.3 Rheological properties of skin during PUVA therapy (▦) are different from normal conditions (▨). For comparison, extensibility of skin (SE) and biological elasticity (BE) are shown for pathological conditions such as cutis laxa (CL), scleromyxoedema (SMO), acroscleroderma (ASD), and in corticosteroid-induced atrophy (CSA)

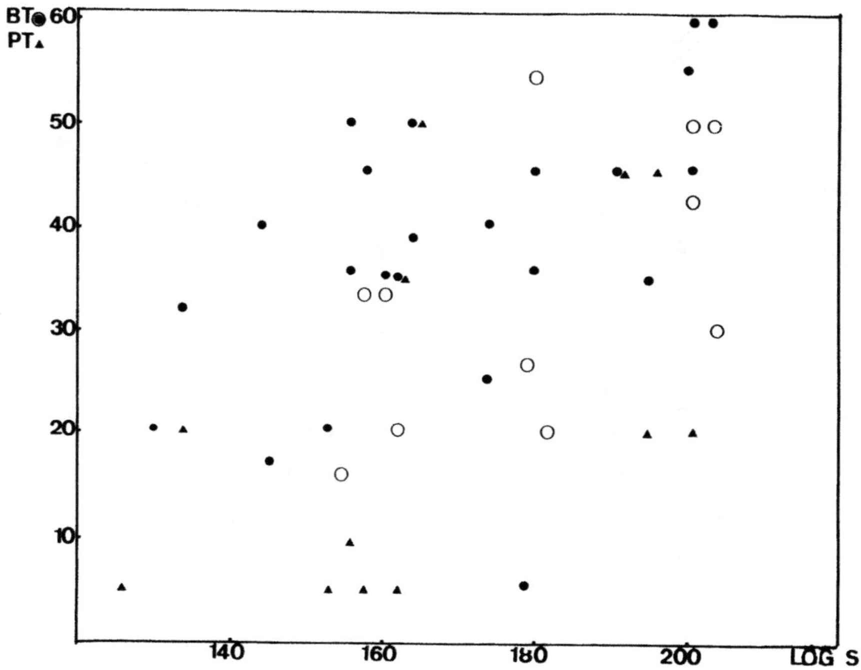

Figure 9.4 Blistering times (circles) and purpura times (triangles) occurring within one hour in normal subjects (black symbols) and in patients receiving PUVA therapy (open symbols) compared with logarithm of skinfold thickness (log S)

treatment. Purpura times were less than 1 hour in 83% of normal subjects but over 1 hour in all patients receiving PUVA treatment. Blistering times and the logarithm of skinfold thickness were linearly related in both groups.

When the total dose of UVA did not exceed 150 J/cm^2, there were linear relationships ($r = 0.9$; $p < 0.001$) between the logarithms of R and the total dose of UVA received, and between the logarithms of R and the mean dose of UVA received in one exposure or in one week. When the total dose of UVA exceeded 150 J/cm^2, R was constant, but had different values in different patients (Figure 9.5). Variations in the angle θ during treatment corresponded to modifications in elasticity and to transient phases of oedema occurring after each exposure to PUVA. Suppression of PUVA therapy did not lead to the recovery of normal values of the rheological parameters.

During the phase of oedema, mean urinary hydroxyproline excretion ($\pm$SE)/24 h (41.2 ± 6 mg) was significantly increased compared with the normal range (7–45 mg) in adults[14,15]. During the phase of established sclerosis, values were normal (24.7 ± 12.3). The ratio between proline and hydroxyproline excretion was normal in all stages of development of photosclerosis (2.6 ± 0.4).

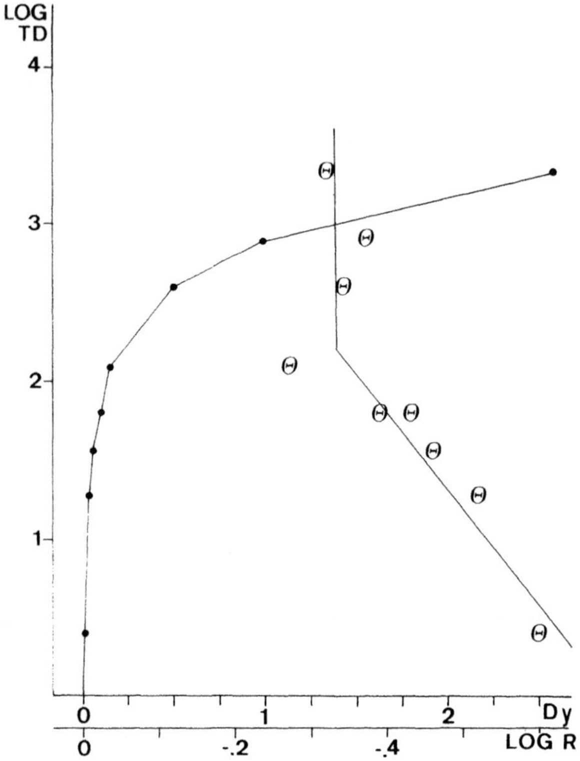

Figure 9.5 Example of relationship between the logarithm of the total dose in J/cm^2 (log TD) administered during a period exceeding 2 years (Dy; ●—●) and the logarithm of modulus *R* (Θ — Θ)

DISCUSSION

Potential risks to the dermis of long-term photochemotherapy with PUVA are expected to be in the spectrum of premature ageing resembling those of actinic damages[4,14]. These effects have not been recognized so far in man, but other side effects are induced[5–12] such as photoatrophy and photosclerosis[6,7,9,12]

Photosclerosis develops mostly early on in the treatment of highly photosensitive young adults requiring a protracted dose of UVA and repeated exposures. It is first associated with an increased hydroxyproline excretion, and oedema in the papillary dermis. This can be prevented by applying a sunscreen[18]. Long-standing oedema is followed by deposit of a homogeneous eosinophilic material at the dermoepidermal junction, within unusual scalloped papillae and around venules[5,8,9,11,12,19]. These abnormal deposits are probably responsible for the increased elastic stiffness of the adventitial dermis, decreased extensibility and variable elasticity.

The increased resistance to bleeding results from the reduced mobility of the skin and from the coating sheaths around vessels. The increased susceptibility to blistering, observed clinically[20,21] and induced experimentally in this study, reflects a decreased dermoepidermal adherence which could be the result of altered basal cells and of fibrosis of the dermis. Such remodellings of the connective tissue are stable as a scar, and the abnormal rheological properties are not normalized even if PUVA treatment is stopped. The exact nature of the deposits in the dermis remains unknown. Colloid amyloid bodies have been observed[10] but are not consistent with our observations. The coating sheaths present around vessels and fibrosis of the superficial dermis could be the combined result of the effect of UVA on microvasculature[22–24], and on collagen fibres[25–30].

References

1. Parrish, J. A., Fitzpatrick, T. B., Tanenbaum, L. and Pathak, M. A. (1974). Photochemotherapy of psoriasis with oral methoxalen and long-wave ultraviolet light. *N. Eng. J. Med.*, **291**, 1207
2. Wolff, N., Fitzpatrick, T. B., Parrish, J. A., Gschnait, F., Gilchrest, B., Honigsmann, H., Pathak, M. A. and Tanenbaum, L. (1976). Photochemotherapy of psoriasis with oral 8-methoxypsoralen. *Arch. Derm.*, **112**, 943
3. Wolff, K., Gschnait, F., Honigsmann, H., Konrad, K., Stingl,, G., Wolff-Schreiner, E. and Fritsch, P. (1977). Oral photochemotherapy: results, follow up, and pathology. In *Psoriasis*, pp. 300–309. Farber, E. M. and Cox, A. J. (eds.) (New York: Yorke Medical Books)
4. Braun-Falco, O., Hofmann, C. and Plewig, G. (1977). Photochemotherapy of psoriasis: clinical and histological findings. In *Psoriasis*, pp. 00–00. Farber, E. M. and Cox, A. J. (eds.) (New York: Yorke Medical Books)
5. Kumakiri, M., Hashimoto, K. and Willis, I. (1977). Biological changes due to long-wave ultraviolet irradiation on human skin: ultrastructural study. *J. Inves. Dermatol.*, **69**, 392
6. Pierard, G. E., de la Brassinne, M. and Lapiere, Ch. M. (1977). Effects of long term photochemotherapy on the dermis. *J. Invest. Dermatol.*, **68**, 249
7. Pierard, G. E. (1978). Iatrogenic alterations on the biochemical properties of human skin. *Br. J. Dermatol.*, **98**, 113
8. Hashimoto, K., Kohda, H., Kumakiri, M., Blender, S. and Willis, I. (1978). Psoralen-UVA-treated psoriatic lesions. *Arch. Dermatol.*, **114**, 711
9. Pierard, G. E. and Ackerman, A. B. (1979). Histopathology of remodelling induced by PUVA in the superficial dermis. *Br. J. Dermatol.*, **100**, 251
10. Hashimoto, K. and Kumakiri, M. (1979). Colloid amyloid bodies in PUVA-treated human psoriatic patients. *J. Invest. Dermatol.*, **2**, 70
11. Worm, A. M. (1980). Microvascular leakage of plasma proteins after short-term PUVA treatment. *J. Invest. Dermatol.*, **74**, 158
12. Niemi, K. M. and Karneva, L. (1980). Electron microscopic observations on photosclerosis of papillary dermis induced by long term PUVA therapy. Abstract 7th European meeting of the Society for Cutaneous Ultrastructure Research, Vienna
13. de la Brassinne, M. (1979). La photochimiothérapie orale en dermatologie. *Méd. Hyg.*, **37**, 2664
14. Wolff, K., Gschnait, F., Honigsmann, H., Konrad, K., Parrish, J. A. and Fitzpatrick, T. (1977). Phototesting and dosimetry for photochemotherapy. *Br. J. Dermatol.*, **96**, 1
15. Piérard, G. E. (1976). Microanatomie et propriétés fonctionnelles du tissu conjonctif de la peau. In *Corticothérapie locale*, Lapière. Ch. M. and Piérard (eds.) pp. 135–148. (Brussels: Brocades)
16. Piérard, G. E. and Lapière, Ch. M. (1977). Physiopathological variations in the mechanical properties of skin. *Arch. Dermatol. Res.*, **260**, 231

17. Franchimont, C. and Piérard, G. E. (1978). Dermoepidermal adherence and stereology of the microarchitecture of the basement membrane zone. In *6th meeting of the European Federation of Connective Tissue Clubs* p. 202
18. Piérard, G. E., Cryns, E. and Franchimont, C. (1978). Perspectives d'évaluation par tonométrie de l'effet de cosmétiques sur les fonctions mécaniques de la peau. *Rev. Inst. Pasteur Lyon,* **11**, 219
19. Piérard, G. E. and de la Brassinne, M. (1979). Dynamics of histopathological changes and of cellular proliferation in psoriasis vulgaris treated by photochemotherapy. Dermatopathology symposium, Amsterdam.
20. Thomsen, K. and Schmidt,, H. (1976). PUVA-induced bullous pemphigoid. *Br. J. Dermatol.,* **95**, 568
21. Robinson, J. K., Baughman, R. and Provost, T. T. (1978). Bullous pemphigoid induced by PUVA therapy. *Br. J. Dermatol.,* **99**, 709
22. Miescher, G. (1957). Zur histologie der lichtbedingten reaktionen. *Dermatologica,* **115**, 345
23. Ramsay, C. A. and Challoner, A. V. J. (1976). Vascular changes in human skin after ultraviolet irradiation. *Br. J. Dermatol.,* **94**, 487
24. Willis, I. and Cylus, L. (1977). UVA erythema in skin: is it a sunburn? *J. Invest. Dermatol.,* **68**, 128
25. Bottoms, E., Cater, C. W. and Shuster, S. (1966). Effect of ultra-violet irradiation on skin collagen. *Nature,* **211**, 97
26. Shuster, S. and Bottoms, E. (1967). The effects of UVR on skin collagen of intact living mice. *Nature,* **214**, 599
27. Raab, W. (1969). Wirkungen von Ultravioletbestrahlung auf dermales Kollagen des Menschen *in vitro. Arch. Klin. Exp. Derm.,* **234**, 36
28. Copper, D. R. (1971). Irradiation of soluble collagen. In Elden (ed.) *Biophysical properties of the skin,* 595–639. (New York: Wiley Interscience)
29. Lovell, W. W. (1973). Ultraviolet irradiation of dermal collagen *in vivo. Transact. St John's Hosp.,* **59**, 166
30. Shuster, S. (1976). The cause of striae distensae. In *Striae Distensae*, Moretti and Rebora (eds.) pp. 97–113. (Milan: Brocades)

10
Skin surface patterns and the directional mechanical properties of the dermis

J. FERGUSON and J. C. BARBENEL

INTRODUCTION

The surface of the skin is covered by intersecting grooves and ridges which produce characteristic skin surface patterns. It has been suggested that these folds provide a reserve of tissue, allowing the skin to stretch during normal function[1–3]. The extensibility of the skin depends on the direction of extension[4], and the skin surface patterns are apparently also directionally disposed[5]. The relationship between the structural directionality of the surface and the directional extensibility of the skin was investigated *in vivo*.

MATERIAL AND METHODS

Mechanical

The *in vivo* load–extension characteristics of the skin of human volunteers were measured using a uniaxial loading device described previously[6]. The apparatus has two load-sensing cantilevered arms, which are driven apart at a constant rate. The device was coupled to the skin by double-sided adhesive tape and two metal tabs 20 mm wide joined by a metal chain. The test site was 20 mm long, and because of the metal chain the extension rate of the metal tabs was 1.56 mm/s. The extension of the skin was measured from photographs of a grid printed on the skin. The skin thickness was not measured.

The skin was stretched in the directions of greatest and least extensibility and in an intermediate direction.

The directional nature of extensibility was investigated using a simpler hand-held suction device[7]. The device consisted of a circular plastic cylinder closed by a concave piston. The open end of the cylinder was placed on the skin surface and the pressure within the device reduced by c. 45 kPa. The skin was sucked into the tube, the shape of the deformed skin surface being limited by the geometry of the piston and cylinder. A ring was printed on the skin around the open end of the cylinder when equilibrium had been reached. When the skin was relaxed the printed ring became an oval. The extensibility and anisotropy was assessed from the difference between the oval and the circle.

Structural

Impressions of the skin surface were taken using a silicone rubber (Silflo, J. & S. Davis, London) and casts made using an epoxy resin. The technique is described in detail elsewhere[5].

The surfaces of the rubber impressions were examined with reflected light microscopy. The casts were used to obtain surface profile and roughness characteristics. A Rank–Taylor–Hobson Talysurf IV instrument was used to obtain the value of the average roughness (R_a), mean peak-to-valley height (R_{tm}), mean spacing between peaks (S_m) and slope of the profile (Δq) in each specimen. A scan length of 10 mm was used, and multiple parallel scans made to estimate both the mean value of the parameters and their variance.

RESULTS

The uniaxial loading device was used to investigate the response of the skin of the volar surface of the forearm. The load–extension curves all displayed the classical non-linear response of the soft connective tissues[8]. An initial complaint phase, during which a large extension was produced by applying a low load, was followed by a progressive stiffening with increased extension (Figure 10.1). The extension at which the stiffening of the skin occurred was dependent on the direction of stretch.

The differences in extensibility of the load–extension curves were characterised by calculating the extensions produced by a load intensity of

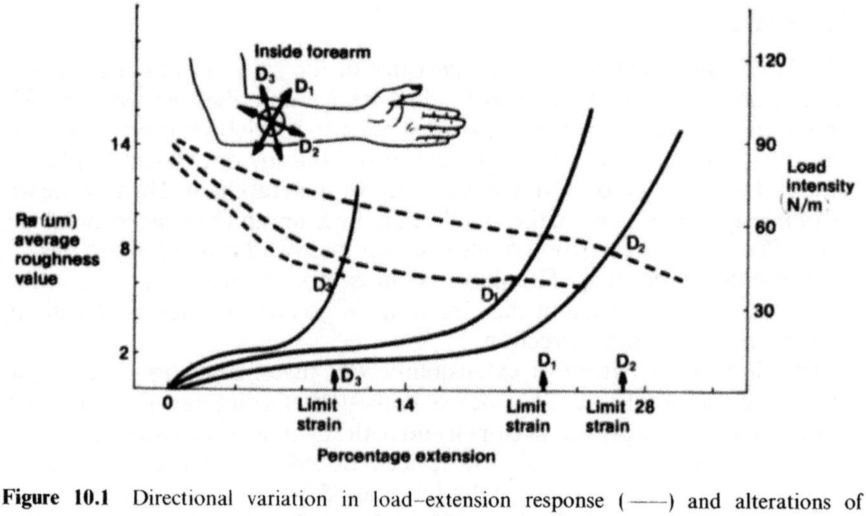

Figure 10.1 Directional variation in load–extension response (——) and alterations of roughness values with extension (---)

60 N/m. This parameter has been used before and the extension called the physiological limit strain[7].

Replicas were made of the skin surface before and during extension. For the unstrained skin there was significant relationship between the limit strain and the values of three of the roughness parameters. These are shown in Figure 10.2a, which displays the results obtained from the test directions in three test subjects.

Both R_{tm} and Δq values were directly related to the limit strains. The mean spacing between the major peaks (S_m) was inversely related to the limit strain.

The contour of the skin surface altered with extension of the skin, and as Figure 10.3 shows the amplitude of the surface undulations decreased to give

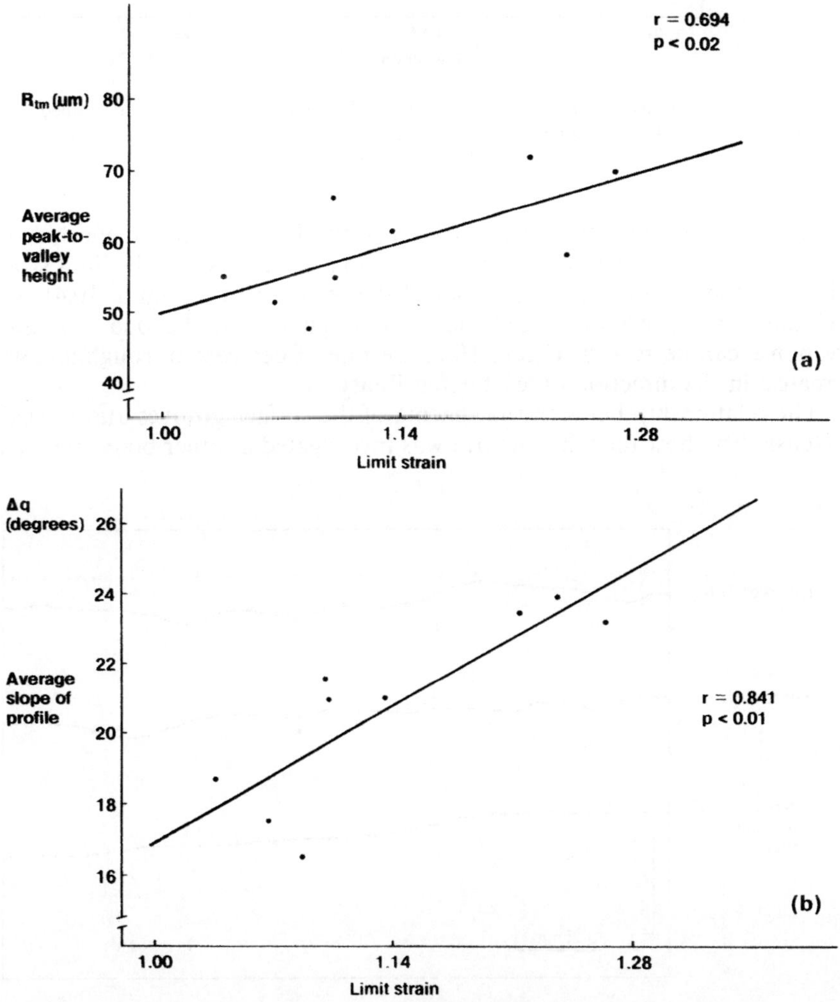

Figure 10.2

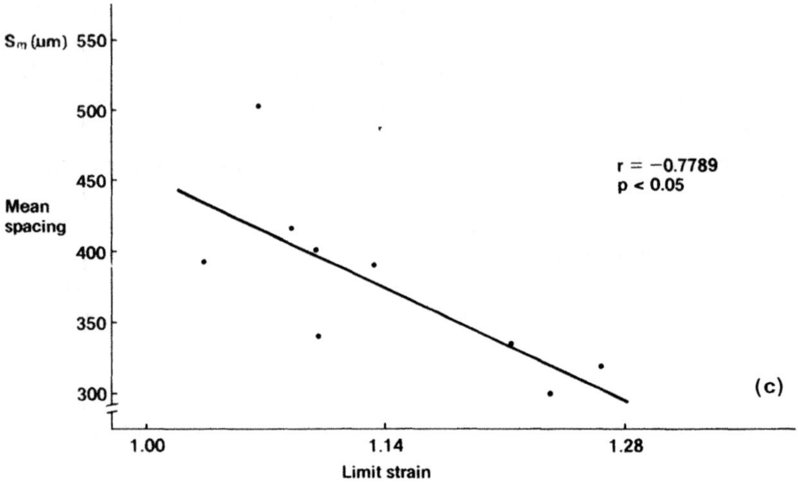

Figure 10.2 Relationship between limit strain and roughness values of unstrained skin. Roughness values: (a) average peak-to-valley height; (b) average slope of profile; and (c) mean spacing between major peaks

an almost complete flattening at the limit strain. The change was reflected as a progressive decrease in the value of the average roughness. At the limit strain the roughness value was about half of the resting value (Figure 10.4). The relationship between the rate of change of roughness and the load–extension response can be seen in Figure 10.1; the rate of decrease of roughness was greatest in the direction of least extensibility.

The relationship between the spacing of the surface grooves and the skin extensibility shown for the forearm was investigated at other body sites. The

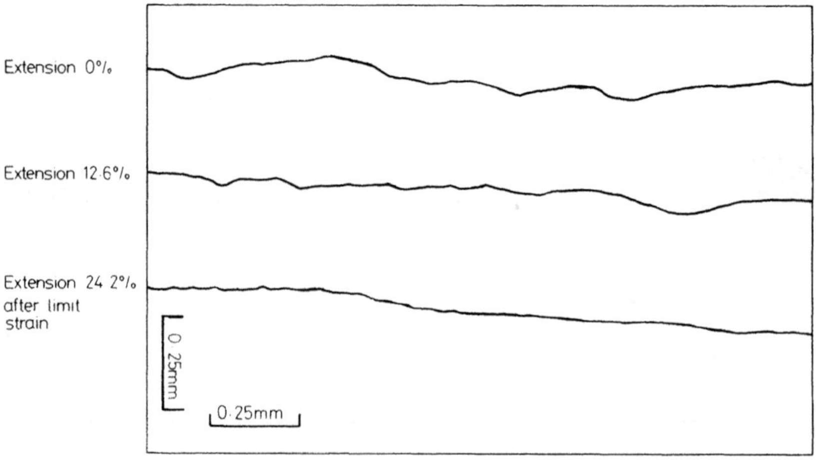

Figure 10.3 Surface profile at different extensions; profiles aligned in direction of loading

Table 10.1 Ranking of anisotropy of test sites

Grossly anisotropic ↓	Thigh
	Thorax
	Abdomen
	Forearm
	Foot
Relatively isotropic	Sternum

skin surface patterns at six body sites were examined in four subjects, and both the degree and anisotropy of the extensibility of the skin assessed using the suction device.

The sites displayed a range of anisotropy and the ranking is shown in Table 10.1. Microscopic examination of the rubber impressions of the test sites, taken in the unstrained position, showed that the major grooves making up the skin surface pattern were aligned perpendicular to the direction of maximum extensibility. The direction of the major grooves could best be displayed by rotating the impression relative to the source of illumination, which also showed the different sets of grooves in the pattern. Figure 10.5a shows the skin surface pattern obtained from the thigh, a grossly anisotropic site. The direction of minimum extensibility is shown by the double-headed arrow and corresponds to the direction in which the spacing of the major grooves is large. This relation between structural and mechanical directionality was a general finding in all the anisotropic sites. The dorsum of the foot was found to be

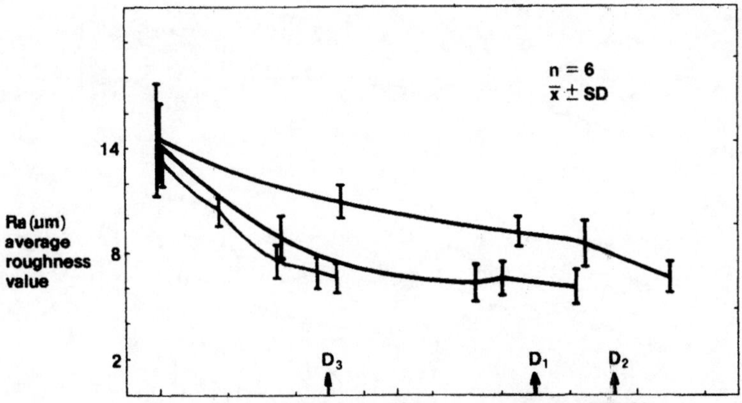

Figure 10.4 Alterations in values of average roughness with extension (results are means ± SD). Rest of key as in Figure 10.1

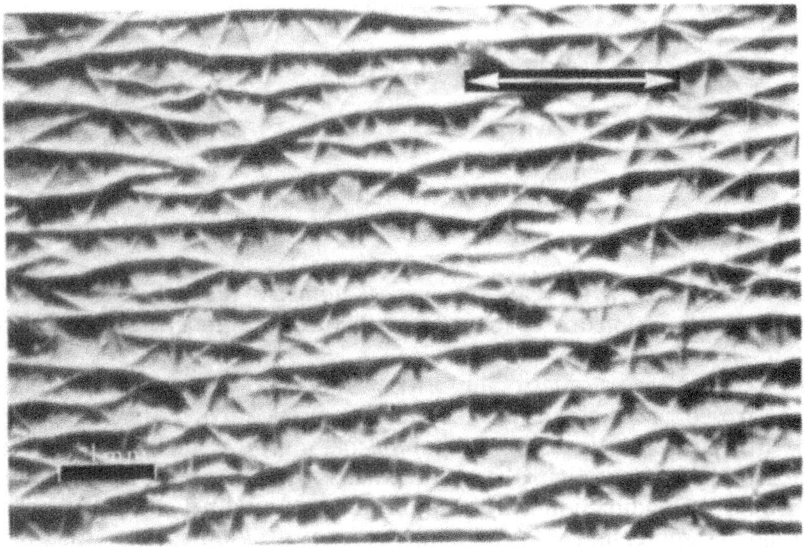

(a)

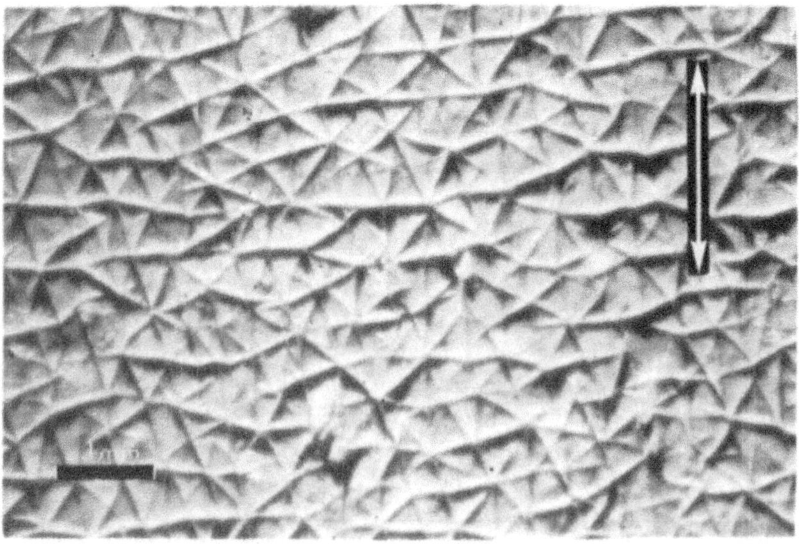

(a)

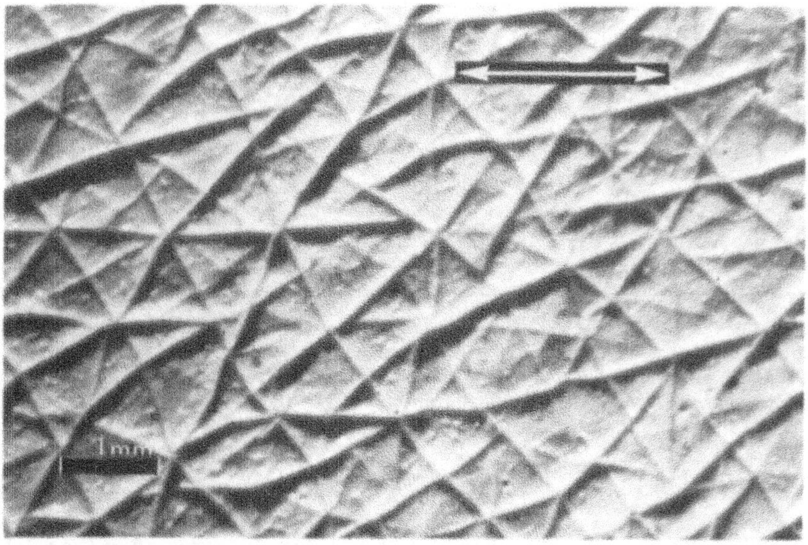

(b)

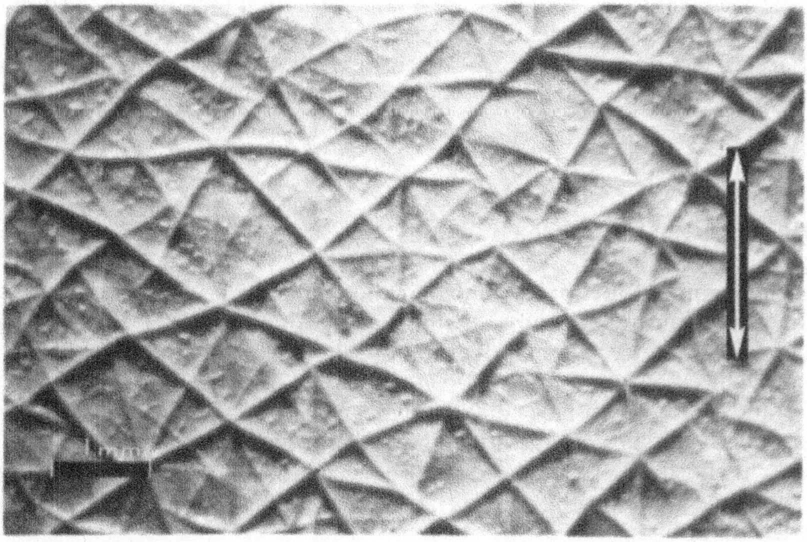

(b)

Figure 10.5 Silicone rubber impressions of surface patterns of unstrained skin of the thigh (a), and dorsum of the foot (b). The double-headed arrow indicates direction of least extensibility

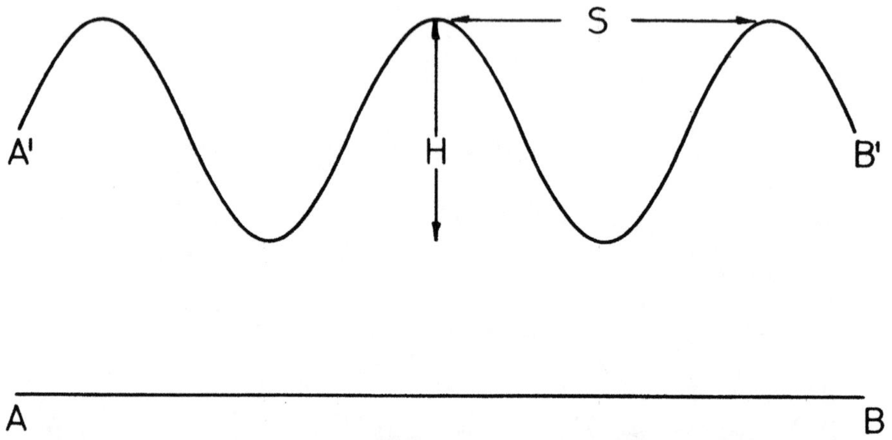

Figure 10.6 Structural uniformity in the dorsum of the foot (a rough surface); the actual profile length (A′ B′) was greater than the nominal scan length (A B). Increasing the peak-to-valley-height (H) and/or reducing the spacing (S) increased the difference between A′ B′ to A B

relatively isotropic, and as Figure 10.6 shows this was reflected in the relative structural uniformity of the site.

The extensibility of the skin of the thigh was greater than that of the dorsum of the foot. This is reflected in the greater spacing of the grooves in the skin surface pattern on the foot. This relationship between the skin-surface patterns and the extensibility of the skin was found in all the sites studied.

DISCUSSION AND CONCLUSIONS

During normal body movement the skin is stretched. *In vitro* studies have shown that during stretching of the skin the undulations of the epidermal surface and dermoepidermal junction are flattened before elongation of the epidermal cells occurs[9]. It has been suggested[1–3] that a similar effect occurs *in vivo* and that the surface folds provide a reserve of tissue, allowing the epidermis to stretch without stretching or disrupting the epidermal cells. If this is true there should be a relationship between the skin extensibility and the grooves and ridges of the skin surface pattern which produce the functional epidermal reserve.

The parameter used to describe the extensibility of the skin was the extension produced by a load intensity of 60 N/m. This is an arbitrary value which has been used because there is no reliable information on the skin extensions which occur during normal function. Nevertheless, the value has a functional importance, being a physiological limit strain at which blanching may occur, and represents an upper limit of functional extension.

Visual examination and measurement showed the roughness of the skin surface, and hence that the actual profile length was greater than the nominal scan length (Figure 10.6). It is this difference which is a measure of the epidermal tissue reserve. Three of the roughness parameters showed a significant relationship with the skin extensibility. Δq is the parameter most clearly related to ratio of profile length to scan length:

$$\frac{\text{actual profile length}}{\text{nominal scan length}} \simeq 1 + \tfrac{1}{2}(\Delta q)^2$$

The direct relationship between Δq and the skin extensibility supports the postulated existence of a functional epidermal reserve associated with the skin-surface folding and the postulation that the magnitude of this reserve is directly related to the magnitude of the local skin extensibility.

The roughness parameters derived from the mean peak-to-valley height (R_{tm}) and the mean spacing between peaks S_m were also related to the ratio of actual profile length to nominal scan length. An increase in the peak-to-valley height and/or a decrease in the mean spacing will increase this ratio. The relationship between the two parameters and the skin extensibility (Figure 10.2) was consistent with the conclusion that the magnitude of the folding associated with the skin-surface patterns is directly related to the skin extensibility.

Functionally the skin surface patterns are best described in terms of the orientation, frequency and prominence of the surface grooves and ridges, and not as geometrical figures[10]. The results of tests using the suction device showed that by viewing surface patterns in this way a clear relationship is shown between the skin extensibility and the degree of folding of the epidermal surface.

Further work remains to be done; in particular the functional relationship between the skin extensibility and the surface patterns might be clarified by investigating the effects of altering the resting skin tensions.

Even without this basic understanding, the experimental result of this study – that there is a relationship between skin surface patterns and skin extensibility – has a practical clinical application in surgical procedures such as the placement and direction of incisions and the design of skin flaps in plastic surgery, in which a knowledge of the directional variation of skin extensibility is required.

ACKNOWLEDGEMENTS

This work was carried out in the Bioengineering Unit, University of Strathclyde (Professor R. M. Kenedi) and the Regional Plastic Surgery Unit, Canniesburn Hospital (Professor T. Gibson), and the authors are grateful for the use of the facilities of both institutions. During the study J. Ferguson was in receipt of an MRC research studentship.

References

1. Gasper, S. and Farkas, L. G. (1970). Transformation of surface relief in skin flaps transplanted to sites with different functional leading. *Folia Morphologica*, **18**, 116
2. Schellander, F. A. and Headington, J. T. (1974). The stratum corneum – some structural and functional correlates. *Br. J. Dermatol.*, **91**, 507
3. Montagna, W. and Parakhal, P. F. (1974). *The Structure and Function of Skin.* 3rd Edn., p. 9. (New York: Acad. Press)
4. Barbenel, J. C., Evans, J. H. and Jordan, M. M. (1979). Tissue mechanics. *Eng. Med.*, **7**, 5
5. Makki, S., Barbenel, J. C. and Agache, P. (1979). A quantitative method for the assessment of microtopography of human skin. *Acta Dermatovener* [*Stockholm*], **59**, 285
6. Evans, J. H. and Siessenopp, W. W. (1967). Controlled quasistatic testing of human skin *in vivo.* Presented at the *7th ICMBE Stockholm*
7. Brereton, W. D. S. (1974). Skin extensibility and its orientational variation *in vivo. MSc thesis*, University of Strathclyde
8. Kenedi, R. M., Gibson, T., Evans, J. H. and Barbenel, J. C. (1975). Tissue mechanics. *Phys. Med. Biol.*, **20**, 699
9. Brown, I. A. (1973). A scanning electron microscope study of the effects of uniaxial tension on human skin. *Br. J. Dermatol.*, **89**, 383
10. Tring, F. C. and Murgatroyd, L. B. (1974). Surface microtopography of normal human skin. *Arch. Dermatol.*, **109**, 223

11
The fibrous structure of the skin and its relation to mechanical behaviour

Y. LANIR

INTRODUCTION

Interpretation of the biophysical behaviour of the skin is often based on the assumption that it is a uniphasic continuum. A better insight into the complex response of the skin to radioactive, thermal, chemical or mechanical stimulii may be obtained if the skin's fibrous structure is considered. Several complex features in the mechanical behaviour of the skin can be accounted for in this way. Furthermore, this structural approach facilitates the use of non-invasive mechanical tests for diagnosing alterations in the skin constituents after exposure to external stimuli, e.g. in determining whether radiation will result in changes of the elastin or collagen and whether the fibers themselves are affected or their structure altered.

To illustrate this method a working mechanical model of the fibrous structure of the skin was developed. It is believed that similar procedures can be used to study other facets of the behaviour of skin.

Skin structure

When examined microscopically, skin is shown to be a fibrous composite material consisting primarily of collagen and elastin fibres embedded in an amorphous matrix (ground substance) of mucopolysaccharides. The collagen and elastin account for 80 % and less than 4 % of the dry weight of the tissue respectively. The collagen is arranged in a primarily planar array of highly undulated fibres in all directions. The collagen and elastin fibres may have cross links with the ground substance and between themselves. The nature and extent of these cross links are not fully established. Mechanically, the skin tissue shows highly non-linear and anisotropic stress–strain–time relationships. This is evident from the non-linear and anisotropic stress–strain relations under constant rate of strain and from the non-linear and anisotropic stress relaxation modes under step change of strain.

MECHANICAL MODELS

Phenomenological models that have been suggested so far have been successfully used to predict the skin's response under specific deformation schemes. A general constitutive model for such complex behaviour, however, has not yet been developed.

An attempt is made to link the observed microstructure with the macroscopic behaviour of the skin. The model is based on the following observations and assumptions.

(1) The skin consists of two primarily planar networks of fibres: highly undulated collagen fibres and straight elastin fibres.

(2) The angular distribution of the numerous collagen and elastin fibres can be expressed by the functions $R_c(\theta)$ and $R_e(\theta)$ respectively.

(3) Upon deformation the fibres stretch and rotate. Collagen fibres gradually straighten with increasing stretch. The distribution of the straightening strains among the collagen fibres is expressed by the function $P(x)$.

(4) The collagen and elastin fibres possess only tensile strength. Their compressive and bending strengths are negligible. The collagen fibre is viscoelastic, having a load–strain relationship expressed by $F_c = F_c(\lambda, t)$, λ being the stretch ratio. The elastin fibre is elastic with $F_e = F_e(\lambda)$.

(5) Based on direct and indirect evidence it is assumed that the contribution of the gelatinous ground substance to the response of the tissue is very small. It will therefore be neglected.

With the above assumptions, the tissue mechanics are characterized by means of five material functions: $R_e(\theta)$, $R_c(\theta)$, $P(x)$, $F_e(\lambda)$ and $F_c(\lambda, t)$. The first three specify the fibres' geometry and the last two are mechanical.

It can be shown[1] that the stress component on an area element ($T_{NS}(t)$) with normal $\bar{n}$ before deformation ($\bar{n} = \bar{i}\cos v + \bar{j}\sin v$) in the direction $\bar{S}$ after deformation ($\bar{S} = \bar{i}\cos\psi + \bar{j}\sin\psi$) is then expressed by:

$$T_{NS}(t) = \int_{-\pi/2}^{\pi/2} W(\lambda, \theta)\cos(\theta - v)\cos(\beta - \psi)\,d\theta, \tag{1}$$

where $\lambda = \lambda(\theta, t)$, $\beta = \beta(\theta)$ and

$$W(\lambda, \theta) = F_e(\lambda)R_e(\theta) + F_c^*(\lambda, t)R_c(\theta) \tag{2}$$

$$F_c^*(\lambda, t) = \int_{1.0}^{\lambda} P(x)/x\left\{\int_0^t F_c[\lambda(z), z]\cdot H[\lambda(z) - x]\,dz\right\}dx, \tag{3}$$

and

$$H[\lambda - x] = \begin{matrix} 0 & \text{if} & \lambda < x \\ 1 & \text{if} & \lambda > x \end{matrix} \tag{4}$$

Here λ and β are respectively the stretch ratio and angle of a fibre of original angle θ; they are functions of the tissue's strain field. H is the step function and ensures that the contribution of slack fibres will not be considered.

The above model is a general constitutive equation for the tissue. If we assume, as a first approximation, that the elastin is linear elastic and the collagen is linear viscoelastic, and after imposing experimental deformation schemes the resulting predicted behaviour agreed qualitatively with the behaviour of the tissue. Non-linear and anisotropic stress–strain–time relationships are obtained in much the same way as the skin exhibits.

For example, if the tissue is suddenly stretched at time $t = 0$ by λ_1 in one direction and by λ_2 in the normal direction, for each fibre where $t \geqslant 0$:

$$\lambda = [\lambda_1^2 \cos^2 \theta + \lambda_2^2 \sin^2 \theta]^{1/2} \tag{5}$$

and

$$\beta = \sin^{-1} [\lambda_2 \sin \theta / \lambda]. \tag{6}$$

The external loads on the edges of the specimen required to maintain this configuration will vary with time for $t \geqslant 0$ as:

$$T_1 = \lambda_1 \int_{-\pi/2}^{\pi/2} [W(\lambda, \theta] \cos^2 \theta \, d\theta \tag{7}$$

and

$$T_2 = \lambda_2 \int_{-\pi/2}^{\pi/2} [W(\lambda, \theta)/\lambda] \sin^2 \theta \, d\theta \tag{8}$$

where $W(\lambda, \theta)$ are as defined in equation (2), and for linear elastic elastin and linear viscoelastic collagen:

$$F_e = K_1(\lambda - 1) \tag{9}$$

and

$$F_c = K_2 G(t) \int_{1.0}^{\lambda} [P(x)(\lambda - x)/x] \, dx \tag{10}$$

where K_1 and K_2 are material constants and $G(t)$ is the reduced relaxation function for the collagen. Bearing this in mind and referring to equation (2) $W(\lambda, \theta)$ may be evaluated and then stresses T_1 and T_2 may be determined as functions of time. Hence the mechanical response (stress–strain–time relations) of the fibrous skin tissue may be analysed.

Reference

1. Lanir, Y. (1980). The rheological behavior of the skin: experimental results and a structural model. *J. Biorheol.* (In press)

12
Mechanical properties of rat skin at high and low loads. Influence of age and desmotropic drugs

H. G. VOGEL

INTRODUCTION

In recent experiments several mechanical and biochemical parameters in animal skin as well as in human skin were studied which depended on the influence of maturation and ageing and treatment with hormones and other desmotropic drugs[1–6]. Some differences were found when the mechanical parameters at low- and medium-extensions were compared with those occurring shortly before rupture[7,8].

METHODS AND RESULTS

In all studies dumb-bell shaped specimens were obtained from skin samples. Skin thickness was measured using calipers and mechanical properties were measured on an *Instron* instrument. Loads at rupture were measured from the stress–strain curves and tensile strengths calculated by dividing appropriate values on these curves by initial cross-sectional area. Furthermore, the ultimate modulus of elasticity was calculated from the straight part of the curve. In rat skin a considerable increase in skin thickness occurred during maturation which reached a maximum at the age of 12 months followed by a slight but significant decrease. These changes paralleled the changes in body weight. When strain at rupture was measured some increase during maturation was found, reaching a maximum much earlier (4 months); afterwards a significant decrease was measured (Figure 12.1). Load at rupture, tensile strength and modulus of elasticity at the straight part of the stress–strain curve showed a rapid increase during maturation, reaching a maximum at 12 months and a decrease during ageing (Figure 12.2).

Similar studies were also performed in samples of human skin. Maximum skin thickness and ultimate strain were reached at the age of about 30 years.

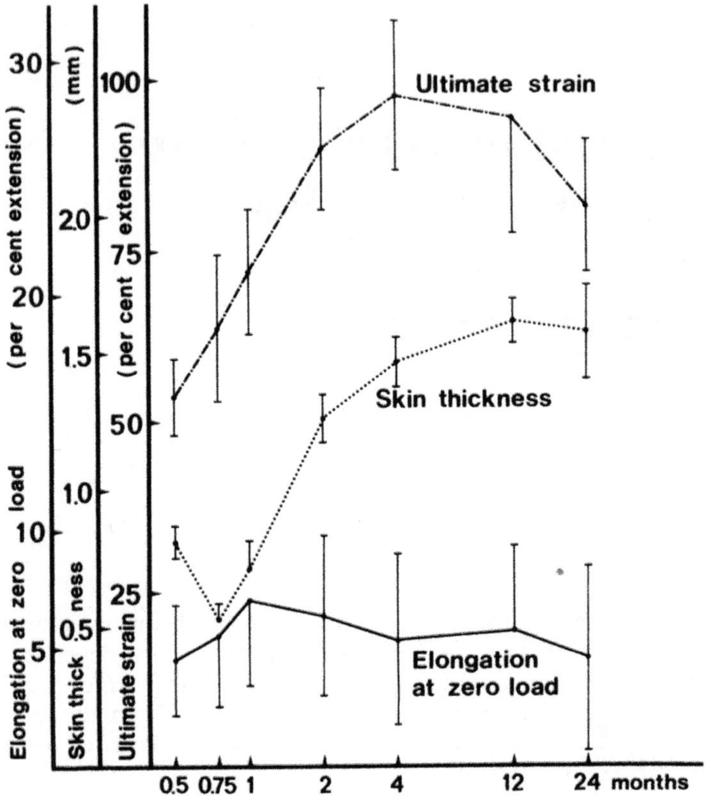

Figure 12.1 Ultimate strain, skin thickness, and elongation at zero load in rat skin at various ages. Results are means $\pm$SE

The maximum of ultimate load occurred earlier (about 20 years), whereas the maximum tensile strength and modulus of elasticity were reached at puberty.

During daily life, skin is extended and loaded much less than at the point of rupture. For this reason the mechanical properties of skin of rats were studied at low loads. The average stress–strain curves in various age groups looked similar, except for the changes in strength already mentioned. When the stress–strain curves were normalized – i.e. for each sample, extension at rupture and stress at rupture were calculated as 100% – the average curves in all age groups looked closely similar except in the 2 month age group. A clear minimum of normalized stress values at the age of 2 months was seen at extensions of between 50 and 90%. To study the stress–strain curves in more detail, the first part was registered with increased magnification. In this way two values for modulus of elasticity characterizing the low part could be calculated. These values showed a minimum in early maturation, whereas the modulus of elasticity at rupture showed a maximum after maturation. In one experimental series, extension was measured at zero load (Figure 12.1) or at

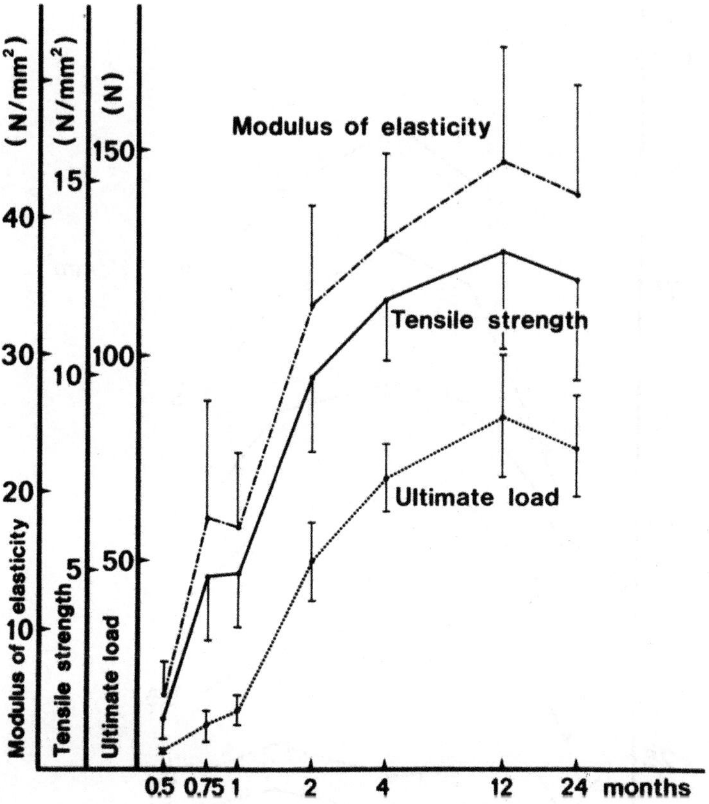

Figure 12.2 Modulus of elasticity, tensile strength, and ultimate load in rat skin at various ages. Results are means ±SE

low loads or low-stress loads. The extension at the lowest stress loads showed a maximum at an age of 4 weeks. This was shifted to 1–4 months with increasing stress loads (Figure 12.3). Closer examination of the stress–strain curves showed the step phenomenon[9,10]. During elongation the measured stress did not increase continuously but dropped down and increased again. Two or three steps occurred before the roughly straight part of the stress–strain curve started. This step phenomenon occurred only in samples obtained perpendicular to the body axis. When the elongation due to the steps or the stress loss was summarized a maximum at 4 months occurred. A further approach to studying the mechanical properties at various degrees of extension was to evaluate the hysteresis phenomenon[11].

The specimen was subjected to a given amount of strain and the stress and the modulus of elasticity measured. The energy input was calculated from the area under the curve. When the specimen was released a hysteresis loop was formed from which the energy dissipation was calculated. When the ratio between energy input was calculated a maximum was found at medium degrees of extension.

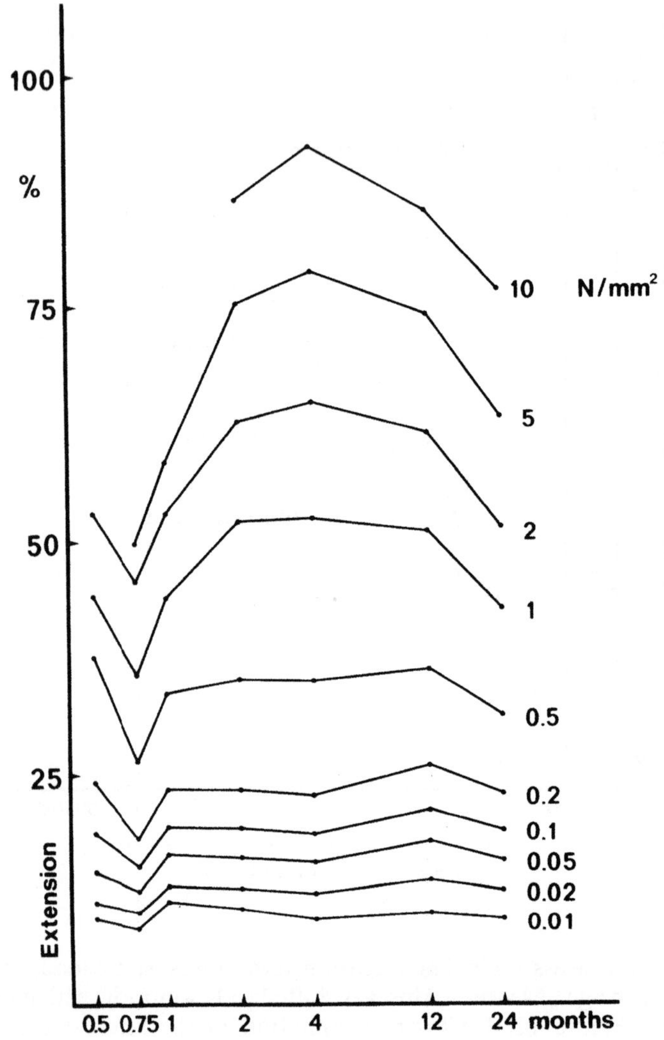

Figure 12.3 Mean extension at various stress loads in rat skin at various ages

However, no dependence on ageing could be detected. When skin samples were strained repeatedly up to 40% extension the stress values decreased with the logarithm of the number of cycles in all age groups. The first as well as all later stress values at 40% extension were lowest at 4 months.

The effects of treatment with hormones and other desmotropic compounds were also studied[1,4,6]. 10 days' treatment with corticosteroids resulted in a dose-dependent increase in strength which was paralleled by an increase in insoluble collagen. The effect was not only dose-dependent, but also time-dependent. Ultimate load reached a maximum after 5 or 10 days' treatment then decreased below control values. Tensile strength reached a maximum between 5 and 10 days. Since skin thickness was diminished, tensile strength did not decrease below control values even after long-term treatment with toxic doses. Likewise other anti-inflammatory drugs such as indomethacin, phenylbutazone and acetylsalicylic acid induced an increase of tensile strength and insoluble collagen[6]. The relative changes, however, were much less than after treatment with corticosteroids. Another compound which considerably influenced mechanical properties of skin was D-penicillamine[3]. This did not influence body weight or skin thickness even at high doses. However, ultimate load, tensile strength and modulus of elasticity were decreased by a large extent. The elasticity modulus at the beginning of the stress–strain curves was influenced by prednisolone and D-penicillamine in the same way as the elasticity modulus at rupture (in contrast with the behaviour depending on age).

CONCLUSIONS

Maturation decreased the values of stress and modulus of elasticity at low degrees of extension but increased them at high degrees of extension. Ageing had the opposite effect. The effects of corticosteroids and D-penicillamine showed a different pattern. These compounds acted antagonistically but the effect of each compound was in the same direction at low and high degrees of extension. One therefore cannot regard the effects of corticoids as an acceleration of ageing or the effects of D-penicillamine as a retardation of ageing.

References

1. Vogel, H. G. (1970). Beeinflussung der mechanischen Eigenschaften der Haut von Ratten durch Hormone. *Arzneimittel-Forsch.*, **20**, 1849
2. Holzmann, H., Korting, G. W., Kobelt, D. and Vogel, H. G. (1971). Prüfung der mechanischen Eigenschaften von menschlicher Haut in Abhängigkeit von Alter und Geschlecht. *Arch. Klin. Exp. Derm.*, **239**, 355
3. Vogel, H. G. (1972). Effects of D-penicillamine and prednisolone on connective tissue in rats. *Conn. Tiss. Res.*, **1**, 283
4. Vogel, H. G. (1974). Correlation between tensile strength and collagen in rat skin. Effect of age and cortisol treatment. *Conn. Tiss. Res.*, **2**, 177
5. Vogel, H. G. (1976). Tensile strength, relaxation and mechanical recovery in rat skin as influenced by maturation and age. *J. Med.*, **7**, 177
6. Vogel, H. G. (1977). Mechanical and chemical properties of various connective tissue organs in rats as influenced by non-steroidal antirheumatic drugs. *Conn. Tiss. Res.*, **5**, 91
7. Vogel, H. G. and Hilgner, W. (1977). Analysis of the low part of stress–strain curves in rat skin. Influence of age and desmotropic drugs. *Arch. Dermatol. Res.*, **258**, 141
8. Vogel, H. G. and Hilgner, W. (1979). Viscoelastic behaviour of rat skin after repeated and stepwise increased strain. *Bioengineering and the Skin, Newsletter*, **1**, 22
9. Vogel, H. G. and Hilgner, W. (1979). The "Step Phenomenon" as observed in animal skin. *J. Biomechanics*, **12**, 75
10. Vogel, H. G. and Hilgner, W. (1979). Influence of age and of desmotropic drugs on the step phenomenon observed in rat skin. *Arch. Dermatol.*, **264**, 225
11. Vogel, H. G. (1978). Age dependence of mechanical parameters in rat skin following repeated strain. *Akt. Gerontol.*, **8**, 601

13
Point impedance characterization of soft tissues *in vivo*

D. E. THOMPSON, H. Mg. HUSSEIN and R. Q. PERRITT

INTRODUCTION

Soft tissues cover the surface of the human body and provide a means of protecting it from the forces and stresses of its environment. They have a structure and thickness which varies from location to location according to the mechanical demand. In this work, the term 'soft tissue' refers to the total tissue thickness which extends from the outer surface of the skin, the epidermis, through the dermal and subcutaneous tissues to the surface of the bone. The role of these tissues is of great interest in rehabilitative medicine and orthopaedics. If these tissues lose their ability to absorb and dissipate mechanical energy, a vicious process of inflammation and autolysis is initiated leading to frank ulceration and, in many cases, amputation of the afflicted part.

The above sequence of events is not well understood, but sensory denervation and changes in nutrition, the so-called trophic changes, are known to be important, as are factors such as the activity levels of the patients, their personal hygiene and many other less obvious variables. Tissues can be traumatized by long-term application of low pressures leading to ischaemia. These are in the order of 4.7–20 kPa (35–150 mmHg) and can lead to necrosis. Trauma may occur after a single application of stress of about 8.8 MPa (1280 p.s.i.) as described by Yamada[1]. Tissue breakdown can also result from many thousands of repetitions of a moderate pressure between the above two extremes. Such pressure ranges from 0.17 to 0.52 MPa (25–75 lbf/in^2) and breakdown occurs as a result of applying mechanical stress to the tissues. This raises the question of how such tissues react to mechanical loads and what we can learn about the tissues from the interaction itself.

The technique of using the mechanical response of soft tissue to characterise the tissues themselves is a comparatively new concept. Many investigators have studied the skin; some have even evaluated certain mechanical properties. Most early studies were simple *in vitro* material property evaluations as typified by the work of Yamada. Hildebrandt *et al.*[2] noted that most of these studies were simple length–tension studies, whereas most soft tissues are loaded compressively.

The above studies were complemented by Franke[3] and Von Gierke[4] who reported on applying mechanical vibrations to human soft tissues *in vivo*. The work of Franke is of particular interest because it is the first application of the concept of mechanical impedance to soft tissues. This concept was introduced theoretically only 50 years ago, and its application to engineering problems is still considered to be advanced[5–7]. Recently, the concept of analyzing the 'signature' of biological specimens in the impedance–frequency domain has been useful in such varied areas as ascertaining the clinical mobility, or looseness, of teeth[8,9], quantifying the degree of healing of fractures[10] and in designing hearing aids using impedance data from the human mastoid[11].

It is the purpose of this study to obtain non-destructive impedance spectral signatures of normal human soft tissues. These will serve as a baseline which can be used to identify abnormal spectra which are associated with pathological changes[12] and to aid in evaluating interface orthotic materials to protect skin used in load bearing[13].

MATHEMATICAL MODEL

Soft tissues may be characterized by their impedance. This is a measure of the resistance the tissues show when forced into motion. This varies with frequency and with the state of stress in the tissues. All materials show a characteristic signature of impedance versus frequency. A simple analytic model of tissues will be presented to aid in interpreting the signatures obtained. This model is attributed to Lord Kelvin (William Thompson, 1824–1907); in its classical form with constant element values, it would fail to portray the non-linear viscoelastic behaviour of soft tissues. If the element coefficients are allowed to vary, however, and only the harmonic behaviour of soft tissues is considered, the model becomes a useful tool.

Impedance is defined as the ratio of applied force to the surface velocity of the tissues. Since both force and velocity are considered as complex harmonic variables having both magnitude and phase, impedance is also a complex quantity. The impedance, Z, is thus:

$$\bar{Z} = \frac{\bar{F}}{\bar{V}} = \frac{F_o e^{j\omega t}}{V_o e^{j(\omega t - \phi)}} = \left(\frac{F_o}{V_o}\right) e^{j\phi} \qquad (1)$$

where F_o and V_o are the amplitudes of the force and velocity, ω is the circular frequency, and ϕ is the phase difference between force and velocity.

The differential equation describing soft tissues as a simple Kelvin element having mass (m), damping (C) and elastic (k) elements can be written for harmonic force and velocity as

$$mj\omega\bar{V} + C\bar{V} + \frac{k}{j\omega}\bar{V} = \bar{F} \qquad (2)$$

Introducing mechanical impedance and simplifying this equation yields:

$$\bar{Z} = C - j(k - m\omega^2)/\omega = Z_c - jZ_k \tag{3}$$

Equation (3) depicts impedance as having real and imaginary components which can be treated vectorially to obtain magnitude (Z_o) and phase (ϕ):

$$Z_o = \sqrt{C^2 + (k - m\omega^2)^2/\omega^2} \tag{4}$$

and

$$\phi = -\arctan\left(\frac{k - m\omega^2}{C\omega}\right) \tag{5}$$

It is apparent from equations (4) and (5) that as the frequency approaches zero, the impedance approaches a value of $Z_o = k/\omega$ and the phase angle approaches $-90°$. At high frequencies the phase angle approaches $+90°$ and the impedance is described by $Z_o = m\omega$. For the same system, with the damping element removed, the impedance vanishes at an important frequency:

$$\omega_n = \sqrt{k/m} \tag{6}$$

defined as the undamped natural frequency. The typical signature of a Kelvin element is summarized in Figure 13.1. The elastic character of the tissues are to be represented by a stiffness element (k) and the viscous behaviour by a damping element (C). Figure 13.1 shows that the elastic element dominates the

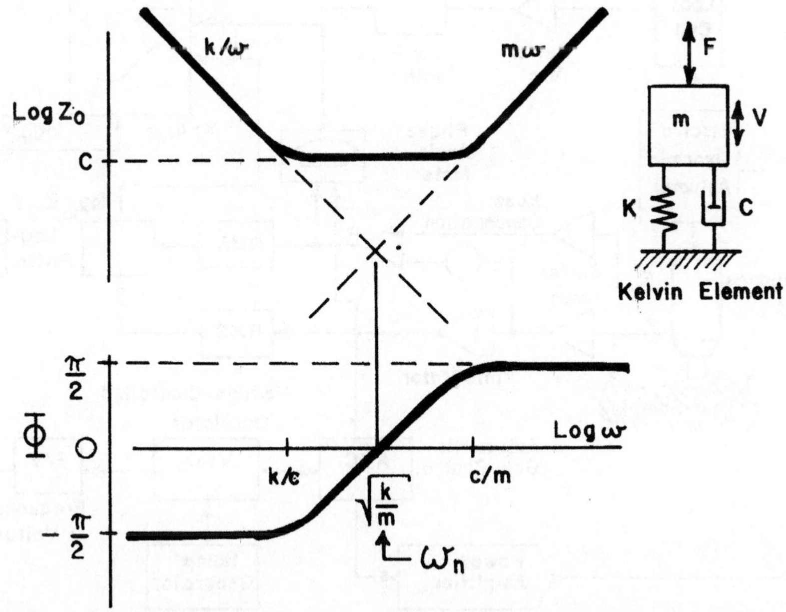

Figure 13.1 Impedance magnitude and phase for a Kelvin element. For key see text

impedance signature at low frequencies where the impedance drops linearly. At intermediate frequencies the impedance is held constant by the viscous element, and at high frequencies inertia becomes the dominant force and impedance rises linearly. Although this is a simplistic model of tissue behaviour, it is sufficient to permit an interpretation of both the impedance and phase-angle signatures.

EXPERIMENTAL PROCEDURE

16 adult subjects (5 women, 11 men) aged between 20 and 46 years were randomly chosen for this study. Tests were performed on the pulp of the terminal digit, third finger of the right hand of each subject. There was no known history of scarring or other tissue damage in any of the subjects. Lateral X-rays of the test fingers were measured and the mean ($\pm$SD) pulp thickness was found to be 7.63 ± 0.67 mm.

A swept-frequency viscoelastometer was constructed to measure the impedance of soft tissues (Figure 13.2). The dynamic force and acceleration signals were obtained from an impedance head (B & K type 8001) which was fitted with a 6.35 mm tip. The impedance head was mounted on an

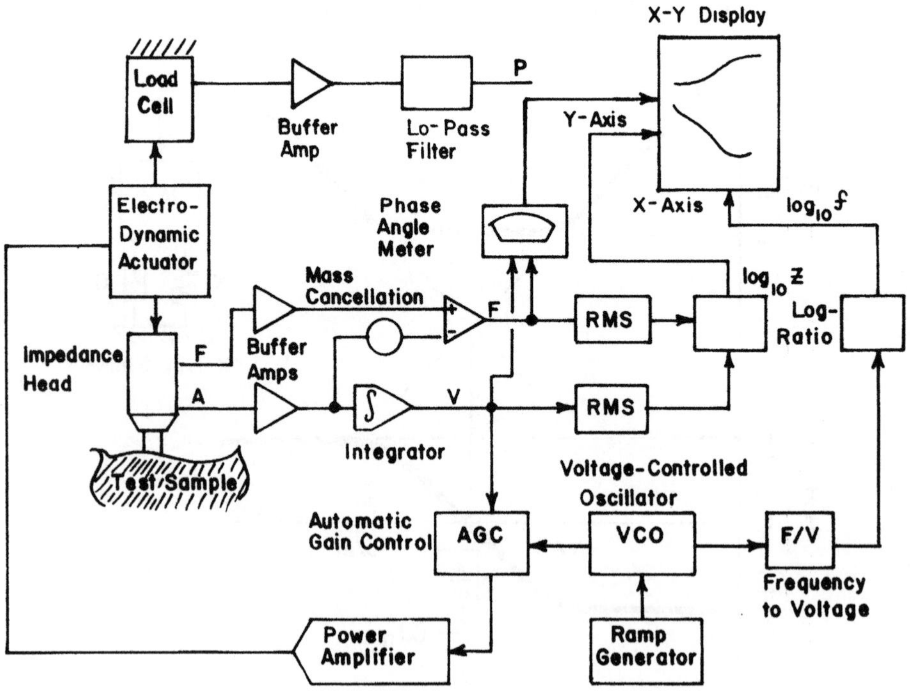

Figure 13.2 Schematic diagram of the swept-frequency point impedance viscoelastometer

electromagnetic actuator or shaker (B & K type 4810) which was driven by a swept sinusoidal voltage. The acceleration signal was integrated to obtain the velocity which was then used in an automatic gain control (AGC) or compressor circuit to control the amplitude of the sinusoidal voltage to the shaker. The velocity was thus constant during testing. The acceleration signal was also used to cancel out that portion of the force signal due to the mass of the impedance head. This corrected force signal and the velocity signal were then input to RMS circuits and then to a log ratio amplifier to obtain an output proportional to the logarithm of the mechanical impedance ($\log_{10} Z$). The phase angle between force and velocity signals was obtained via a phase meter (B & K type 2971). Both phase angle and log-impedance were used as vertical drive signals through a multichannel display multiplexor to an $X-Y$ storage display oscilloscope (Tektronix 4701, 613). The horizontal drive was obtained by using a frequency-to-voltage converter and log-amplifier. In this manner real-time plots of log Z and ϕ versus log of frequency (f were obtained.

A separate load cell was used with a low-pass filter-amplifier (0–5 Hz) to obtain the mean static force, or preload force. This was necessary since the impedance head uses piezoelectric sensors and thus cannot be used for measuring the DC component of force.

The test frequency range was chosen as 30 Hz–1 kHz. Preload forces were tested over the range 0.5–5.6 N. These ranges were selected to stay within the operational limitations of the hardware used. Earlier experiments determined that frequencies below 30 Hz did not yield any additional information on tissue properties.

The tests were carried out by placing the finger on the test table with the dorsum of the finger rigidly supported. The tip of the impedance head was then positioned on the centre of the fingerprint swirl and the desired preload applied. Frequency sweeps were taken and calibration lines added before producing a hard copy of the screen image. The impedance and phase spectra were obtained by placing the hard copy on a digitizer (Tektronix 4956) connected to a computer graphics system (Tektronix 4051). The digitized data were then stored on magnetic tape for each subject and each preload for later recall and averaging.

RESULTS

The impedance and phase-angle spectra of the 16 subjects were averaged at discrete frequencies for each preload. Figures 13.3 and 13.4 show these averaged results for all of the preload values. Several features are readily apparent and will be interpreted with the Kelvin model. The impedance spectra clearly depict a linear region at low frequencies, where elastic effects dominate. At higher frequencies this gives way to viscous effects. As preload increases, the impedance also increases with concomitant increases in both elastic and viscous contributions. The ratio of these is not constant, however, as the first corner frequency, shown to be k/C by the Kelvin model, rises slightly with increasing preload.

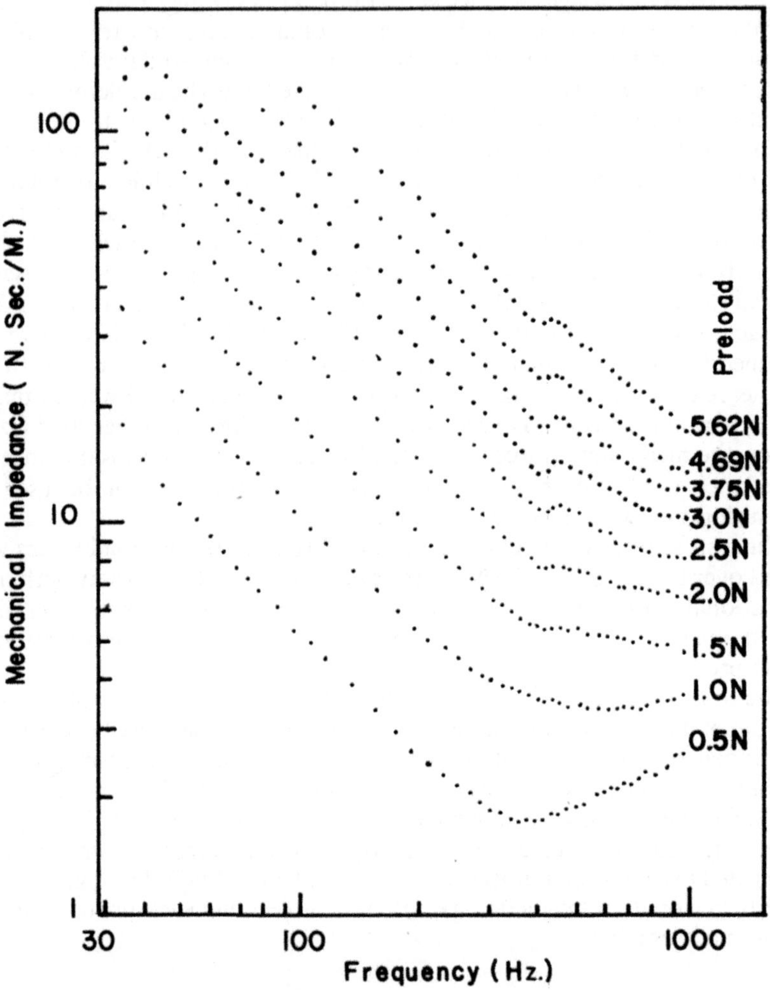

Figure 13.3 Mechanical impedance spectra of human finger pulp tissues in 16 subjects at various preloads

These observations are borne out by the phase-angle measurements. The model predicts that increasing the viscous damping parameter would tend to broaden the phase-angle transition from −90 to +90 degrees; this effect is clearly seen in Figure 13.4 as the preload increases. Increasing the elastic modulus, however, tends to raise the 'undamped natural frequency' of the system where the phase shift is zero. This effect is also evident in Figure 13.4 as the preload increases. Artifacts in the impedance and phase spectra were also observed at about 90 Hz and 500 Hz. These have been identified as resonance

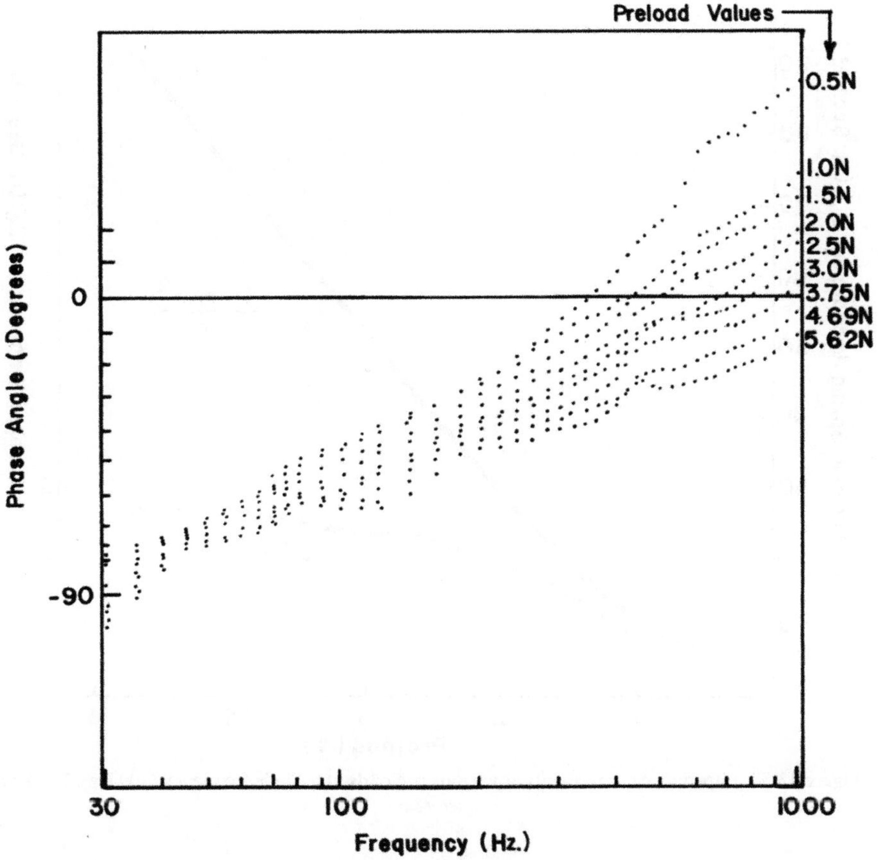

Figure 13.4 Impedance phase-angle spectra of human finger pulp tissues in 16 subjects at various preloads

frequencies of the device, and were not due to osseous structures in the soft tissues.

The influence of preload on soft tissues can also be studied at a single frequency. Figure 13.5 shows the relative contributions of the two components of impedance (Z_c and Z_k).

DISCUSSION

The increase in the elastic and viscous moduli with preload can be understood by considering the structure of the tissues. As the tissues are compressed the collagen network and the cellular structures become more compact and the solid elements offer an increasing resistance to compression. This is typical of cross-linked polymeric materials. In addition, the movement of both the

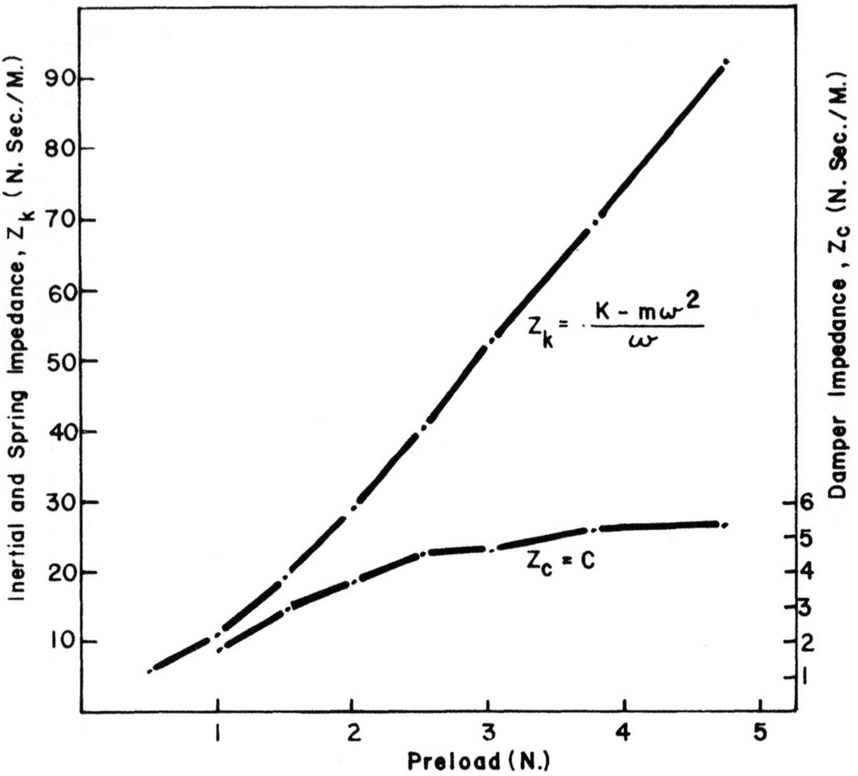

Figure 13.5 Impedance components at various preloads at a single frequency (300 Hz). For key see text

mobile solid and the liquid elements from high-pressure to low-pressure regions gives rise to a resistance caused by viscous and frictional stresses. This effect is shear-rate dependent and increases with frequency. It also increases with preload because of the restriction of the collagen network and a narrowing of interstitial spaces as the tissues are compressed. As shown in Figure 13.5, this increase in viscous resistance diminishes as the tissues become tightly packed, and the resulting viscous response becomes primarily a frictional effect that is independent of the preload. These forces are well above the occlusion level for the circulation, and there is no significant viscous contribution from blood motion into and out of the affected region at the higher preloads.

In summary, the characteristic signature of soft tissue can be described by an essentially elastic response at frequencies up to about 200 Hz. The characterizing elastic parameter (k) increases with tissue compression. The viscous parameter (C) also increases with compression and its contribution to impedance is a dominant factor throughout the 200 Hz to 1 kHz range.

ACKNOWLEDGEMENTS

This research was supported by grants from the United States Public Health Service (PHS 240-79-002, PHS 258-78-0044). We thank the Rehabilitation Branch, Carville USPHS Hospital, for their co-operation and assistance in this programme.

References

1. Yamada, H. (1970). *Strength of Biological Materials.* (Baltimore: Williams and Wilkins)
2. Hildebrandt, J., Fukaya, H. and Martin, C. J. (1969). Completing the length–tension curve of tissue. *J. Biochem.*, **2**, 463
3. Franke, E. K. (1950). Mechanical impedance of the surface of the human body. *J. Appl. Physiol.*, **3**, 582
4. Von Gierke, H. E., Oestreicher, H. I., Franke, E. K., Parrack, H. O. and von Wittern, W. W. (1952). Physics of vibrations in diving tissues. *J. Appl. Physiol.*, **4**, 886
5. *Fundamentals of Mechanical Impedance Analysis.* (1974). Technical Publication M-Z, 11-74. (San Diego: Spectral Dynamics Corporation)
6. Kerfoot, R. E. (1966). *Solutions for Mechanical Impedance Measurement Problems.* Technical Publication No. M-1, 9-66. (San Diego: Spectral Dynamics Corporation)
7. Broch, J. T. (1976). *Mechanical Vibration and Shock Measurements*, pp. 235–246. (Bruel & Kjaer Instruments)
8. Noyes, D. H., Clark, J. W. and Watson, C. E. (1968). Mechanical input impedance of human teeth *in vivo*. *Med. Biol. Eng.*, **6**, 487
9. Noyes, D. H. and Salt, C. W. (1972). Relationship between tooth mobility and mechanical mobility of teeth. *J. Peridontol.*, **43**, 301
10. Cambell, J. N. and Jurist, J. M. (1971). Mechanical impedance of the femur: A preliminary report, *J. Biomechanics*, **4**, 319
11. Bednin, F. V., Sagalovich, B. M., Vodolazskii, L. A. and Golikov, V. A. (1975). Mechanical impedance of tissues of the human mastoid. *Biofizida*, **20**, 298
12. Hussein, H. M.-G., Thompson, D. E. and Perritt, R. Q. (1979). Dynamic response of human soft tissue with surface scar. In *Proceedings of the 32nd ACEMB.*
13. Thompson, D. E. and Hussein, H. M.-G. (1977). Characterization of orthotic materials by mechanical impedance method. In *Proceedings of the 30th ACEMB.*

14
The gas bearing electrodynamometer (GBE) applied to measuring mechanical changes in skin and other tissues

C. W. HARGENS

INSTRUMENT CAPABILITIES

The gas bearing electrodynamometer (GBE) (Figure 14.1) has now reached a state of perfection and proved itself of sufficient value in several medical and biological applications to deserve wider publication. It has filled an important gap in the spectrum of measuring devices.

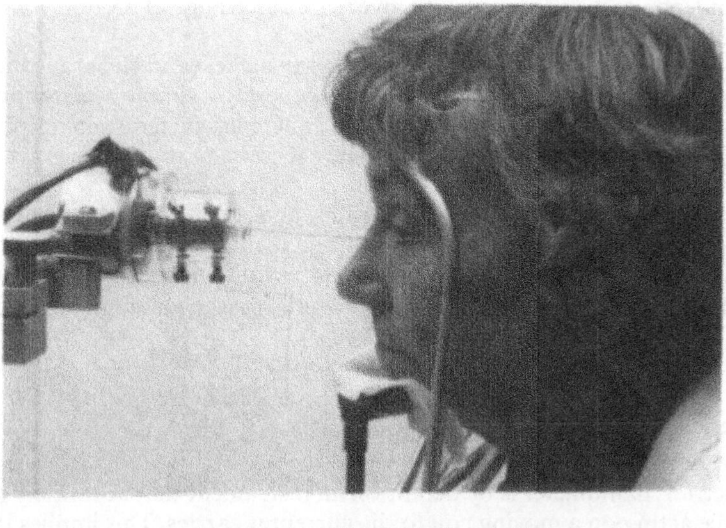

Figure 14.1 GBE attached to facial site

The GBE solves the problem of measuring small forces acting in soft, highly compliant biological materials or organs where the intervention of a stiffer transducer would mask the modulus or effect being studied. For example, strain guages and load cells operating with a sensitive electric bridge can be used to measure small forces but they themselves, or a flexural element upon which they are mounted, will be inherently quite stiff under any appreciable deflection. The spring rate of such a guage or combination is high. The GBE, on the other hand, can have a spring rate which approaches zero and may be neglected in comparison with that of the substance measured.

The force and displacement signals of the GBE can be readily registered electrically. The displacement can be several millimeters or more without the specimen undergoing more than a minute stress. This is important in studying biological materials and organs which are soft and when the effects to be observed without imposing large forces require that considerable displacements occur. For example, soft tissue may show non-linear stress–strain characteristics only after experiencing larger and more rapid oscillations than a conventional testing machine can achieve. Moreover, the cumbrousness of such equipment makes its use in tests on delicate living systems impractical.

An essential difference between the GBE as a transducer and other types is its ability to exert a large range of forces, down to a fraction of a gram, over a considerable range of displacements. In comparison the strain guage has minimal strain capability and can only become useful for measuring sizable strains by attaching it to a large flexible cantilever element. The latter requires considerable force to deflect it, and thus cannot be used where such magnitudes of force are not available or tolerated.

Most force-measuring transducers have this problem i.e. they possess inherently large built-in restoring forces, like a spring, which require a high level of potential energy input to deflect them over their useful ranges. The GBE has a negligible internal restoring force and can exert a widely adjustable force, independent of displacement, over its entire range of motion which can be adjusted as required.

Any time function of applied force is readily achieved in this transducer by external electrical means such as a function generator, simple oscillator or DC (battery) source. A combination of AC and DC bias currents makes possible operation of the GBE in any position, the DC serving to oppose gravity and 'float' the armature.

The armature's mass is only a few grams, and it can therefore be driven at high oscillatory rates if desired. We have tried to minimize inertial effects. This makes the determination of dynamic moduli quite simple and permits the determination of viscous or loss parameters in viscoelastic materials.

PHYSICAL DESCRIPTION

An electrodynamometer is by definition an instrument that accurately relates the force acting on a moving coil to the current it carries. This implies that its conductors move in a steady and uniform magnetic field produced by a fixed-coil electromagnet or permanent magnet. Thus it is basically a current-to-force transducer, and it can be used in many ways.

For the moving coil to undergo displacement it must be mounted on some kind of compliant structure such as a spring or a bearing. Accurate electrodynamometers, such as those used to give electrical standards of current, are usually constructed similarly to an analytical balance using precision pivots to minimize friction. Such an arrangement needs mechanical components of such mass that rapid response is impossible. However, they are used as static balances quite successfully.

The introduction of a gas bearing (Figure 14.2), on the other hand, provides the moving coil with a suspension of the highest compliance. Its mechanical components can be made light in weight (a few grams) so that it can have rapid movements in response to alternating currents.

This ability to respond rapidly permits many applications of the electrodynamometer which would not exist for the conventional balance form of the device.

Especially important is its use as an instrument to test the compliance of soft and delicate materials. Since many soft substances, like biological tissue and organs, are viscoelastic, their compliance can only be brought out by stressing them rapidly. The viscous response of the specimen causes a phase lag between strain and stress if the cyclic process is carried out at an appropriate rate. The frequencies we find convenient for these effects are in the order of or greater than one Hz.

The value of the GBE is enhanced for the above application by adding a sensitive displacement measuring device, the LVDT (linear variable differential transformer). In the present design a thin magnetic sleeve, which forms the core of the LVDT, rides coaxially with the coil-carrying armature on the gas bearing. Thus the position of the armature is precisely indicated at any instant during the oscillatory cycles. This is the strain signal, and it can be recorded and compared with the stress signal in real time to yield a hysteresis loop on an oscillograph or other type of $X-Y$ recorder.

Connection to an anatomical test site is usually made as shown in Figure 14.1 or in Figure 14.3 by an attachment wire or rod and adhesive-coated button. A small removable collet (Figure 14.4) designed to have a taper fit within the armature sleeve provides a convenient means of changing probes. In practice the button is attached to the skin and remains in place for the test period, which may be up to a whole day. The probe is pressed into a mating hole in the button and can then be removed to disconnect instrument and test subject.

Operation of the LVDT requires an electronic signal conditioner. These circuits provide a 2500 Hz (or higher for rapid motion) excitation current for the LVDT primary coil. The conditioner also contains the demodulator, which compares the signals from the two secondary coils and produces a rectified DC signal proportional in sign and magnitude to the displacement of the LVDT core. In this case a special hollow core fits coaxially over the GBE's armature. The core is centered on the LVDT and secured to the armature

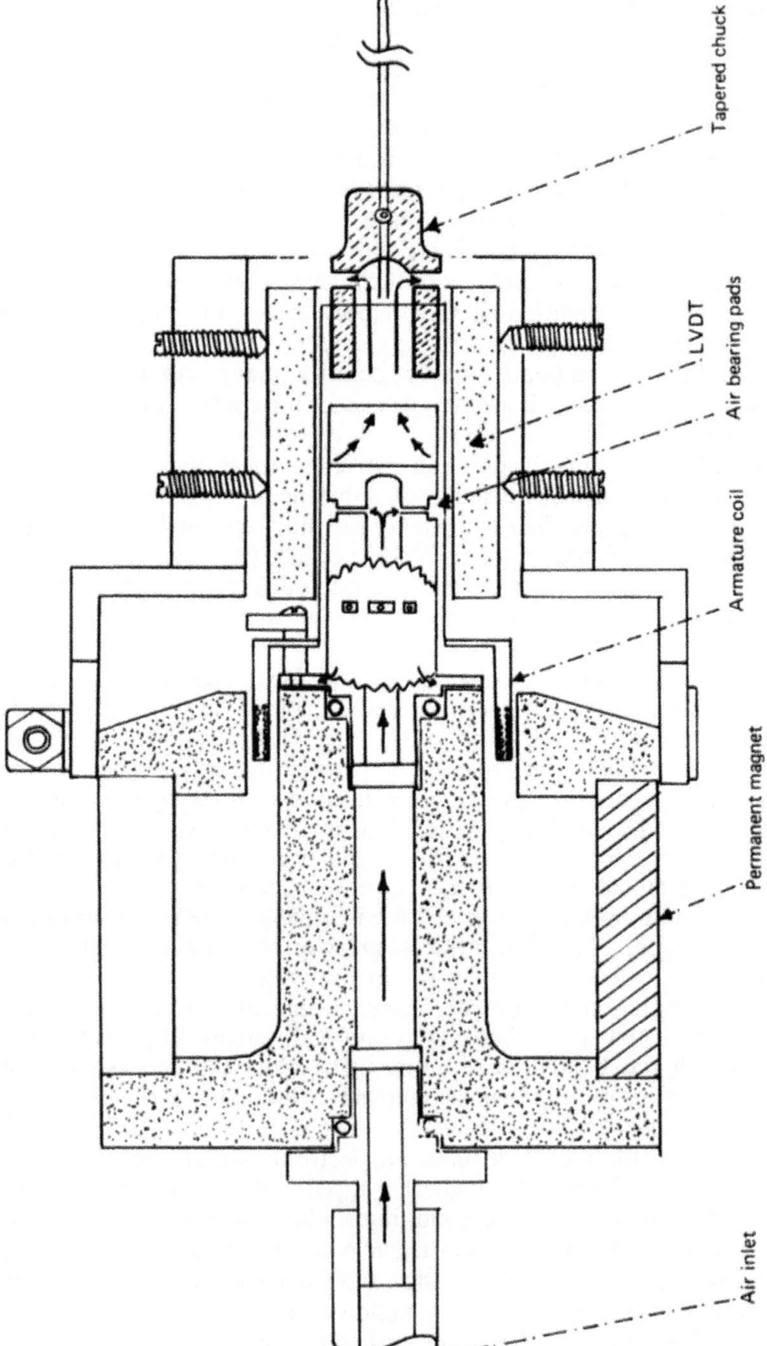

Figure 14.2 Sectional sketch of GBE with LVDT displacement output (× 1.4 actual size)

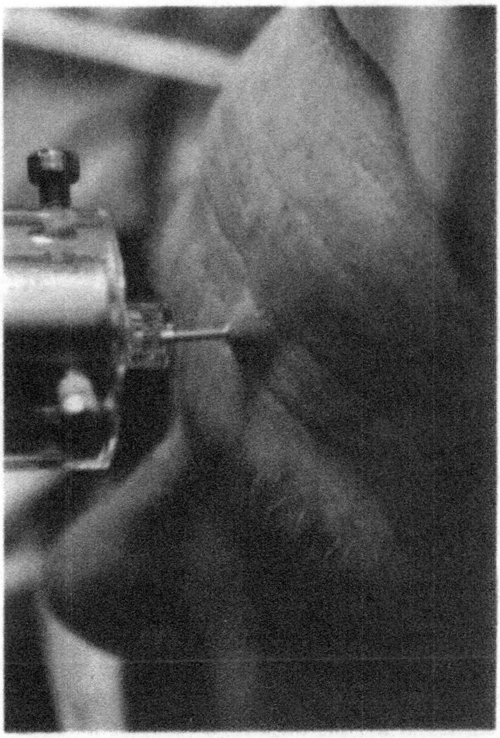

Figure 14.3 Tension–compression test on forehead skin

when it is at its midpoint of travel, so that the electrical displacement signals represent movement about this neutral position.

The linearity and positional accuracy of the LVDT and signal conditioner used here is 0.05% for movement up to ±2 mm and 0.1% for greater excursions out to the limits of armature travel. This is more than adequate for a measurement system using an oscillographic display which can only be, at best, accurate to within 1%.

The accuracy of the force co-ordinate is again as good as the electronic display. The transducer will obey absolutely the law defining the force on a current-carrying conductor in a magnetic field. As long as the field is constant and uniform in the space through which the coil moves fidelity of the force–current relationship is dependable. The force signal that is displayed is simply derived from the voltage across a resistor in the current circuit in series with the armature coil.

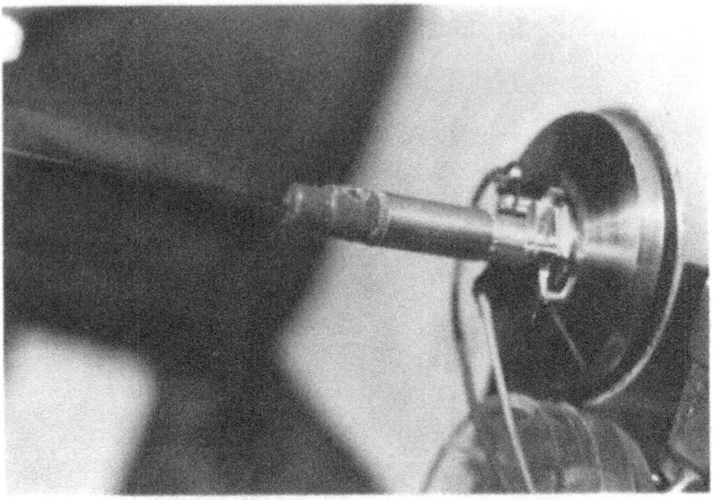

Figure 14.4 Probe collet

TYPICAL DATA

The advantage of applying a rapidly oscillating force to a specimen and of the resulting dynamic stress–strain diagram (hysteresis loop) which appears on an oscilloscope screen lies in the instantaneous disclosure of several mechanical characteristics such as stiffness, lossiness and any non-linearities. This is illustrated in the loops in Figure 14.5 which are given for comparison of variety. The first typifies a stiff material with minimal losses, such as a dry human hair subjected to oscillatory elongation in tension. The second is for the same hair treated with shampoo making it less stiff (note less steep slope of loop) but lossier (more open loop). The third shows the compression characteristics of sponge rubber showing greater stiffness (steeper slope) but more loss (and more viscous behavior) than the other two materials. All of these display rather linear, Hookean, elastic tendencies with some degree of loss factor.

Figure 14.5 Dynamic hysteresis loops – comparison of hysteresis loops for (from left to right): stiff (low loss); less stiff; and viscous (lossy) materials

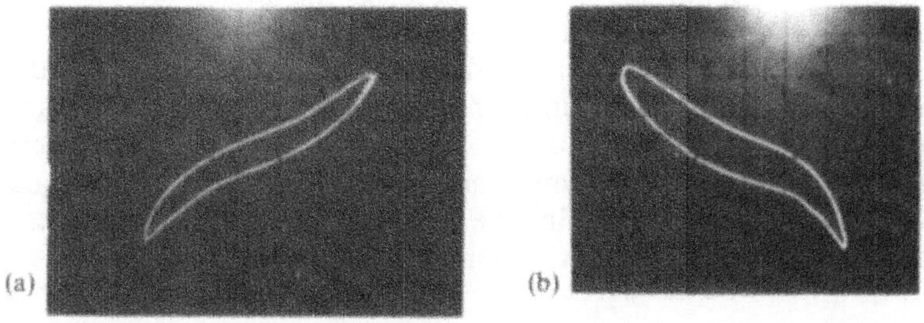

Figure 14.6 Shear and compression–tension moduli for dermatologic surfaces: (a) for facial skin; (b) for forehead skin

Figure 14.6 shows a shear modulus of a test subject's facial skin (as in Figure 14.1), showing that the area was stressed beyond the linear range but symmetrically at both ends of the stroke. Figure 14.6b shows the results of a tension–compression test on the forehead (as in Figure 14.3), where the abrupt compressive stiffness of the right-hand part of the curve results from the bony underlying skull structure. Both were made at the 1-second cycle rate, i.e. one second was the time required to complete the loop.

The speed with which the loops are drawn allows an experimenter to observe rapid changes in a specimen that is actively undergoing some alteration in its characteristics. This is a great advantage over bulky testing machines whose time to complete a test and draw the stress–strain characteristic might occupy the entire period of interest, when the transient phenomenon was occurring. In some equipment the test cycle is so protracted that the specimen will have changed appreciably during the test.

An example of this rapid change is in tests of single hair fibres where the drying rate after treatment is extremely rapid and the accompanying internal processes for restoration of the stiffness modulus can be missed unless a new loop can be plotted nearly every second. Figure 14.7 shows at the left a stored series of such cycles (like the single loop at the right) one second apart, representing the first moments of a wet hair fibre's recovery. Interestingly, the

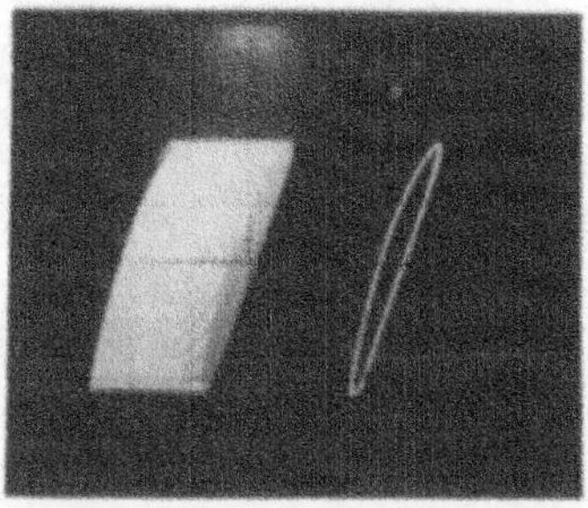

Figure 14.7 Recording the time history of rapid alterations in dynamic modulus

strain axis is self-calibrating in time because the cycles are of 1-second duration. Thus there is a time history of the changing dynamic modulus as well as the slow creep in the specimen as it dries and is worked by the cyclic stress routine. The slow creep is evidenced by the gradual movement of the loop sequence progressing to the left and producing a kind of Moiré pattern.

In this figure we have an example of a material whose dynamic modulus does not change appreciably during the period of observation, and yet there is obvious creep or elongation of the fibre.

THE INSTRUMENT MECHANISM

All of the electromechanical characteristics of the GBE are readily determined by simple measurements. First, the current-to-force conversion efficiency is derived from the force calibration. Figure 14.8 shows the unit mounted vertically over a platform balance so that the weight of the air-floated armature could be determined. In the unit shown this is about 4.5 g. The armature current to achieve this balance is recorded and thereafter becomes the basis of force calibration. Since the $F = ilB$ law* states that the force acting on the conductors is directly proportional to the current carried, we can assume that a linear relationship between current and armature force exists. The armature conductors must always remain within a uniform, steady field for this rule to apply. This must be guaranteed by the dimensional design, which restricts armature movement relative to the magnetic pole pieces. The field is produced by a strong permanent alnico ring magnet.

Figure 14.8 Two-inch diameter GBE mounted vertically on balance for initial force calibration

* F = force; i = current; l = conductor length; and B = flux density.

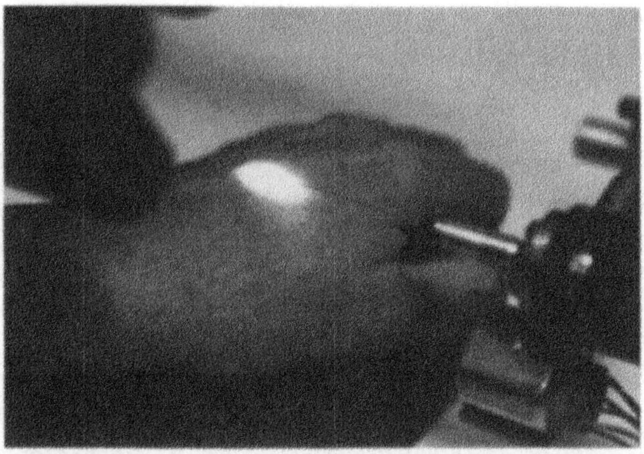

Figure 14.9 One-inch diameter GBE for ophthalmic studies

Three sizes of GBE have been made, the one shown in Figure 14.8 (2 inches in diameter) being the medium sized. Maximum thrust is limited by the amount of power that can be tolerated in the armature coil. A 1 inch diameter instrument (Figure 14.9) was developed for ophthalmic exploration of the cornea; and a large, 3 inch version (Figure 14.10) was used in our own and another laboratory for materials studies. This was mounted in a heavy stand and could therefore be operated throughout most of the audio spectrum without encountering problems of structural resonance. It did not use a strain-measuring LVDT but functioned by virtue of an AC bridge capable of determining the mechanical impedance of the specimen by its relationship to the reflected electrical impedance.

The air for the gas bearings can be either the laboratory bench supply or a compressed-gas cylinder with pressure reduction to 5 p.s.i. The latter is satisfactory because of the small volume flow (about 0.05 cubic feet per minute).

OTHER ASSOCIATED EQUIPMENT

The storage type of oscilloscope affords a convenient and flexible medium for the display of dynamic hysteresis loops. By making the stress and strain signal circuits direct-coupled, plotting can be either rapid or very slow. If a trace is not satisfactory, as is often the case due to movement of living subjects, a touch of a button erases it and provides another one second later. One thus has the ability to select physiological data and eliminate artifacts.

The stress or force function applied to the specimen is the independent variable in this case; the displacement or strain is the resultant, which depends on the material's mechanical impedance. Thus the choice of a suitable function

Figure 14.10 Heavy-duty GBE used for materials testing

generator to drive the GBE is important. It should have a variety of wave shapes (especially sine and ramp functions) with a wide range of very low to medium frequencies. An adjustable positive and negative offset voltage mixed with the AC output is extremely important; this is used as a DC bias to offset gravity when the transducer is in any position but horizontal. It can also be a valuable means of applying a steady bias force – e.g., to prevent lift-off of a probe during a sinusoidal compression series. The same biasing is needed when testing a hair filament with an oscillatory tension force. A Polaroid camera attachment for the oscilloscope and a convenient means of clamping and adjusting the position of the transducer relative to the test subject or specimen are also needed. The GBE requires average experimental competence to yield reliable results. Several investigators have gained sufficient skill in its operation after a few days spent using it in our laboratories while participating in specific test procedures of interest to them.

15
International standardization of instruments used to determine mechanical properties of human skin

T. H. COOK

The formulation of standards for and endorsement of procedures, devices and units used in investigations of the mechanical behaviour of human skin will require the definition of desired mechanical properties and evaluation of existing and future technology in terms of its ability to generate this information.

The objective of this report is to identify issues which must be resolved before standards can be established for measuring and reporting mechanical properties of skin. The discusion will be confined to *in vivo* investigations.

Table 15.1 contains a list of investigators who are currently conducting research on the mechanical behaviour of intact skin. It also contains the type of measurement system used and the measured quantities reported by each investigator. This table shows that there are many different devices used to study the mechanical behaviour of skin and a variety of measured quantities are reported. Unfortunately, a comparison of the results of these studies is impossible. This inability to compare results and possibly corroborate findings is one of the primary reasons why standardization is needed. However, the proliferation of measurement devices is not the only reason for the incomparibility of data. Interpreting the mechanical characteristics is made extremely difficult by the unknown stress and strain states existing in the skin before testing.

To demonstrate their effects and show how various deformation states imposed on the skin by different measurement devices affects the mechanical response, mechanical tests were performed with three devices: (1) a circular suction cup; (2) a race-track suction cup; and (3) a strip biaxial extensometer. Each of these devices imposes a different type of deformation to the skin.

The deformation state applied to the skin by the circular suction cup is non-homogeneous. If the skin was deformed from the undeformed state, an equal biaxial state ($\lambda_1 = \lambda_2 > 1$) would exist at the centre of the inflated skin membrane and a state of pure shear ($\lambda_1 > 1$, $\lambda_2 = 1$) would exist at the

Table 15.1 Methods currently used for investigating the mechanical behaviour of skin

Investigators	*Measurement system*	*Measured quantities*
Alexander, H. and Cook, T. H.[1]	Race-track Suction cup	Material Constants
Barbenel, J. C. and Evans, J. H.[2]	Torsional	Torque and phase angle
Gunner, C. W., *et al.*[3]	Extensometer	Slope of stress–strain curve at a fixed strain
Christensen, M. S., *et al.*[4]	Electrodynamometer	Dynamic spring rate and loss angle
Cook, T. H.[5,6]	Strip biaxial extensometer, race-track suction cup	Slope of terminal portion of stress–strain curves, material constants
Hussein, H. MG., *et al.*[7]	Compressive	Mechanical impedance
Landel, R. F., *et al.*[8]	Skin transducer	Qualitative description
Pierard, G. E. and Lapiere, Ch. M.[9]	Tonometric	Index of deformability, skin extensibility, biologic elasticity
Thacker, J. G., *et al.*[10]	Extensometer	Material constants
Tosti, A., *et al.*[11]	Ballistometer	Coefficient of restitution
Wijn, P. F. F., *et al.*[12]	Extensometer, torsional	Young's modulus, shear modulus

boundary. λ_1 and λ_2 are the extension ratios in the meridional and circumferential directions of the deformed skin respectively. Generally, *in vivo* testing does not permit the deformations to be applied to skin in the undeformed state. The deformations applied to the skin are superimposed on an unknown existing state. Therefore to apply the data generated with the circular suction cup the existing state is ignored, the stress and strain states are taken to be homogeneous and the deformed skin is assumed to be a sector of a sphere. Under these conditions, equations for stress and strain in the skin can be obtained from force equilibrium and geometrical considerations:

$$\sigma = \frac{P}{2t}\left(\frac{d^2}{8\delta} + \frac{\delta}{2}\right) \tag{1}$$

and

$$\lambda = \frac{1 + 4\left(\frac{\delta}{d}\right)^2}{4\left(\frac{\delta}{d}\right)} \arcsin\left[\frac{4\left(\frac{\delta}{d}\right)}{1 + 4\left(\frac{\delta}{d}\right)^2}\right] \tag{2}$$

where σ is the true membrane stress (force per deformed cross-sectional area), λ is the extension ratio (deformed length per undeformed length), P is the vacuum in the cup, t is the deformed skin thickness, d is the diameter of the suction cup and δ is the centre height of the deformed skin.

Equations (1) and (2) are gross simplifications of the actual stress and strain states in the skin. They have been used here because this form of data reduction has been used by others[13] and their use raises the issue of validity of data-reduction techniques while showing the effect of the existing state on data reduction as well as data acquisition.

The deformations applied to the skin by the race-track suction cup are homogeneous. The deformed shape of the skin has been shown to be a sector of a cylinder[6]. It is found from geometrical considerations and force equilibrium that the equations for stress and strain are:

$$\sigma_c = \frac{P}{t}\left(\frac{L^2}{88} + \frac{\delta}{2}\right) \tag{3}$$

and

$$\lambda_c = \frac{1 + 4\left(\frac{\delta}{L}\right)^2}{4\left(\frac{\delta}{L}\right)} \arcsin\left[\frac{4\left(\frac{\delta}{L}\right)}{1 + 4\left(\frac{\delta}{L}\right)^2}\right] \tag{4}$$

where σ_c is the true membrane stress in the circumferential direction of the inflated skin cylinder (the direction of stretch), λ_c is the extension ratio in the circumferential direction, P is the vacuum in the cup, t is the deformed skin thickness, L is the cup width and δ is the skin centre displacement. These equations are simplifications for the present case since the deformations applied by the device are perturbations on the existing state.

The strip biaxial extensometer applies a homogeneous plane-strain state to the skin[5]. Again, if the existing state in the skin is ignored, equations for stress and strain in the direction of stretch can be obtained:

$$\sigma_1 = \frac{F}{W} \tag{5}$$

and

$$\lambda_1 = \frac{\ell_1}{\ell_0} \tag{6}$$

where σ_1 is the true stress in the direction of stretch, λ_1 is the extension ratio in the direction of stretch, F is the force developed in the skin between the pads of the device, W is the width of the pads, t is the deformed skin thickness, ℓ_1 is the deformed length of skin between the pads and ℓ_0 is the initial separation of the pads.

Equations (1) to (6) are used here to present the results of constant-strain rate tests. Since an accurate measurement of skin thickness is not possible, tension ($T = \sigma t$) was used in place of stress.

Two tests were performed on the volar forearm using each device, one with the forearm fully extended (at 0°) and one with the forearm unextended (at 90°). These two positions established different initial stress and strain states in the skin.

The results of these tests showed that the mechanical response is specific to each test. This becomes apparent after inspecting the six curves in Figure 15.1. Each of these tension versus extension ratio curves has been obtained from the same material, yet the response characteristics are very different in each case. This is expected and is a result of the existing state and the different deformations imposed on the skin by the three devices. Any method used to quantify these curves, other than a valid constitutive equation, does not yield results representative of intrinsic material properties of skin.

Until a method is developed for determining the existing states of stress and strain in the skin and a tractable constitutive equation is uncovered, the use of

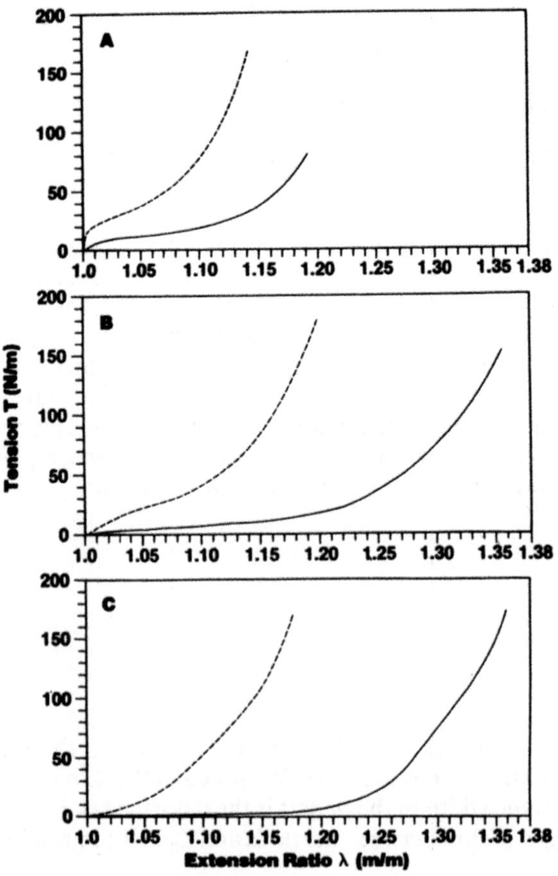

Figure 15.1 Tension versus extension ratio curves obtained with: (A) a circular suction cup; (B) a race-track suction cup; and (C) a strip biaxial extensometer. (---) = Arm extended; (— —) = arm unextended

validated testing devices and standardized quantifying techniques will help in interpreting and comparing the data of mechanical properties reported by various investigators.

The following mechanical properties are suggestions which may be useful when assessing changes in the skin's mechanical behaviour: (1) secant modulus; (2) tangent modulus; (3) hysteresis; (4) the slope of a straight line curve fit to the data; and (5) constants in a curve fit other than a straight line. The deformation state applied to the skin by the test device, its dimensions and the subject's position (existing state) should also be reported for the mechanical properties thought to be most valuable.

This discussion emphasizes the importance of knowing the state imposed on the skin by a particular testing device and its effect on data reduction. It has also shown the pronounced effect that changes in the existing state of stress and strain have on the results of skin tests. It has questioned the use of terminology, specifically the use of the terms material property, mechanical property and mechanical response or behaviour. Terms like Young's modulus have been used to describe the mechanical behaviour of skin. They are well-defined properties of a material and should not be used to describe the non-linear viscoelastic behaviour of skin. The use of the term material property should be reserved for a property which is independent, over a large range of the material's behaviour, of the measurement system used to obtain that property. However, several defined mechanical properties can be used to measure the skin's response to standardized or validated mechanical testing devices and testing procedures.

References

1. Alexander, H. and Cook, T. H. (1977). Accounting for natural tension in the mechanical testing of human skin. *J. Invest. Dermatol.*, **69**, 310
2. Barbenel, J. C. and Evans, J. H. (1977). The time dependent mechanical properties of skin. *J. Invest. Dermatol.*, **69**, 318
3. Gunner, C. W., Hutton, W. C. and Burlin, T. E. (1979). The mechanical properties of skin *in vivo* a portable hand-held extensometer. *Br. J. Dermatol.*, **100**, 161
4. Christensen, M. S., Hargens, C. W., Nacht, S. and Gans, E. H. (1977). Viscoelastic properties of intact human skin: instrumentation, hydration effects and the contribution of the stratum corneum. *J. Invest. Dermatol.*, **69**, 282
5. Cook, T. H. (1978). Mechanical properties of human skin. *Proceedings of the 3rd International Congress of Biorheology*. August 28 (La Jolla)
6. Cook, T., Alexander, H. and Cohen, M. (1977). Experimental method for determining the 2-dimensional mechanical properties of living human skin. *Med. Biol. Eng. Comput.*, **15**, 381
7. Hussein, H. MG., Thompson, D. E. and Perritt, R. Q. (1978). Impedance of human soft tissue under various static preloads. Presented at the *31st Annual Conference on Engineering in Medicine and Biology*. October 21–25 (Atlanta)
8. Landel, R. F., Peng, T. J. and Brody, G. S. (1979). *In vivo* measurement of stress–strain and stress relaxation behavior of human skin. *J. Rheol.*, **23**, 113
9. Pierard, G. E. and Lapiere, Ch. M. (1977). Physiopathological variations in the mechanical properties of skin. *Arch. Dermatol. Res.*, **260**, 231
10. Thacker, J. G., Iachetta, F. A., Allaire, P. E., Edgerton, M. T., Rodeheaver, G. T. and Edlich, R. F. (1977). *In vivo* extensometer for measurement of the biomechanical properties of human skin. *Rev. Sci. Instrum.*, **48**, 181

11. Tosti, A. T., Compagno, G., Fazzini, M. L. and Villardita, S. (1977). A ballistometer for the study of the plastoelastic properties of skin. *J. Invest. Dermatol.*, **69**, 315
12. Wijn, P. F. F., Brakkee, A. J. M., Stienen, G. J. M. and Vendrik, A. J. H. (1976). Mechanical properties of the human skin *in vivo* for small deformations; a comparison of uniaxial strain and torsion measurements. In Kenedi, R. M. and Cowden, J. M. (eds.) *Bedsore Biomechanics*, pp. 103–108. (Baltimore: University Park Press)
13. Grahame, R. (1970). A method for measuring human skin elasticity *in vivo* with observations on the effects of age, sex and pregnancy. *Clin. Sci.*, **39**, 223

16
Changes in the mechanical properties of intact guinea pig skin resulting from ultraviolet irradiation

T. H. COOK, R. MITCHELL and D. DARR

INTRODUCTION

Skin, by virtue of its composition and location, is an important system in the study of extracellular ageing and the effects of radiation on both cellular and non-cellular compartments. Indeed, skin is of prime importance for protecting underlying tissues, water and salt balance, temperature regulation and sensation. Although much is known about the histology of normal skin (both epidermal and dermal layers), there are still many unanswered questions concerning the effects of normal ageing on structure and function of skin and, more specifically, how ultraviolet radiation influences the rate of skin ageing.

Most of the form, strength and 'elasticity' of skin is determined by the amount and nature of the collagen, elastic tissue and ground substance of the dermis. All of these structural elements play major roles in both the structural and functional integrity of healthy skin. Many of the changes to the skin which occur during normal ageing or in many pathological conditions are a result of alterations in content or conformation of these dermal constituents. The only way to *directly* measure ageing- or disease-related changes in dermal constituents is to sample by biopsy. However, biopsies are inconvenient, time-consuming, expensive and often not possible. Hence there is much interest in the development of non-invasive mechanical testing devices to be used on the surface of skin. The mechanical properties of each of the skin components and their mutual interaction establishes the response of whole skin to deformations and loads. Data of mechanical properties, therefore, coupled with a multidimensional constitutive relation for skin and a correlation with skin constituents, provides a quantitative means of monitoring and evaluating changes in dermal microstructure. Relating age and ultraviolet exposure to the mechanical properties measured *non-invasively* on the surface of skin can provide a useful, quantitative means of assessing degenerative changes in skin without the need for excising skin samples. This knowledge is also a necessary prerequisite for establishing a sensitive, non-invasive method for evaluating the effectiveness of treatments for various diseases associated with ageing skin.

The objective of this pilot study was to determine whether the effects of repeated exposure to ultraviolet radiation of known wavelengths and intensity on the mechanical properties of guinea pig skin could be measured non-invasively and to separate these effects from intrinsic changes, due to ageing. Biochemical analyses of irradiated and non-irradiated skin samples were carried out to correlate changes in the mechanical properties with corresponding changes in the skin's constituents.

MATERIALS AND METHODS

Abdominal skin in four groups of two, 6-month-old albino guinea-pigs was irradiated with varying known doses of UV light over a 5-month period. One additional group of two animals served as controls. Mechanical tests were performed on all animals before irradiation and again after each group had received its total dosage. Biochemical tests were done on skin samples to assess soluble and insoluble collagen, type I to type III collagen ratios, and total glycosaminoglycans.

Two Westinghouse FS20 sunlamps and two Sylvania Black/light Blue tubes were used as a UV source. These fluorescent tubes were housed in an aluminium reflector 15 cm from the bottom of the animal cage. The bottom of the cage was made from a UV transmitting, acrylic sheet (3 mm Acrylite OP-1). The spectral characteristics of the fluorescent tubes (Figure 16.1) were determined with a monochromator and the total irradiance was measured with an Eppley thermopile. The total intensity was 1325 $\mu W/cm^2$; 62 % of the energy was contained in the UV-A range and 38 % in the UV-B range.

The abdomen of each animal was initially sheared with clippers and then shaved with an electric shaver. The strip biaxial extensometer developed by Cook[1] was used to determine three baseline stress–strain curves for the abdominal skin of each animal. All animals were anaesthetized (during mechanical testing) to prevent movement.

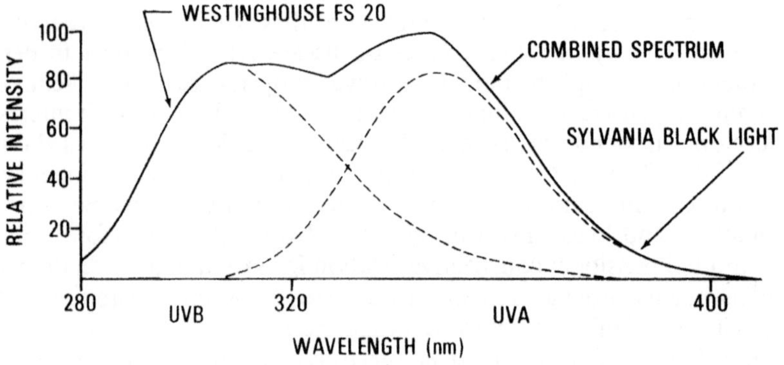

Figure 16.1 Relative intensity versus wavelength spectrum for UV source

For irradiation the animals were placed in an exposure cage about 15 cm from the ultraviolet source. One half-hour exposure was given every Tuesday, Thursday and Saturday. Toward the end of the 5-month period, one group of animals received 1-hour exposures. Two animals were removed from the study after 3 hours (14.3 J/cm^2), 7.5 hours (35.8 J/cm^2), 21 hours (100 J/cm^2) and 39 hours (186 J/cm^2) of total exposure (total dose).

Three or four mechanical tests were performed with the strip biaxial extensometer on the irradiated abdominal skin in each group. These tests were run over a 3–4 day period after the last day of exposure for the particular group. The animals were killed with an overdose of sodium pentabarbitol and skin harvested for biochemical analysis. The control group was tested and killed at the same time as the last group, which received 186 J/cm^2. Therefore, these animals were the same age.

Skin was removed in small pieces (1 cm^2) from the animals' abdomen and back and quick frozen in a dry ice-alcohol-n-hexane bath, followed by storage in a freezer at −20 °C. When samples were needed, pieces were thawed, blotted of excess water, and weighed.

For total protein and collagen analysis, 100–300 mg of tissue (wet weight) were taken up in 25 ml of 0.5 mol/l acetic acid and homogenized by a Polytron homogenizer. The homogenate was left for 18 hours at 4 °C and was then centrifuged at 100 000 *g* for 1 hour. The supernatant was kept and the pellet redissolved in additional acetic acid and again left for 18 hours at 4 °C. This procedure was repeated twice. After the fourth centrifugation a negligible amount of protein and collagen was found in the supernatant. The protein in the acid-soluble and insoluble fraction was determined by the method of Lowry[2]. The total collagen was determined by assaying for hydroxyproline according to the method of Parekh and Jung[3].

For measuring total glycosaminoglycans (GAGs), about one gram of tissue was retrieved and homogenized in 20 mol/l Tris–HCl, pH 8. The GAG extraction procedure used by Conrad *et al.*[4] was followed, except that papain was used instead of pronase. Uronic acid was analysed according to the method of Bitter and Muir[5]. This gives a fair estimate of the total glycosaminoglycan in tissue.

For type I and type III collagen extraction, about one gram of skin was taken from the freezer and homogenized in 1.0 mol/l NaCl–0.05 mol/l Tris–HCl, pH 7.5. The homogenate was left for 4 days, after which the sample was centrifuged and placed in 0.5 mol/l acetic acid and left for another 4 days. Insoluble type I and type III collagen was then extracted by the method of Chung and Miller[6]. The amount of collagen was determined by assaying for the amount of hydroxyproline according to the procedure of Parekh and Jung[3].

RESULTS AND DISCUSSION

To measure changes which had taken place in the stress–strain curves the terminal portion (quasilinear portion) of each curve was fitted to a straight

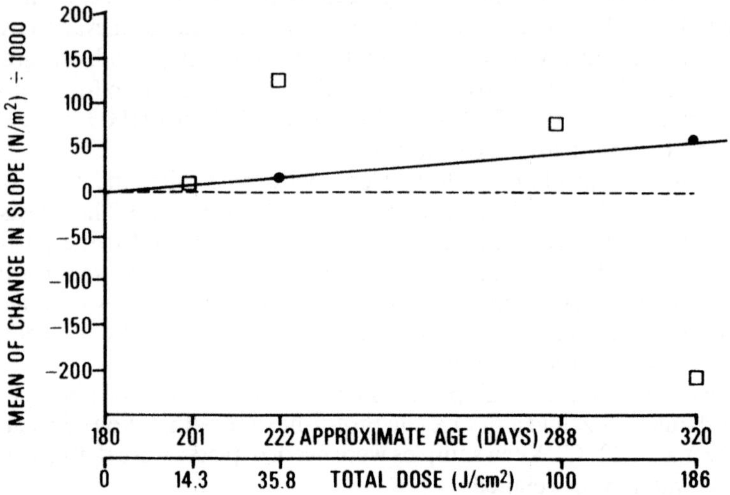

Figure 16.2 Mean changes in skin stiffness (as mean change from mean slope of baseline stress–strain curves) with age and UV dose. ● = Control animals. □ = Irradiated animals

line. The slope (stiffness) of the straight line was used as an index of change in the stress–strain curves.

The slopes of the baseline stress–strain curves for each group of two animals were pooled and a mean slope computed for each group. Each of the slopes computed for the post-irradiation curves were subtracted from the appropriate baseline mean slope. The mean group value of these differences was calculated.

Figure 16.2 presents the mean of the differences for each group and the control group plotted against the animals' age and the total dosages. Points on this plot which lie above the broken line represent an increased stiffness and those below the line represent a decreased stiffness. This data shows an increased stiffness of animal skin in the control group with increased age. With increasing doses of radiation, the data again shows a progressive increase in skin stiffness, but this is followed by a dramatic loss in stiffness in those animals which received the highest UV dose. Alpermann and Vogel[7] reported that the magnitude of some mechanical parameters showed an increase during maturation and diminished during senescence. The preliminary data presented in Figure 16.2 may represent an acceleration of this process.

Three different biochemical variables were assayed to correlate changes seen in mechanical properties with changes in dermal constituents responsible for those mechanical properties. Soluble and insoluble collagen, type I to type III collagen ratios, and total glycosaminoglycans were measured. Multiple assays were performed in each skin sample to establish reproducibility of testing procedures and to determine a mean value for assayed material in individual animals. These mean values were pooled for each group and the group means were compared. Because the number of animals in each group was small and the variation between animals large only trends will be discussed here. The biochemical results are presented in detail elsewhere[8].

Changes in dermal collagen composition correlate with and partly account for mechanical property changes. With both age and UV exposure dermal collagen because increasingly insoluble, and with large doses of radiation the type I to type III ratio of collagen decreased. A small but insignificant increase was detected in dermal glycosaminoglycan content with large doses of UV radiation.

CONCLUSION

A major objective of this study was to test the sensitivity of the Cook strip biaxial extensometer for *in vivo*, non-invasive measuring of the stress–strain properties of skin. This device was shown to be sensitive to changes in the mechanical properties of skin caused by age and ultraviolet radiation. Its value as a diagnostic tool is enhanced by virtue of its non-invasive, *in vivo* design. More detailed information on changes in collagen types and structure and knowledge of changes in glycosaminoglycan content and elastic tissue alterations will help in increasing the use of this device in the diagnosis, and monitoring or treatment, of skin diseases.

Acknowledgement

Parts of this work were taken from a thesis submitted by one of the authors (D. D.) to the Pennsylvania State University in partial fulfilment of the requirements of the M.S. degree.

References

1. Cook, T. H. (1978). Mechanical properties of human skin. *Proceedings of the 3rd International Congress of Biorheology*, August 28 (La Jolla)
2. Lowry, D. H., Rosebrough, N. J., Farr, A. L. and Randall, R. J. (1954). Protein measurement with the folin phenol reagent. *J. Biol. Chem.*, **193**, 265
3. Parekh, A. C. and Jung, D. H. (1970). An improved method for determination of total hydroxyproline in urine. *Biochem. Med.*, **4**, 446
4. Conrad, C. W., Hamilton, C. and Haynes, E. (1972). Difference in glycosaminoglycans synthesized by fibroblast-like cells from chick cornea, heart, and skin. *J. Biol. Chem.*, **252**, 6861
5. Bitter, T. and Muir, H. M. (1962). A modified uronic acid carbazole reaction. *Anal. Biochem.*, **4**, 330
6. Chung, E. and Miller, E. H. (1974). Collagen polymorphism: Characterization of molecules with chain composition $(\alpha 1(III))_3$ in human tissues. *Science*, **183**, 1200
7. Alpermann, H. and Vogel, H. G. (1978). Effect of repeated ultraviolet irradiation on skin of hairless mice. *Arch. Dermatol. Res.*, **262**, 15
8. Darr, D. J. (1979). Effects of age and repeated ultraviolet exposure on the mechanical, biochemical and histochemical properties of female, albino, guinea-pig skin. *MS Thesis*, Pennsylvania State University

17 The alinear viscoelastic properties of human skin *in vivo* related to sex and age

P. F. F. WIJN, A. J. M. BRAKKEE, J. P. KUIPER and A. J. H. VENDRIK

INTRODUCTION

From the published data on the mechanical properties of the human skin, it seems that authors using different measuring techniques obtain different data[1,2]. We examined the mechanical properties of the human skin of the calf *in vivo* for small deformations in several subjects of both sexes and different ages with two different measuring techniques, uniaxial strain and torsion, and compared the results. The aim of this study was to clarify apparent discrepancies on the basis of an improved mechanical skin model in which the fibrous structure of the skin has been taken into account and to determine the normal values of these skin properties.

METHODS

In uniaxial strain studies two rectangular tabs were attached to the skin with a double-sided adhesive tape (Scotch 400). One of these tabs was fixed to the body of the apparatus (Figure 17.1) and the other was connected to a cylindrical magnet on air bearings which moved freely along the axis of a cylindrical coil. A force was generated by an electric current in the coil. Changes in the distance between the tabs were detected by a displacement transducer (Philips PR9314/10). By varying the width of the tabs and their original distance apart, the width and effective length of the piece of skin were defined. Displacement and force were converted into strain (ε, relative increase in length) and stress (σ, force divided by cross-sectional area) using a constant value for the skin thickness[3] of 1.3 mm. From the obtained values of ε and σ coefficients of elasticity (E) were calculated.

In torsion experiments a disc and a concentric guard ring (Figure 17.2) were taped to the skin. The disc was mounted on the rotation axis of a coil which

Figure 17.1 Uniaxial strain apparatus attached to the skin of the calf in the direction of Langer's lines

moved freely within the homogeneous field of a fixed magnet[4]. A twisting moment was generated by a current in the coil. The rotation of the disc was detected by a rotation transducer (WT1 Cam. Bauer AG). The effective radii of the disc and the guard ring in the skin were determined by comparing the results found with different sized discs and guard rings.

Assuming homogeneity and isotropy of the medium, Vlasblom[4] has presented an equation with which the shear modulus (G) can be calculated from the relationship between the torsion angle (ϕ) and the twisting moment (M). Since in the theory for homogeneous isotropic media the relationship between E and G is known, comparing the results of both techniques checks the applicability of that theory to the skin.

MEASUREMENTS

Stepwise loads were used to study the time-dependent behaviour of the skin. In the response three regions were distinguished: an instantaneous, a delayed and a permanent deformation; U_A, U_B and U_C respectively (Figure 17.3). These three regions can be considered as the results of three different viscoelastic

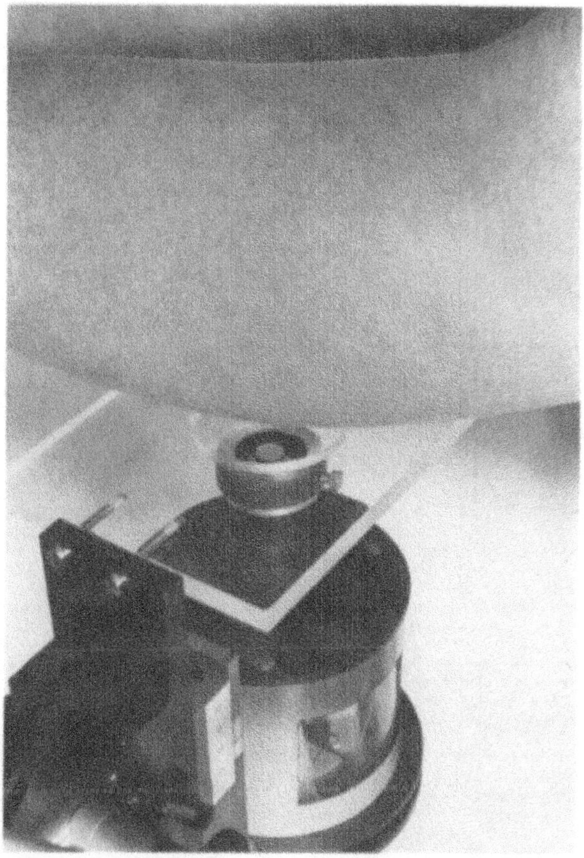

Figure 17.2 The torsion device underneath the calf ready for attachment

processes in the skin: a purely elastic, a delayed elastic and a purely viscous process. Each process was studied by comparing the amplitudes of the specific region of the response as a function of the magnitude of the load. In this paper only the purely elastic process will be discussed.

Apparently even for the smallest deformations the skin behaved alinearly. Furthermore, a high anisotropy was found (Figure 17.4); the stiffest direction coincided with the dominant direction of the fibres in the skin as deduced from the course of the so-called Langer's lines. When the stress to strain ratios were plotted as a function of the stress (Figure 17.5, using the data of Figure 17.4) two fairly straight and strikingly parallel lines were found.

In torsion measurements the shape of the M–ϕ curve was similar to the curves presented in Figure 17.4. The M–ϕ ratios, when plotted as a function of the twisting moment, again showed a fairly straight line.

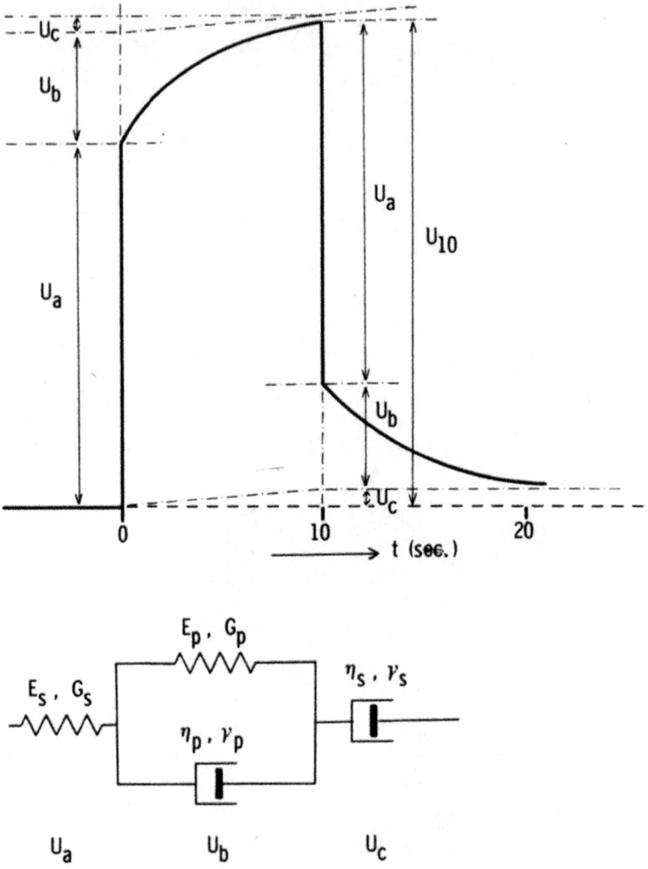

Figure 17.3 Schematic representation of the response of the skin on a stepwise load with a duration of 10 seconds, and a basic model of the three viscoelastic processes. For key see text

DESCRIPTION OF THE ALINEAR BEHAVIOUR

According to Fung[5] the alinear behaviour of various soft tissues can be described by the equation:

$$\frac{d\sigma}{d\varepsilon} = E_o + k \cdot \sigma \tag{1}$$

This means that the tangent modulus of elasticity $\left(E' = \frac{d\sigma}{d\varepsilon}\right)$ depends in a linear way on the stress. E_o is the coefficient of elasticity for zero stress and $k \cdot \sigma$

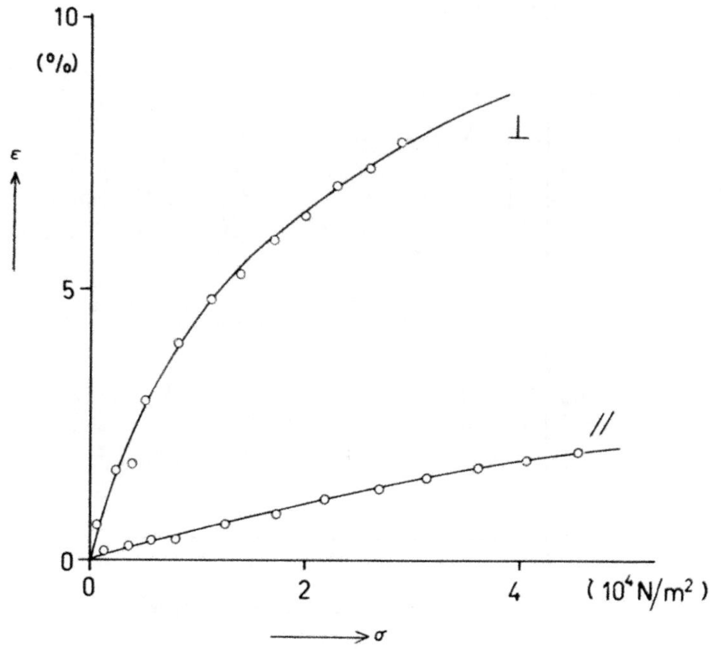

Figure 17.4 Stress–strain relationships of the instantaneous response of the skin on the calf in the direction along (direction along (||) and across (⊥) Langer's lines

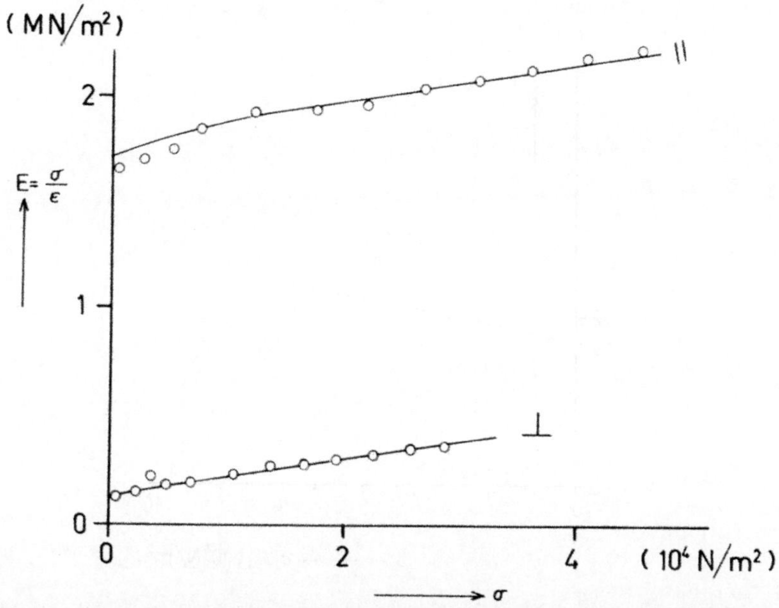

Figure 17.5 Stress–strain ratio's (coefficient of elasticity, E) of the instantaneous responses of the skin on the calf as a function of the stress. The same data are used as in Figure 17.4

$$E = \frac{d\sigma}{d\varepsilon} = E_0 + k.\sigma$$

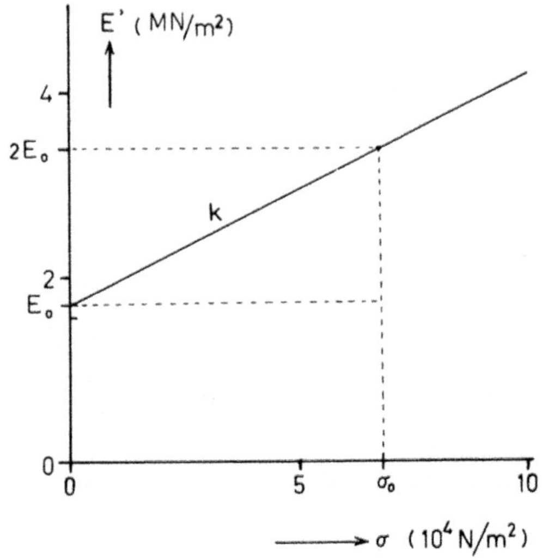

$$E = \frac{\sigma}{\varepsilon} = \frac{k}{\ln.(1 + \frac{k\sigma}{E_0})} \cdot \sigma$$

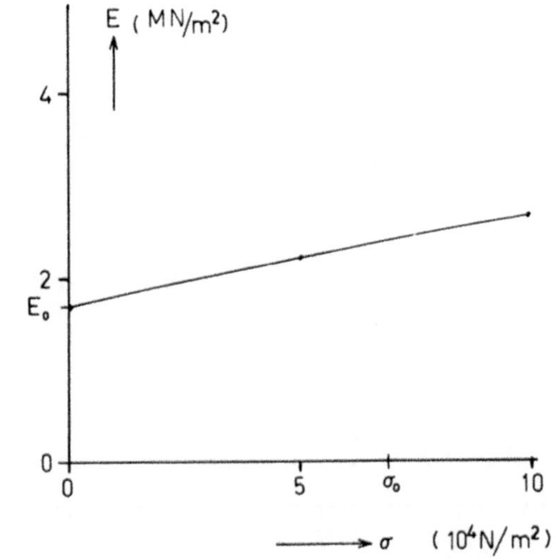

Figure 17.6 Stress dependency of (a) tangent modulus and (b) stress–strain ratio according to Fung's equation. For key see text

coefficient of alinearity. Using Fung's equation the stress to strain ratio becomes:

$$E = \frac{\sigma}{\varepsilon} = \frac{k \cdot \sigma}{\ln\left(1 + \frac{k \cdot \sigma}{E_0}\right)} \tag{2}$$

Both relations are shown in Figure 17.6. The values of E_0 and k in this figure were obtained from the results of experiments along Langer's lines. A comparison of Figure 17.6(b) and Figure 17.5 showed the validity of Fung's equation in describing the alinear behaviour of the skin. The corresponding stress–strain equation is:

$$\varepsilon = \frac{1}{k} \cdot \ln\left(1 + \frac{k \cdot \sigma}{E_0}\right) \tag{3}$$

In all our investigations (in over 50 subjects) we found an excellent correlation between the experimental points (from 12 loads) and the optimal fit according to equation (3) ($r > 0.998$). The parallelism of the two lines in Figure 17.5 corresponds with the result that the k values in both directions are equal. Thus the anisotropy was only reflected in the different E_0 values for different directions. $E_{0\parallel}$, k and the maximal anisotropy A, defined as the ratio of $E_{0\parallel}$ to $E_{0\perp}$, characterize the anisotropic and alinear behaviour of the purely elastic process in the skin.

CONSIDERATIONS WITH RESPECT TO SKIN STRUCTURE

In a previous paper we[6] showed that, assuming homogeneity and isotropy, the results of uniaxial strain experiments do not agree with those of torsion measurements. Even with a modified theory, supposing orthotropy[7], and taking into account the anisotropy which is found experimentally, no agreement can be obtained. For this reason a new model, a fibre model[8], has been developed which is based upon the morphology of the skin[9] and upon our experimental results (a high anisotropy of the coefficients of elasticity with an alinearity which is isotropic). The fibre model is based mainly on the following two assumptions:

(1) the stress–strain relationship of the participating skin in a specific direction is a result of the summated properties of the elastin fibres;
(2) only elastin fibres which make a very small angle with the direction of the applied force participate in the straining process.

With these assumptions an improved description of a torsion experiment is found with which torsion results can be understood from the findings of uniaxial strain measurements[8]. A direct consequence of the validity of the model is that the parameters E_0, k and A have a specific meaning with respect to the morphology of the skin: E_0 is a measure of the amount of elastin fibres participating in a specific direction; k is a measure of elastin fibre quality; and A is a measure of the angular distribution of the elastin fibres.

NORMAL VALUES RELATED TO SEX AND AGE

We have determined the normal range of each parameter in 49 subjects (25 males; 24 females) in four age groups (of about 20, 35, 50 and 65 years) who appeared normal. Because of the large differences between subjects, age and sex dependencies were studied with a twofold analysis of variance[10]. When significant dependencies were found they were taken into account in calculating the normal range. The inner and outer limits of the percentiles P_5 and P_{95} (determined with a reliability of 90%[11]) were used to indicate the borders of the normal range. Individual results inside the inner limits (dotted lines) were considered normal, those between the inner and outer limits (thin lines) suspicious and those outside the outer limits abnormal. In Figures 17.7, 17.8 and 17.9 the results of $E_{o\parallel}$, $k_{\perp}$ and A are presented respectively. In these figures the subscripts on the parameters mark the region of the viscoelastic

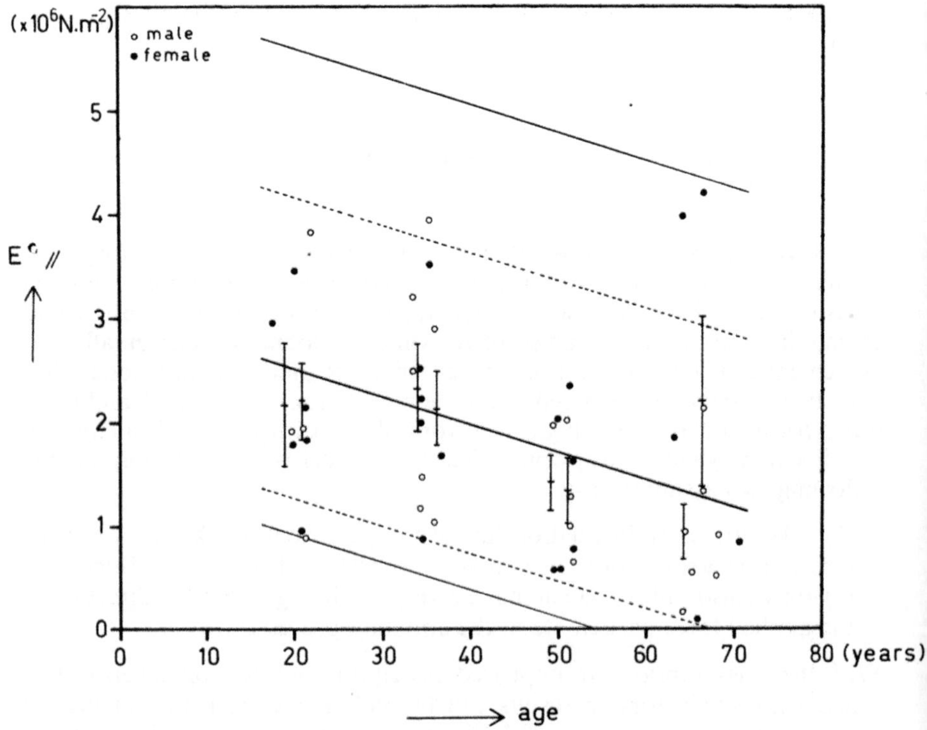

Figure 17.7 The normal behaviour of the coefficient of elasticity for zero stress ($E_{o\parallel}$) related to sex and age. ○ = Result in male subject. ● = Result in female subject. For further details see text

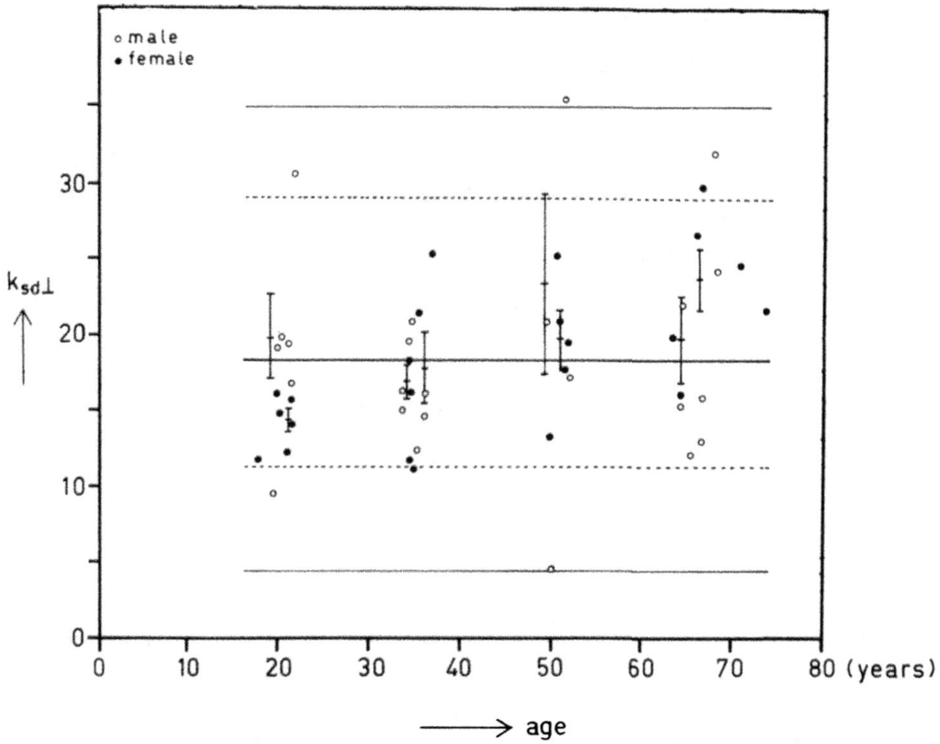

Figure 17.8 The normal behaviour of alinearity ($k_\perp$) related to sex and age. ○ = Results in male subjects. ● = Results in female subjects. For further details see text

response: s, serial spring (purely elastic process); d, 'down' (the response after loading). For each age group and both sexes the mean value and the SEM are shown separately (results in males on the left). The mean of all measured values or the result of linear regression is shown with a thick line. Figure 17.7 shows that $E_{o\parallel}$ is independent of sex and decreases between 20 and 65 years by a factor of two. A similar decrease in the amount of elastin fibres has been reported[12]. The alinearity $k_\perp$ (Figure 17.8) did not show any relationship with sex and/or age. The anisotropy A (Figure 17.9) was also independent of age. On average females showed a somewhat higher anisotropy than males.

PSEUDO XANTHOMA ELASTICUM

As an example of the potential clinical value of the parameters the results in five patients suffering from pseudoxanthoma elasticum (PXE) will be discussed. PXE is a hereditary disease resulting in a generalized dystrophy of the elastic tissue[13]. The hypothesis is that the elastin fibres degenerate and fragment in the presence of calcium deposits. In Table 17.1 the individual and

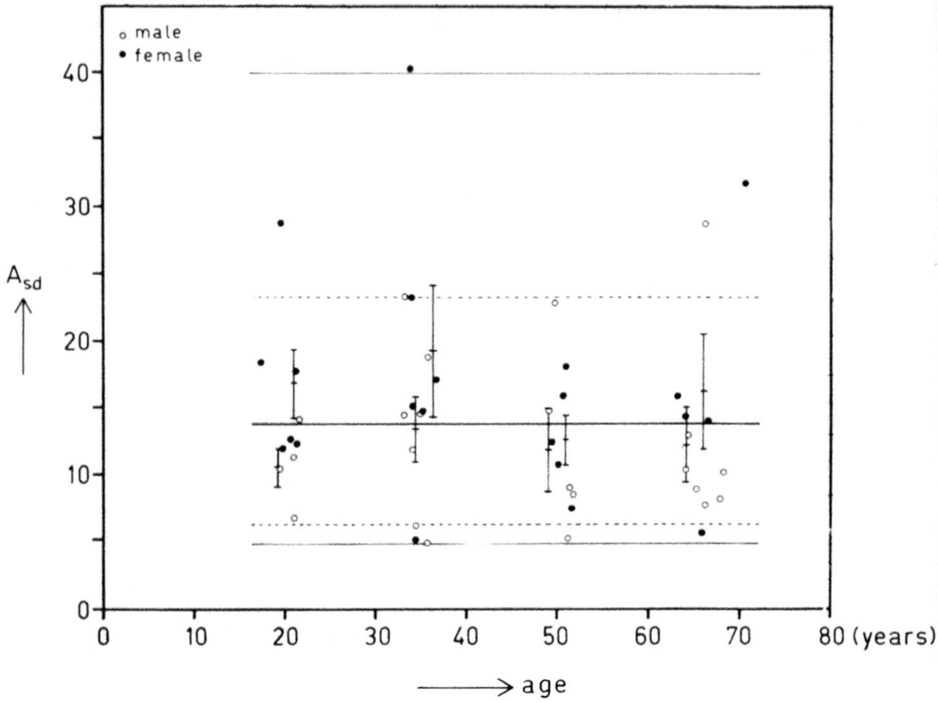

Figure 17.9 The normal behaviour of anisotropy (A) related to sex and age. ○ = Results in male subjects. ● = Results in female subjects. For further details see text

mean results are presented as percentages of the normal values for that age and sex. From Table 17.1 it can be concluded that the individual subjects do not show significantly abnormal results but that the five patients as a group show that in PXE the $E_{o\|}$ value is reduced to about half the normal value ($P < 0.001$, normal distribution), $k_{\perp}$ is normal and A is about 60% of the normal value ($P = 0.004$). Combining the first and the third result one can conclude that $E_{o\perp}$ is about 80% of its normal value; this difference is not significant. Harvey *et*

Table 17.1 Individual results and mean values of skin parameters in five subjects with PXE. Results are given as percentages of normal values

Parameter	*Subject*					*Mean*
	I	II	III	IV	V	
$E_{o\|}$	28*	42*	40*	77†	50*	47
$k_{\perp}$	86†	117†	62†	175*	102†	108
A	40*	38*	115†	53*	61*	61

* Suspicious result.
† Normal result.

al.[14] reported a decrease in skin stiffness of the same magnitude for the 'autosomal dominant' types of PXE. The present results suggest that the fibre quality is normal but the number of elastin fibres is decreased mainly in the direction along Langer's lines. We believe that this gives new additional information on the functional aspects of the disease.

References

1. Sanders, R. (1973). Torsional elasticity of human skin *in vivo. Pflüg. Arch.*, **342**, 255
2. Vogel, H. G. (1977). Untersuchungen über die Mechanischen Eigenschaften der Haut. *Lymphology*, **1/2**, 28
3. Finlay, B. (1971). The torsional characteristics of human skin *in vivo. Bio-med. Eng.*, **6/12**, 567
4. Vlasblom, D. C. (1967). *Skin Elasticity. PhD Thesis*, University of Utrecht, The Netherlands
5. Fung, Y. C. B. (1972). Stress–strain-history relations of soft tissues in simple elongations. In Fung, Y. C. B., *et al.* (eds.) *Biomechanics, its Foundations and Objectives*, pp. 181–208. (Englewood Cliffs: Prentice Hall)
6. Wijn, P. F. F., Brakkee, A. J. M., Stienen, G. J. M. and Vendrik, A. J. H. (1976). Mechanical properties of human skin *in vivo* for small deformations; a comparison of uniaxial strain and torsion measurements. In Kenedi, R. M., *et al.* (eds.) *Bed Sore Biomechanics*, pp. 103–108. (London: Macmillan)
7. Lanir, Y. and Fung, Y. C. (1974). Two-dimensional mechanical properties of rabbit skin – II. Experimental results. *J. Biochem.*, **7**, 171
8. Wijn, P. F. F., Brakkee, A. J. M. and Vendrik, A. J. H. (1978). The alinear viscoelastic properties of the human skin *in vivo* for small deformations. In Reul, H., (ed.) *Conference Digest of the 1st ICMMB, Aachen 1978*, pp. VI–207–210. (Baden-Baden: Witzstrock)
9. Gibson, T. and Kenedi, R. M. (1967). Biomechanical properties of skin. *Surg. Clin. North Am.*, **47**, 279
10. Scheffé, H. (1959). *The Analysis of Variance.* (New York: Wiley)
11. Rümke, C. L. and Bezemer, P. D. (1972). Methoden voor de bepaling van normale waarden. *II*–nieuwe methoden. *Ned. T. Geneesk.*, **116/35**, 1559
12. Dick, J. C. (1951). The tension and resistance to stretching of human skin and other membranes with results from a series of normal and oedematous cases. *J. Physiol.*, **112**, 102
13. Ross, R. (1973). The elastic fibre. A review. *J. Histochem. Cytochem.*, **21**, 199
14. Harvey, W., Pope, F. M. and Grahame, R. (1975). Cutaneous extensibility in pseudoxanthoma elasticum (PXE). *Br. J. Dermatol.*, **97**, 679

18
The rheological properties of human skin and scar tissue

G. S. BRODY, T. J. PENG and R. F. LANDEL

INTRODUCTION

We have described the rationale and methodology for *in vivo* strip biaxial testing of human skin[1–4]. In summary, we have designed a force–displacement transducer which is bonded to the forearm skin using cyanoacrylate glue (Eastman 910). The transducer's configuration constrains the skin sufficiently in its lateral extent to prevent any appreciable change in its transverse dimension. Thus changing the longitudinal dimension provides information on two dimensions.

Skin stretch is obtained by the subject flexing and extending the elbow so that the range remains within physiological limits. A mechanical stretcher was also used to range the same skin and comparable data was obtained.

Most tissue when stretched and relaxed without exceeding the elastic limit produces a simple hysteresis loop. This effect can be modelled as a non-linear spring dash pot diagram. Repetitive cycling generates a great deal of information.

The viscoelastic properties of normal human flexor forearm skin have been characterized. We have confirmed the observations of others that human skin is anisotropic and rate-independent, at least over a decade range.

THE FIRST CURVE EFFECT

When the strip biaxial transducer is glued to the proximal volar forearm, the elbow is cyclically flexed and extended to generate typical viscoelastic curves as seen in Figure 18.1, for normal skin and in Figure 18.2 for early burn scar. A phenomenon dubbed 'the first curve effect' has been seen throughout all our testing (Figure 18.1) and is characterized by a different shape of the initial curve. This effect has been noted by Fung[5] but he deferred analysis. When viscoelastic material such as a nylon thread is stretched and released the first cycle will show a different initial 'preconditioning' curve from subsequent cycles, which generate relatively repetitive properties. The nylon will not recover its prestretched condition for very long periods of time, if ever. It must

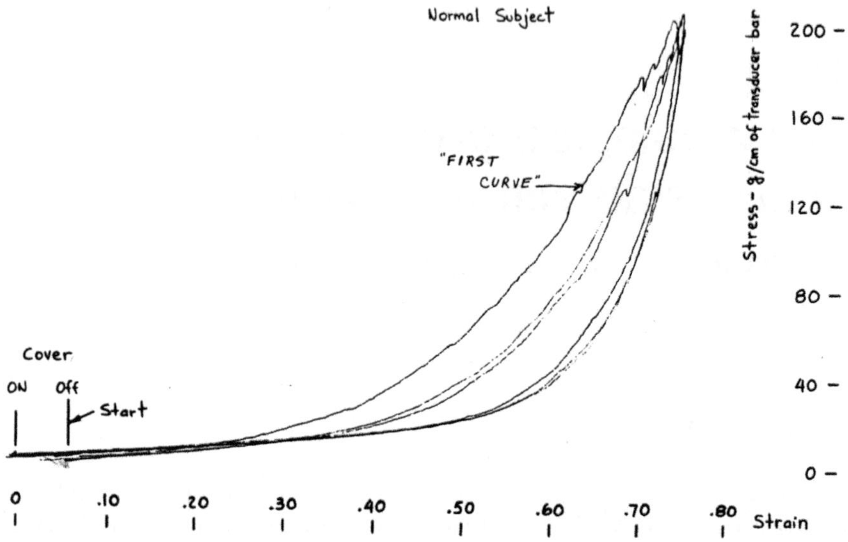

Figure 18.1 Three cycles of extension and flexion of normal forearm skin yielding typical stress–strain curves

be noted, however, that skin is not virgin and has always been almost continuously stretched and relaxed. Therefore, human skin does not 'precondition' in a material sense as it is restored to its original state, even after a brief rest. This does not seem to be an artifact, as it can be reproduced by resting a few minutes between cycles and is seen to some degree in all test modes.

THEORETICAL EXPLANATION OF FIRST CURVE EFFECT

A tentative heuristic phenomenological explanation for this effect has been developed as follows.

When we examine the repetitive cycling in normal skin, we see a drawing together of the load and unload curves with time, suggesting a lessening of the viscous effect and a tendency for the elastic properties to predominate (Figure 18.3). Thus skin can be characterized (Figure 18.4) as a nonlinear spring in series with two valved dash pots. The first dash pot contains a low viscosity fluid (water?) and the second a high viscosity material (other contents of the ground substance). The 'valve' permits a different rate of flow of the viscous material on extension and compression. The lymphatic system may be the mediator of this effect as it is unidirectional and normally under negative pressure and thus could function as the 'valve'. Water passes into the lymphatics more rapidly than the other substances that constitute the lymph.

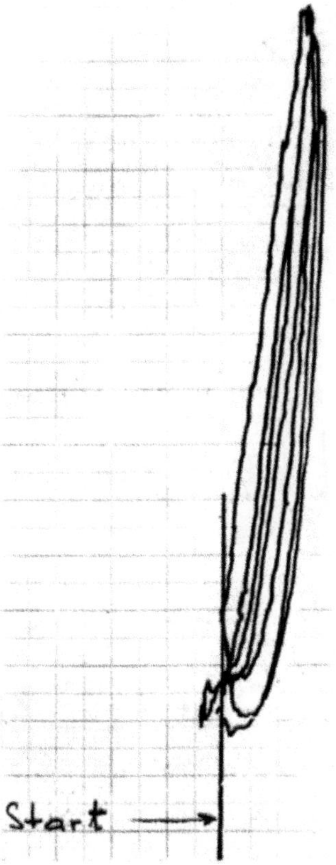

Figure 18.2 Stress–strain curves produced by cycles of extension and flexion of early burn scar tissue

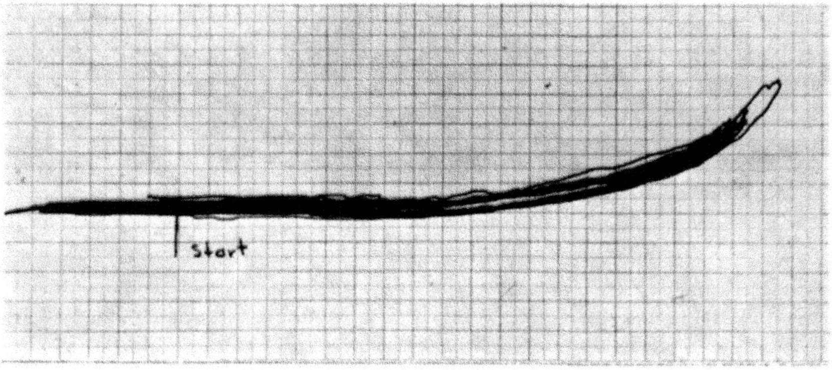

Figure 18.3 Stress–strain curves produced by 25 cycles of extension and flexion of normal forearm skin

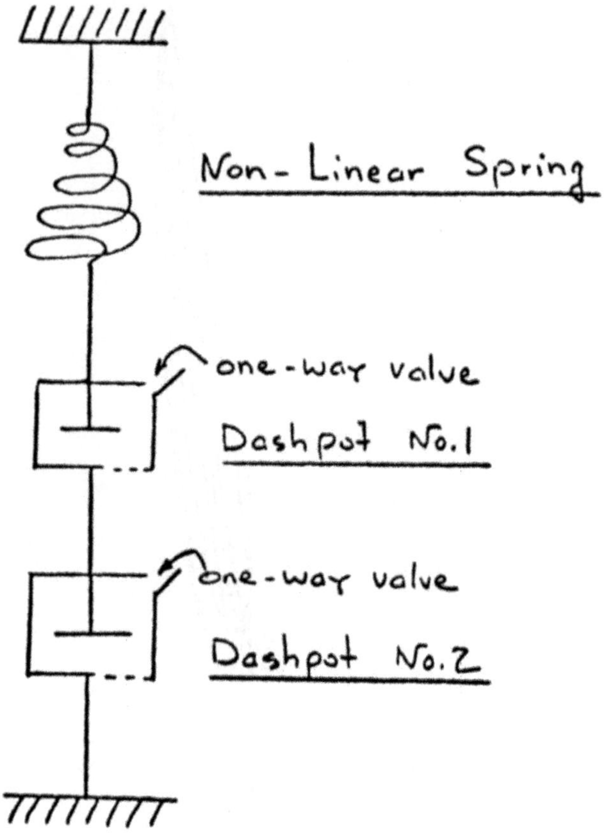

Figure 18.4 Schematic representation of skin

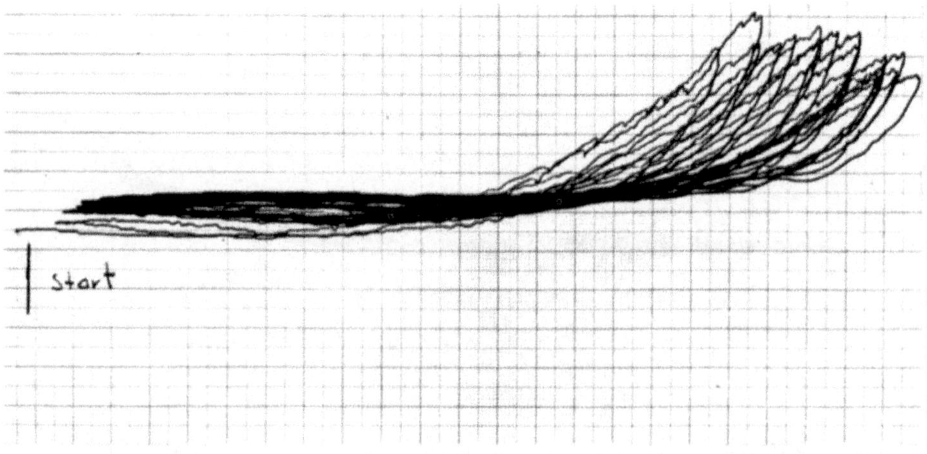

Figure 18.5 Stress–strain curves produced by 25 cycles of lymphoedematous skin

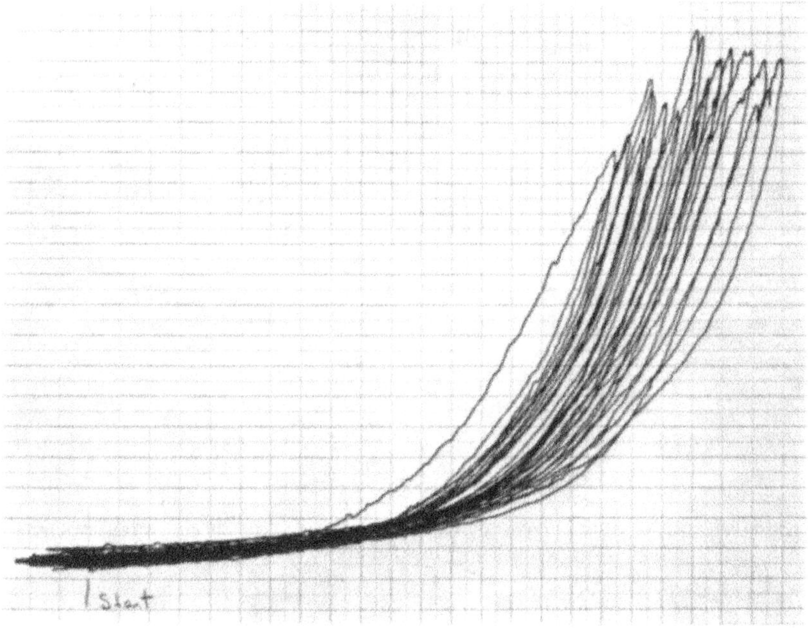

Figure 18.6 Stress–strain curves produced by 25 cycles of mature burned scar skin

A preliminary test in a subject with unilateral lymphoedema showed normal mechanics in the normal arm and a 'softening' pattern in the lymphoedematous limb closely similar to that seen in burn scar (Figures 18.5 and 18.6). The same effect is seen in a normal arm whose circulation is obstructed by a tourniquet (Figure 18.7). Maturing scar has no demonstrable lymphatics, thus it seems reasonable to assume that the dermal–vascular mechanisms are incompetent, at least in the early healing phase, mimicking lymphoedema. Therefore, 'the first curve effect' suggests that the first dash pot is emptied of water with the first stretch. The subsequent trend to elastic predominance implies gradual emptying of the more viscous material in the second dash pot. Even after a brief rest the 'first curve effect' recurs, suggesting the rapid restoration of water to the system.

CROSS-OVER EFFECT

A most unusual 'cross-over' effect has been noted in most subjects tested (Figure 18.8). When the forearm is extended minimally, the skin behaves like a typical viscoelastic material in that the unloading or down curve is always below the loading or up curve. However, if the forearm skin is stretched beyond a critical point (usually 30°–50° extension), the down stroke crosses *above* the upstroke. Thus after high strains there is greater force developed

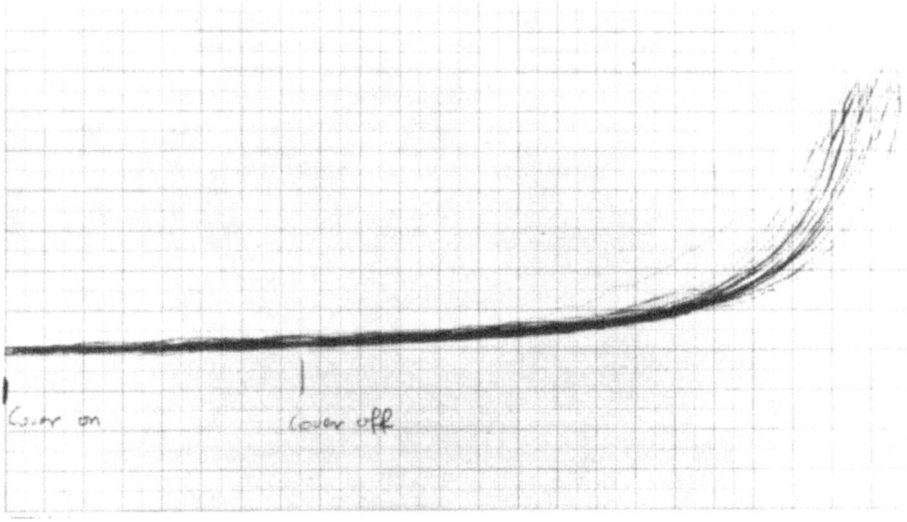

Figure 18.7 Stress–strain curves produced by 25 cycles of extension and flexion on a normal arm with a tourniquet inflated above systolic pressure

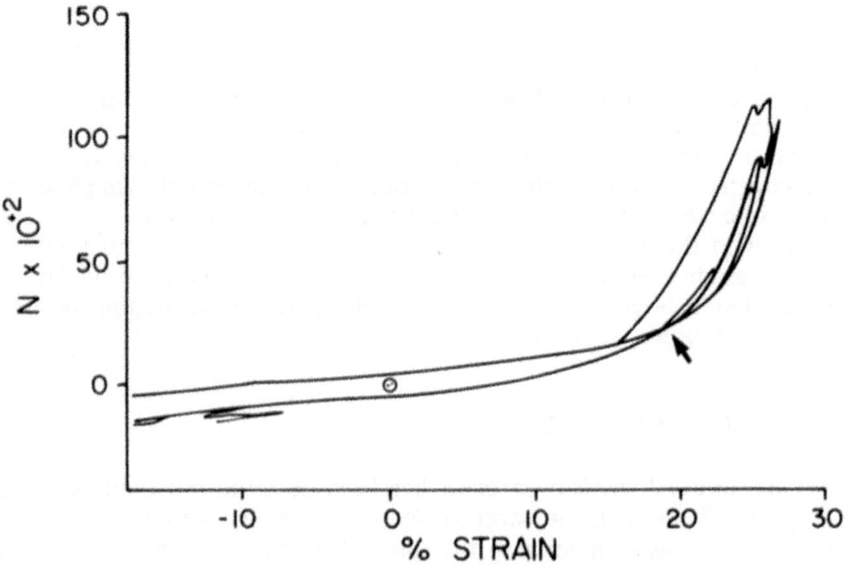

Figure 18.8 Stress–strain curves produced by 3 cycles of extension and flexion of normal forearm skin (arrow at 'cross-over' point)

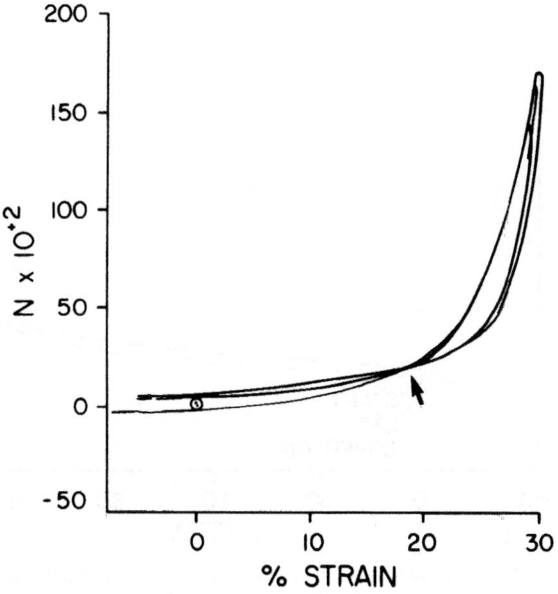

Figure 18.9 Stress–strain curves produced by 3 excursions of extension and flexion of normal skin with arm tourniquet inflated to 200 mmHg (arrow at 'cross-over' point)

during the first stages of unloading than during stretch. This remarkable phenomenon can occur only if there is an energy *input* (i.e. a net energy gain) when the skin is compressed. We postulate that the energy must arrive via the microcirculation powered either by the arterial pressure or the underlying musculature. Recent studies by McCraw *et al.*[6] have shown that the main circulatory supply of much of the skin is from the underlying muscle. To try to demonstrate that the circulation is indeed the source of this energy input, the following experiments were performed:

(1) injecting xylocaine to eliminate axonic reflexes, or epinephrine to produce vasospasm both diminished but did not eliminate the cross-over;

(2) injecting saline as a control produced the same, therefore non-specific, effect;

(3) a tourniquet inflated between systolic and diastolic or above systolic pressure diminished but did not eliminate the cross-over (Figure 18.9);

(4) ranging of the elbow while attempting to contract the wrist flexors (whose muscle mass is directly beneath the transducer, and therefore supplies blood to the overlying skin) produced a larger cross-over than when an attempt was made to relax these muscles (Figures 18.10 and 18.11); and

(5) ranging the skin with a mechanical stretcher with the forearm relaxed eliminated the cross-over.

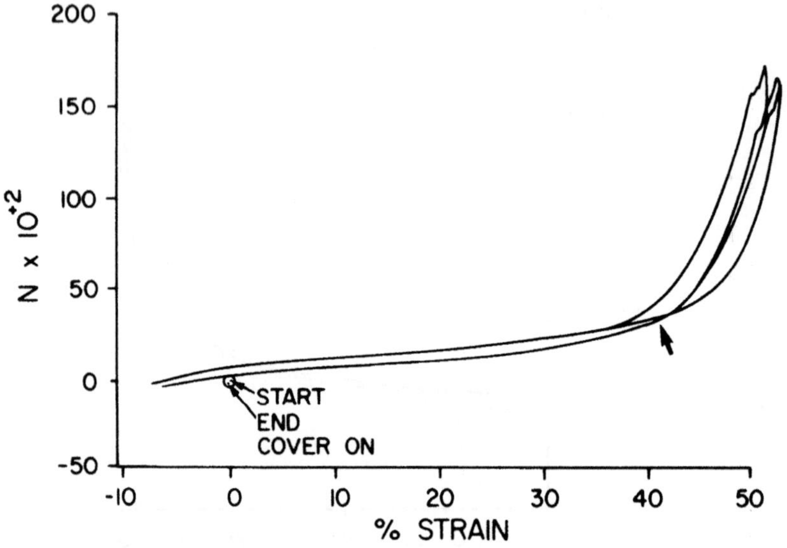

Figure 18.10 Stress–strain curves produced by 3 cycles of extension and flexion of normal forearm skin while attempting to contract wrist flexors

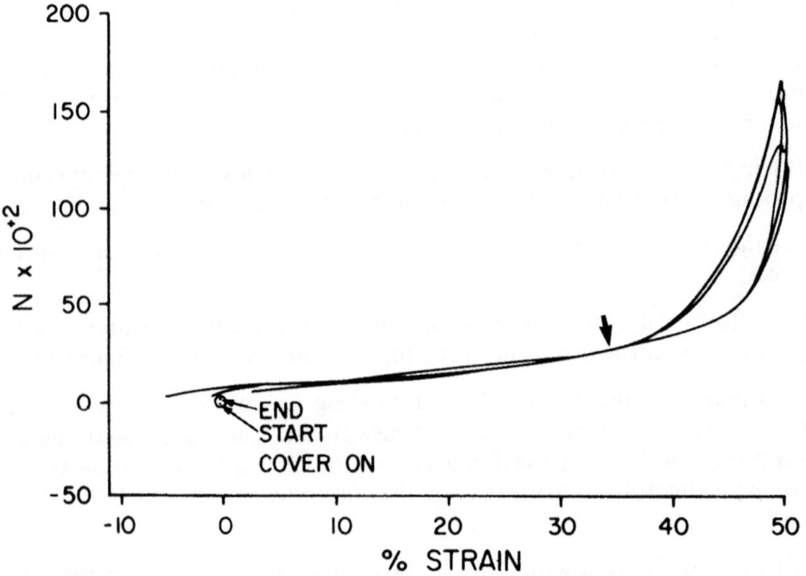

Figure 18.11 As in Figure 18.10, except that attempt being made to relax wrist flexors

Thus we postulate that when the forearm is triggered beyond a certain extension there is a fluid shift (probably mediated through the microcirculation) powered by both the muscle lying immediately beneath the skin and the blood pressure. The forearm skin with normal elbow flexion and extension will strain up to 50%. Most materials will permanently deform under such strain. Thus this mechanism may be the physiological way in which the original dimensions are restored by physically or biologically imparting an increased tension to retract the skin. Thus, homeostasis is preserved over the almost infinite repetitive stretching encountered during a lifetime. Clearly these assumptions are quite new and must be considered conjectural until more definite evidence is produced. Rheologically, therefore, skin must be thought of as a complex structure rather than as a simple material.

STRESS RELAXATION

Stress–relaxation experiments were performed in a 'step-up' and 'step-down' mode. Living human skin behaves as expected on step-up experiments but on step-down experiments the stress relaxation (i.e. the viscosity) almost disappears. Too little data is available to formulate any theory, but the phenomenon is reported here as it seems useful to exploit this effect therapeutically (Figures 18.12 and 18.13).

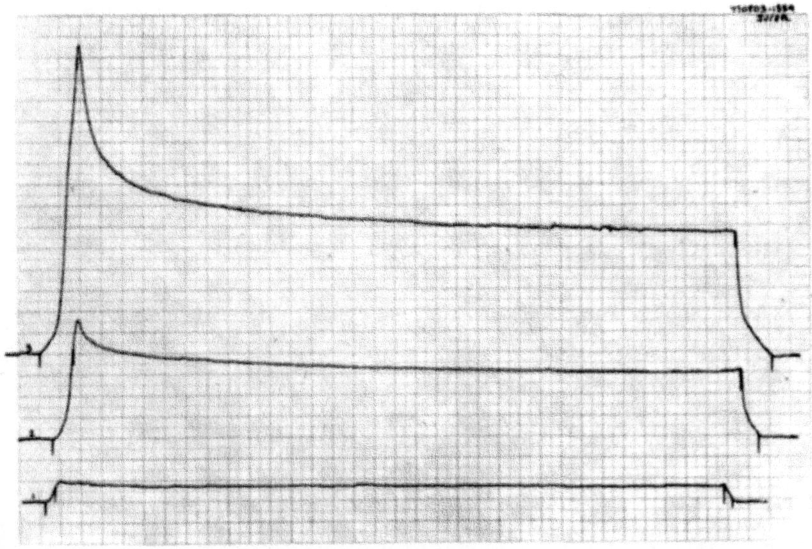

Figure 18.12 Stress–strain curves produced in three-stage 'step up' stress–relaxation test of normal forearm skin

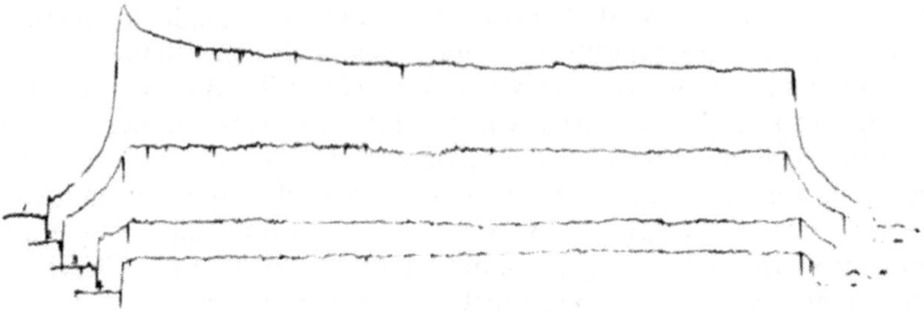

Figure 18.13 Stress–strain curves produced in three-stage 'step down' stress–relaxation test of normal forearm skin

RHEOLOGY OF BURN SCAR

Burn scar, in contrast with the behaviour of normal skin, more closely follows the rheological behaviour of known materials. Early burn scar is very stiff and does not 'cross-over' (Figure 18.2). Repetitive ranging produces continued softening which does not stabilize even after 75 cycles (Figure 18.6). This continued softening is also seen in lymphoedema and in a normal subject with circulation obstructed by tourniquet. If the curve of the first cycle were extrapolated to match the extension of the fifth or subsequent cycles enormous

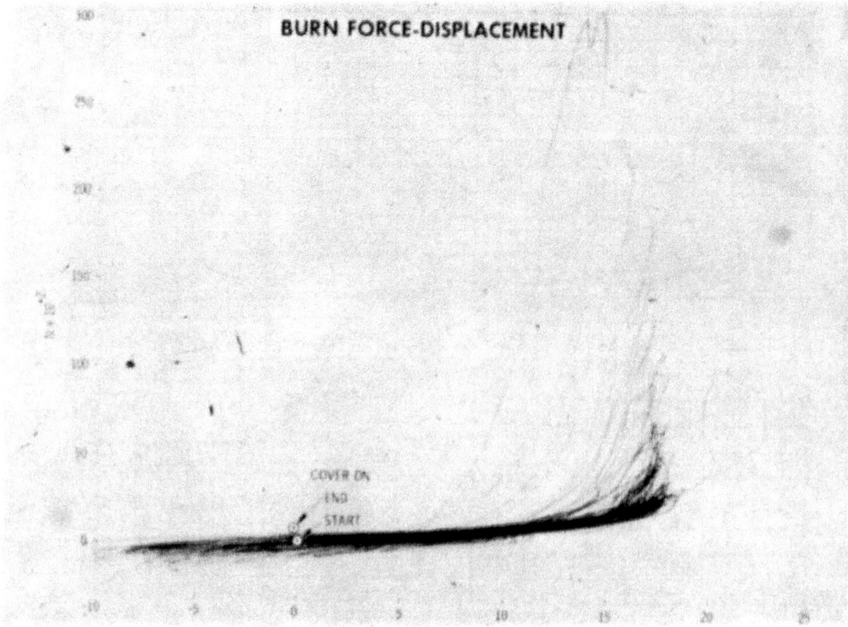

Figure 18.14 Stress–strain curves produced during 25 cycles of extension and flexion of immature scarred skin

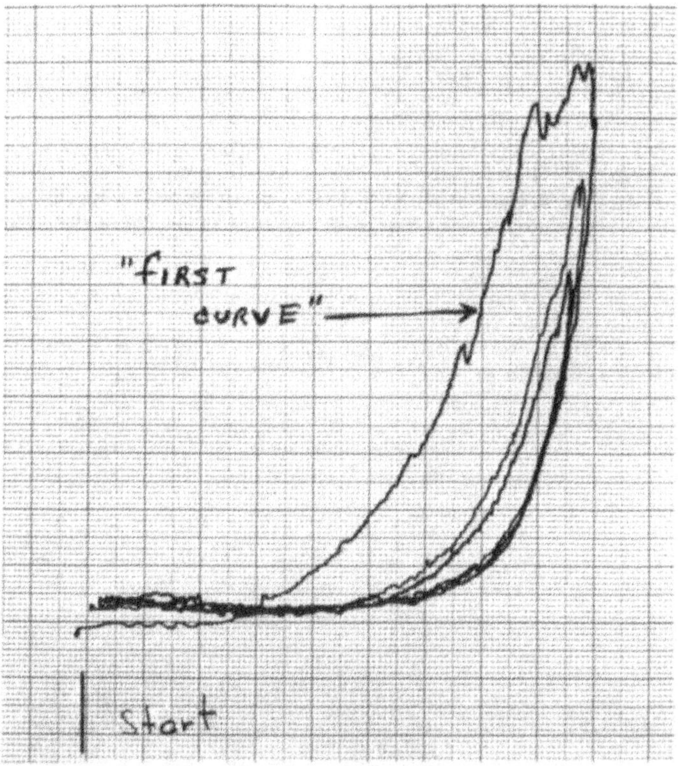

Figure 18.15 Stress–strain curves, as in Figure 18.2 after one year

forces would have been generated. Yet with repeated cycling greater skin stretch (permitting greater elbow range) is obtained with much less force. The 'yield' effect seen at the top of the burn–scar curve (Figure 18.14) when the limb is extended could represent the unfolding and 'ungluing' of the tangled and kinked collagen fibres. Baur *et al*[7] have documented this reorientation of scar under stretch ultrastructurally.

As the scar matures the rheological properties begin to approach normal until cross-over eventually recurs (Figure 18.15). We theorize, therefore, that early burn scar is incapable of maintaining its rheological homeostasis, at least partly because of incompetence of the circulatory shifting mechanisms described above. With maturation, competency returns coincident with softening and decreased swelling (hypertrophy) of the scar.

This correlates well with the known clinical ability to affect softening and lengthening of early burn scar with drug therapy, splints etc. The mature scar is more fixed and resistant to non-surgical treatment.

References

1. Chu, B. M. and Brody, G. S. (1975). Nondestructive measurements of the properties of burn scar. *Med. Instrum.*, **9**, 139
2. Peng,, T. J., Landel, R. F. and Brody, G. S. (1978). *In Vivo* Study of Human Skin Rheology. In *Proceedings of the Sixth New England Biomedical Conference, University of Rhode Island, March 1978* (New York: Pergamon Press)
3. Landel, R. F., Peng, T. J. and Brody, G. S. (1980). A strip biaxial stress–strain transducer for *in vivo* measurements of human skin. *Ann. Biomed. Eng.* (In press)
4. Landel,, R. F., Peng, T. J. and Brody, G. S. (1977). *In Vivo* Measurement of Stress–Strain and Stress-relaxation Behaviour of Human Skin. Presented at the *40th Annual Meeting of the Society of Rheology Bio-Solid Symposium, October 23–27*
5. Fung, Y. C. (1972). *Biomechanics*, p. 181. (Englewood Cliffs: Prentice-Hall)
6. McCraw, *et al.* (1977). Clinical definition of independent myocutaneous vascular territories. *Plast. Reconstr. Surg.*, **60**, 341
7. Baur, P. S., *et al.* (1978). Wound contractions, scar contractures and myofibroblasts: a clinical case study. *J. Trauma.* **18**, 8

Measurement of Function and Dimension

19
Indirect measures of transepidermal water loss

D. L. MILLER, ANNE M. BROWN and E. J. ARTZ

SUMMARY

The common method of measuring transepidermal water loss (TWL) using a flowing gas stream is subject to errors, many of which can be avoided in the applications where convective removal of skin surface water is minimized with respect to diffusive mechanisms. Simple closed-cup and open-cup TWL measurement systems are made possible by applying modern electronic humidity sensors. These two indirect methods afforded rapid measurement of TWL. However, modelling by computer simulation techniques showed limitations at high ambient humidity and high TWL. The process of healing of wounds inflicted by stripping skin with adhesive tape as shown by TWL appeared somewhat different using the indirect-static TWL technique than with the air-flow technique. These differences are related to the limitations of both methods.

INTRODUCTION

Transepidermal water loss (TWL) refers to the rate at which water migrates from the viable dermal tissues through the layers of the stratum corneum to the external environment. In the absence of profuse sweating, the TWL is predominantly controlled by the diffusion of water vapour in the stratum corneum under the influence of the difference in vapour concentration between the 'inside' and 'outside' surfaces.

TWL measurements have been used in studying the restoration of barrier function in wounded skin[1–3], as well as in evaluating occlusive properties of topical preparations[4].

METHODS OF TWL MEASUREMENT

Flow hygrometry

Direct measurement of TWL with flow hygrometry has been used by many investigators[1–7]. This approach is illustrated in Figure 19.1. The flux of water

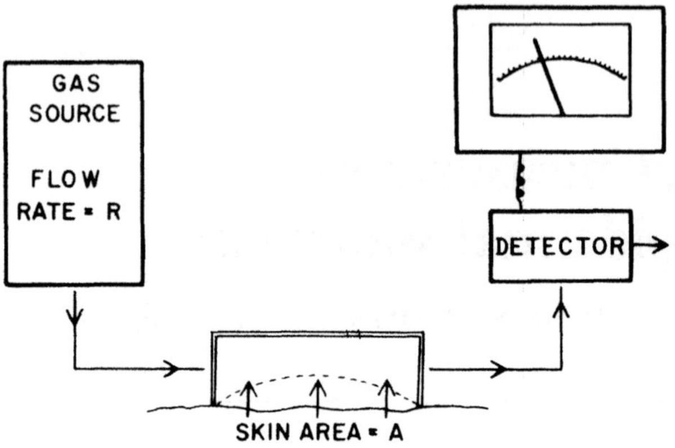

Figure 19.1 General schematic of a flow hygrometer. Dashed line suggests still air boundary location

vapour out of a fixed area of skin is determined by measuring the increase in water vapour concentration in the flowing gas stream. The increase in humidity is read at the detector output once steady-state conditions have been reached. TWL is calculated using the formula:

$$\text{TWL} = \frac{K \times V \times R}{A} \tag{1}$$

where K is an instrument constant, V the increase in detector output voltage, R the gas flow rate and A the skin area isolated by the measuring chamber. There are numerous possible sources of error in the flow hygrometry system, which include uncertainty in the gas flow rate, uncertainty in the actual area of skin exposed to the flowing gas, absorption of water vapour in tubing connecting the skin chamber with the detector and leaks in the seal of the chamber to the skin site. These errors can be minimized by careful design of a system. On the other hand, measurement artefacts from a boundary air layer of unknown thickness which exists between the flowing gas and the skin surface are difficult to account for (Figure 19.1). The contribution of boundary layer phenomena can be significant when the TWL is high and/or the flow rate of gas is low. Lastly, if the moisture content of the analysing gas is significantly different from that of the skin test site before measurement, an artefact resulting from the rapid change in environment of the skin surface can result.

In the direct method just described, water vapour transfer to the detector is by convection, which normally requires gas tanks or pumps, valves and tubing in addition to the detector and skin chamber. In contrast, indirect techniques to be described require little, if any, auxiliary equipment.

The indirect measurement of TWL relies on establishing a boundary air layer over the skin of known geometry. The perturbation of the initial humidity conditions by the water vapour escaping from the skin surface is measured using high-speed solid-state humidity sensors which have recently become available*.

Description of the closed-cup method

In the closed-cup method, the detector is sealed into one end of a cylinder of known length. The cylinder is placed on the skin, trapping a volume of unstirred air between the skin and the detector as shown in Figure 19.2. Within a few seconds, the detector voltage begins to rise steadily followed by a gradual decrease in rate. The application of diffusion principles[8] to the water vapour in the volume of trapped air predicts this behaviour and shows that the TWL is directly proportional to the slope of the transient linear portion of the detector output curve (Figure 19.2). The TWL is found from:

$$\mathrm{TWL} = K \times l \times \frac{\mathrm{d}v}{\mathrm{d}t} \qquad (2)$$

where K is an instrument calibration constant, l is the distance between the detector and the skin surface, v is the detector voltage and t is time.

Description of the open-cup method

In the open-cup method, the detector is mounted in the wall of a cylindrical chamber. Both ends of the chamber are open; one end is placed against the skin for measurement. This configuration is illustrated in Figure 19.3. As in the closed-cup method, the cylindrical chamber defines a volume of undisturbed air in which the transfer of water vapour is controlled by diffusion in air. A few minutes after applying the chamber to the skin, the detector reaches a new equilibrium in which the rate of flow of vapour out of the open end of the cylinder is equal to the rate of flow of vapour out of the skin, the TWL. In this

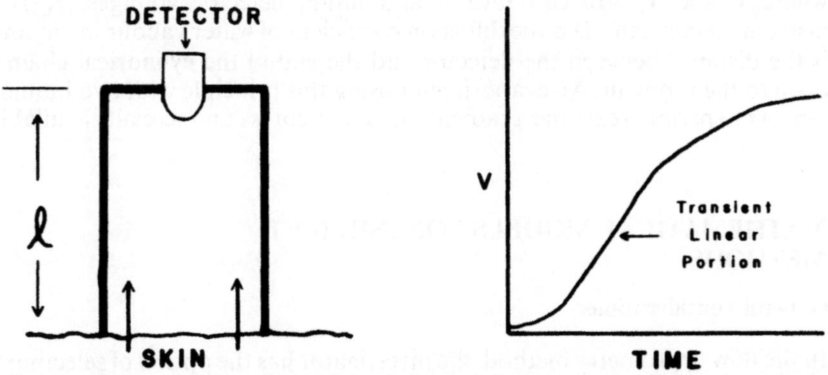

Figure 19.2 The closed-cup method: (a) configuration of measuring cup; and (b) typical detector output curve. For key see text

* For example, detectors available from Thunder Scientific, Alberquerque, USA; and Vaisala, Helsinki, Finland.

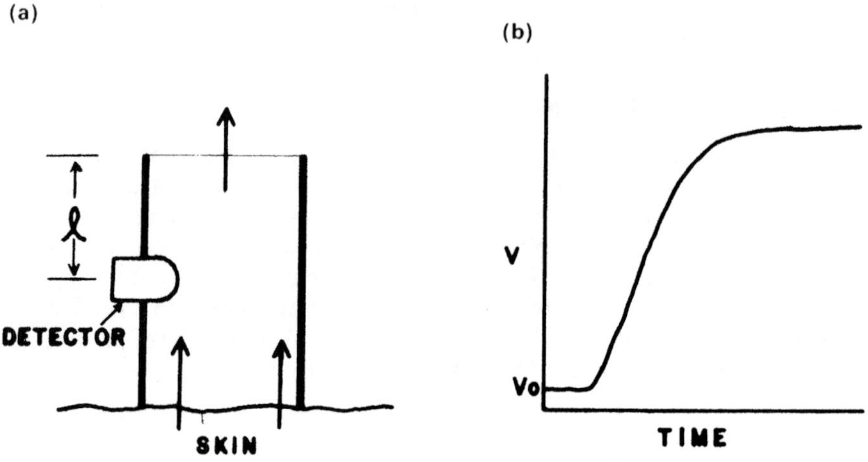

Figure 19.3 The open-cup method: (a) configuration of measurement cylinder; and (b) typical detector output curve. For key see text

condition the rate of flow can be calculated, according to the principles of diffusion[8], from the product of the diffusion coefficient of water vapour in air and the concentration gradient. For the configuration shown in Figure 19.3 the detector is mounted at a distance l from the end of the cylinder where the humidity is determined by the room condition. Thus the increase in detector response divided by the length determines the humidity gradient.

The equation for TWL measurement with the open-cup method is then:

$$\text{TWL} = \frac{(V - V_0) \times K \times D}{l} \quad (3)$$

where V and V_0 are equilibrium and initial detector voltages, K is an instrument constant, D is the diffusion coefficient of water vapour in air, and l is the distance between the detector and the end of the cylindrical chamber open to the room air. An evaporimeter using this principle with two humidity sensors to permit 'real time' gradient measurement is commercially available†.

MATHEMATICAL MODELS FOR INDIRECT METHODS

General considerations

In the flow hygrometry method, the investigator has the option of selecting the water content of the sweep gas within the limits of his moisture detector. For the indirect methods, unless measurements are made in a humidity controlled

† ServoMed AB, PO Box 110, S-162 12, Stockholm - Vallingly, Sweden.

room, the variable ambient condition can have a profound effect on the resulting measurements.

The interaction between the initial TWL, the room humidity and the quantity measured by both indirect methods was investigated by applying computer-generated models for each method based on diffusion mathematics. The details of the computer programs for the models are beyond the scope of this paper and will be dealt with separately. The general approaches are described below.

Model for the closed-cup method

The closed-cup mathematical model is based on the general solution of the diffusion equation for the case where there is a constant flux of water vapour at one surface (the skin) with the other surface impermeable (the detector)[8].

The following two equations are derived from the general solution to the diffusion equation:

$$\frac{\mathrm{d}C'}{\mathrm{d}t} = F/l\left[1 + 2\sum_{n=1}^{\infty}(-1)^n \exp(-Dn^2\pi^2 t/l^2)\right] \tag{4}$$

and

$$\frac{\mathrm{d}C}{\mathrm{d}t} = F/l\left[1 + 2\sum_{n=1}^{\infty}\exp(-Dn^2\pi^2 t/l^2)\right] \tag{5}$$

where C' and C are vapour concentrations at the detector and skin interfaces respectively, t is time, F is flux or TWL, l is the length of the diffusion path from skin to detector and D is the diffusion coefficient of water vapour in air. Moisture is continually introduced to this closed system. This has a suppressing effect on the TWL and is accounted for by

$$F = F_0[1 - (C - C_0)/(C^* - C_0)], \tag{6}$$

where F_0 is the initial flux or TWL and C_0 and C^* are the initial (ambient) and saturation vapour concentrations respectively.

Beginning with the initial constants F_0 and C_0, the computer program solves equations (5) and (6) for a given increment in time by a numerical iteration procedure. The resulting value of F is substituted in equation (4) to determine the detector response at the same increment in time. Hence the program calculates the time course of the detector response.

Figure 19.4 shows the computer output – a family of curves for a given ambient humidity and various values of F_0 or initial TWL. Each curve in this figure is normalized to its initial TWL. The curves for small F_0 show a broad flattened maximum close to unity which accounts for the instrumental observation described in the section dealing with the closed-cup method (p. 157. As TWL increases the maxima become sharper and much less than unity as a result of the overriding influence of the moisture build-up in the closed cup.

Figure 19.5 shows a TWL correction chart. Each curve on this graph represents the maximum flux (or TWL) at the detector plotted against the

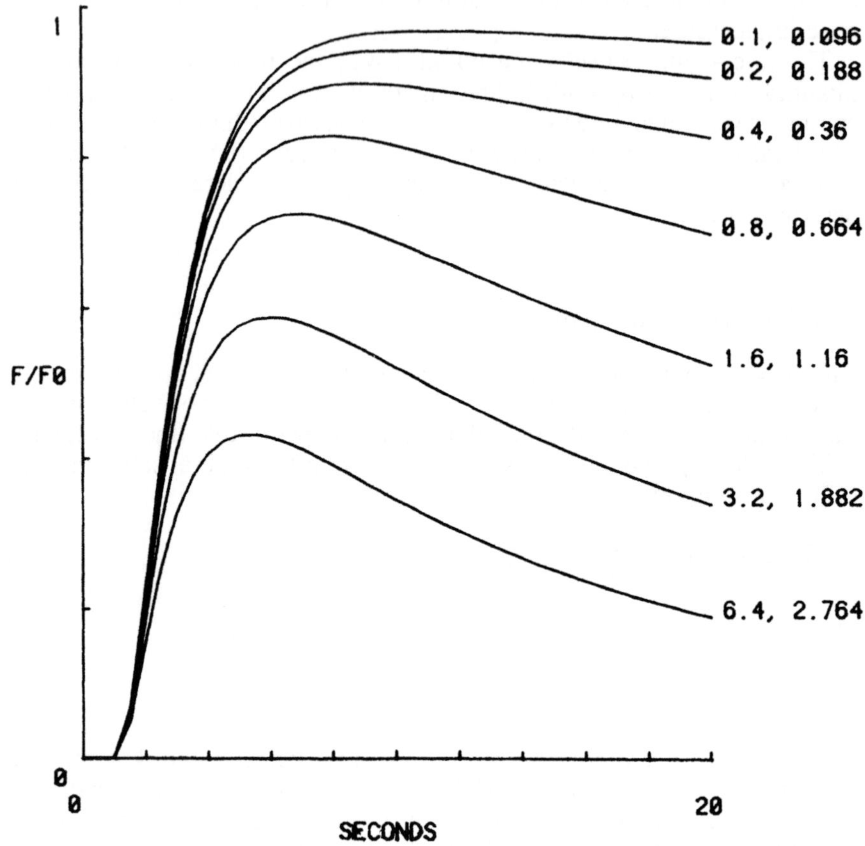

Figure 19.4 Computed curves of water vapour flux relative to initial flux (F/F_0) versus time for 50% ambient RH. The numbers to the right of each curve are the initial and maximum observed TWL (in/mg/cm^2/h) respectively

initial TWL for a given ambient humidity. From this chart it can be concluded that the closed-cup method is most accurate at low ambient humidities and for TWL values under 1. This TWL range includes most observations on intact skin.

Model for the open-cup method

The open-cup mathematical model is based on a set of water vapour sorption and diffusion properties in human stratum corneum measured *in vitro*[9]. Numerical expressions for the sigmoidal shaped sorption isotherm and for the concentration dependence of the water diffusion coefficient were incorporated into a computer program.

The water concentration profile with one surface of the skin in contact with water and the other at ambient humidity is generated by the computer using

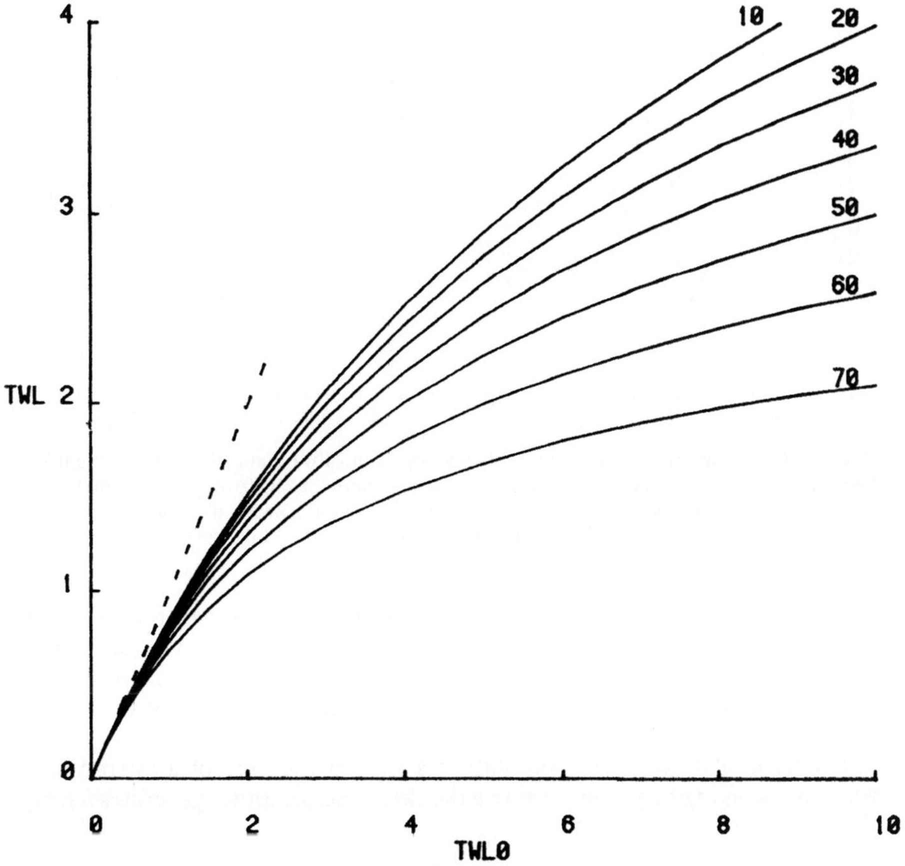

Figure 19.5 Computer correction curves relating initial TWL (TWLϕ) and observed TWL for the closed-cup method. The dashed line indicates the ideal (1 to 1) correlation. Numbers on each curve are relative humidity (%)

the Schmidt finite difference method (Crank, J., p. 178)[8]. The initial TWL is then computed from the water concentration gradient at the skin–air interface using a Fick's law expression:

$$\text{TWL} = F_0 = -D(c)\,\delta C/\delta x \qquad (7)$$

where $D(c)$ is the magnitude of the water diffusion coefficient at the skin surface and δC is the finite difference of concentration computed for the increment of distance δx.

The process of obtaining the TWL measurement itself is modelled by 'sandwiching' the skin with a layer of still air. The thickness of the air layer is set equal to the height of the open cup measuring capsule. One side of the layer should be in equilibrium with the skin surface moisture, the other maintained at ambient humidity. Because the water vapour leaving the skin must now move through this still air at a rate dependent on the diffusion coefficient of

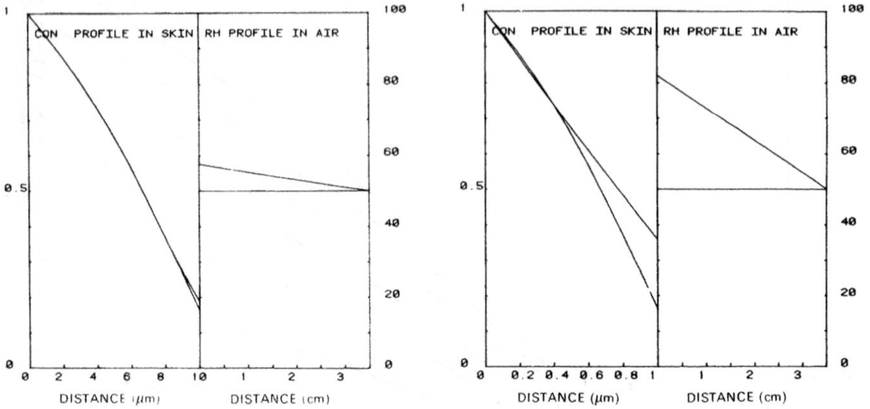

Figure 19.6 Computed concentration profiles for open-cup TWL simulation, 50 % ambient RH. Left hand side: stratum corneum thickness of 10 μm; initial TWL 0.6 $mg/cm^2/h$; equilibrium TWL 0.5 $mg/cm^2/h$. Right hand side: stratum corneum thickness of 1 μm; initial TWL 6 $mg/cm^2/h$: equilibrium TWL 2 $mg/cm^2/h$. Only initial and final curves are shown

water vapour in air, a new equilibrium condition is generated by the computer program. The finite difference method is applied to both the skin and the air layer with the requirement that, at the skin–air interface, the magnitudes of moisture concentration in the skin and the relative humidity of the air are consistent with the sorption isotherm.

The result of the computer simulation is a plot of the concentration profile of water in the skin and in the air above the skin. The computer generates a set of

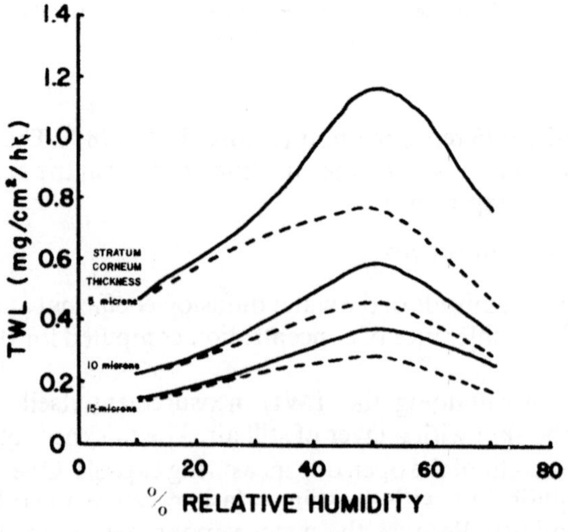

Figure 19.7 Influence of ambient humidity on TWL computed by the open-cup simulation. = Initial TWL. = Equilibrium TWL observed instrumentally

profiles for each increment of time, proceeding until the change in profile produced by further increments can be neglected; this is the point of new equilibrium. In Figure 19.6, two simulations are compared to illustrate the effect of stratum corneum thickness. For clarity, only the initial and final profiles are drawn. It is evident from this comparison that at 50% ambient relative humidity (RH) the 'open-cup' TWL measurement perturbs the pre-existing water concentration profile, particularly when the barrier is thin, and thus the TWL high. As a result the observed TWL is predicted to be one third of the unperturbed levels for 1 μm skin thickness at 50% RH. From computer simulations generated by varying the RH over a range of values at a given stratum corneum thickness, it was found that the disparity between observed and true TWL is a strong function of the RH. This result is illustrated in Figure 19.7. Errors in TWL values are apparently minimized when operating at ambient conditions of 20% RH or less. The maximum observed in the TWL–RH function is determined by the net effect of the diffusion coefficient, which increases with RH, and the overall gradient of moisture activity across the skin (which decreases with RH). Similar maxima in TWL measurements have been observed *in vivo* by other workers[5].

BARRIER REPAIR MEASUREMENT

Figure 19.8 illustrates the results of two separate experiments[3,10] in which the return of barrier properties was measured using TWL. Wounds were created in each case by adhesive tape stripping on day zero. TWL data after stripping is represented as a percentage of the TWL measured immediately after stripping on day zero. Although the relative TWL levels of air-exposed control and sites occluded with Saran-Wrap are comparable in both air-flow and static (open-cup) techniques, there are two obvious differences in the results of the two methods.

First, the range of 'percent damage' observed in the air-flow technique is an order of magnitude greater than for the static (open-cup) method. This difference can be accounted for by examining the data in Figure 19.8. The day zero reference TWL values for the air-flow method are as much as ten times larger than those measured by the static method: a direct result of convection enhancement of the evaporation rate from an essentially wet surface. In the static system, day zero TWL values are fixed at the upper limit imposed by the rate of diffusion of water vapour in still air under a gradient of RH going from nearly 100% RH at the skin surface to the ambient condition.

Second, while the air-flow technique indicates that the largest decrease in damage occurs during the first 2 days, the static method yields a nearly constant high TWL during the same period, suggesting a possible delay in the healing process. Once some form of barrier is present, the TWL measurement by air-flow drops from those levels determined by convection-enhanced evaporation to levels determined by the barrier resistance; thus, the *mechanism* of water loss, as imposed by the measurement technique, changes. In the static-measurement method, the diffusional resistance of a sufficiently thin barrier will be small compared with the resistance of the much thicker layer of air

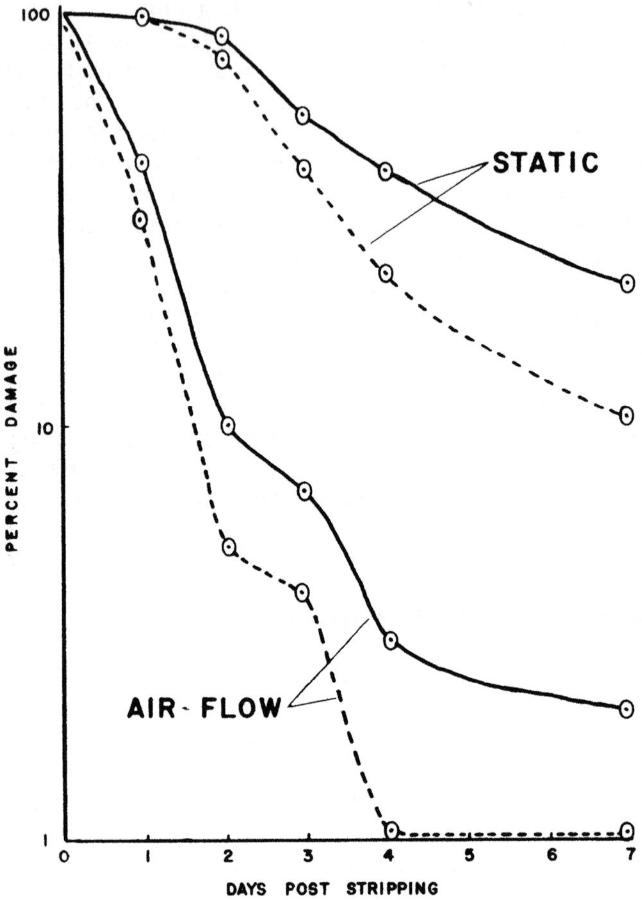

Figure 19.8 Assessment of wound healing with TWL measurements. —— = Air-exposed strip wound. --- = strip wound occluded with Saran Wrap. Comparisons of static (open-cup) and air-flow hygrometry measurements. Damage (%) = TWL/TWL (day 0) × 100

above the skin. Therefore the skin damage, measured in a static system, will decrease only after the resistance of the barrier becomes comparable in magnitude with the resistance of the static air layer.

CONCLUDING COMMENT

It should be evident from this discussion that none of these methods for measuring TWL are free from error or artefact. Each method has its advantages and disadvantages. In addition to the critical comparison of the techniques, evidence has been presented which shows that making the measurement has a significant effect on the local skin condition and thus the actual TWL value.

References

1. Spruit, D. and Malten, M. D. (1968). Estimation of the injury of human skin by alkaline liquids. *Berufsdermatosen*, **16**, 11
2. Malten, K. E. and Thiele, F. A. J. (1973). Evaluation of skin damage II. *Br. J. Dermatol.*, **89**, 565
3. Rovee, D. T. *et al.* (1972). Effect of local wound environment on epidermal healing. In Maibach, H. I. and Rovee, D. T. (eds.) *Epidermal Wound Healing*, pp. 159–181. (Chicago: Year Book Medical Publishers)
4. Berube, G. R. and Berdick,, M. (1974). Transepidermal moisture loss II. The significance of the use thickness of topical substances. *J. Soc. Cosmetic Chem.*, **25**, 397
5. Grice, K., Sutton, H. and Baker, H. (1972). The effect of ambient humidity on transepidermal water loss. *J. Invest. Dermatol.*, **58**, 343
6. Goodman, A. B. and Wolf, A. V. (1969). Insensible water loss from human skin as a function of ambient vapor concentration. *J. Appl. Physiol.*, **26**, 203
7. Aly, R., Shirley, C., Cunico, R. and Maibach, H. I. (1978). Effect of prolonged occlusion on the microbial flora, pH, carbon dioxide and transepidermal water loss on human skin. *J. Invest. Dermatol.*, **71**, 378
8. Crank, J. (1967). *The Mathematics of Diffusion*, pp. 42–61. (London: Oxford University Press)
9. Miller, D. L. (1974). Unpublished results
10. Brown, A. M. (1974). Unpublished results

20
Dermatological applications of the Magiscan image analysing computer

G. L. GROVE

SUMMARY

It is hoped that the range of applications covered by this brief survey provide some idea of how valuable the Magiscan system can be in studying skin structure and function. By coupling a TV camera to a computer-assisted image processor, we were able to count, measure and classify features according to a wide range of densitometric, geometric and textural criteria. In addition to allowing objective measurements to be made, this powerful instrument also increased speed of analysis and ease of operation. Since image analysis is a continually expanding field, we will probably see even more intensive use of this method in the next few years.

INTRODUCTION

The human eye and associated regions of the brain can serve quite well in certain forms of image analysis. This would include such tasks as pattern recognition of various objects and discerning the topographical features and spatial relationships these objects may have. On the other hand, the unaided human is extremely poor at extracting quantitative information from images. Unfortunately, although the genuine need for increased objectivity and accuracy in investigative dermatology has long been appreciated, 'eyeballing' still persists as the major method of measurement in this field.

During the past decade, the development of quantitative image analysis techniques has advanced rapidly. This growth, facilitated especially by space-age advances in electronic and computer technologies, has resulted in powerful instruments that can often aid and sometimes even supplant the eye in extracting quantitative information from images. In this paper we will briefly describe one such system, the Magiscan[1,2] (Joyce Lobel) and its application to dermatological research.

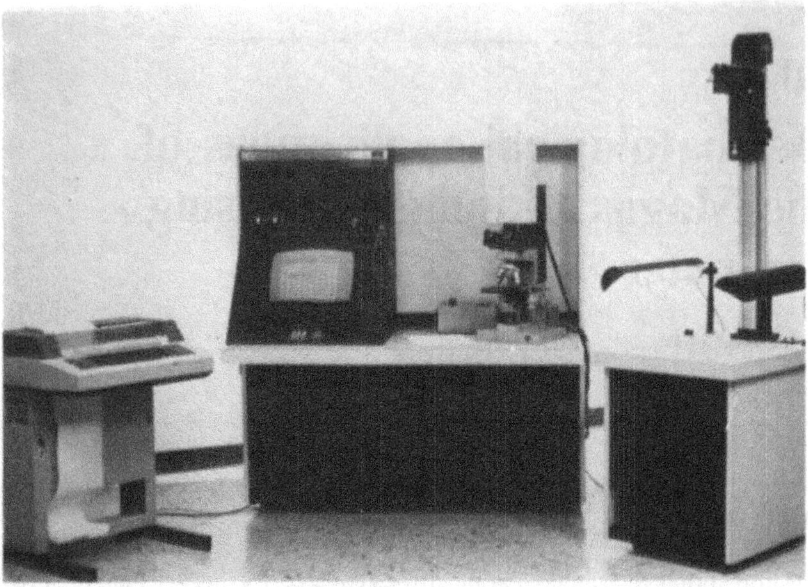

Figure 20.1 General view of the Magiscan image analyser system from left to right: minicomputer, image processor and video monitor, light microscope with TV scanner attached, and macroviewer table

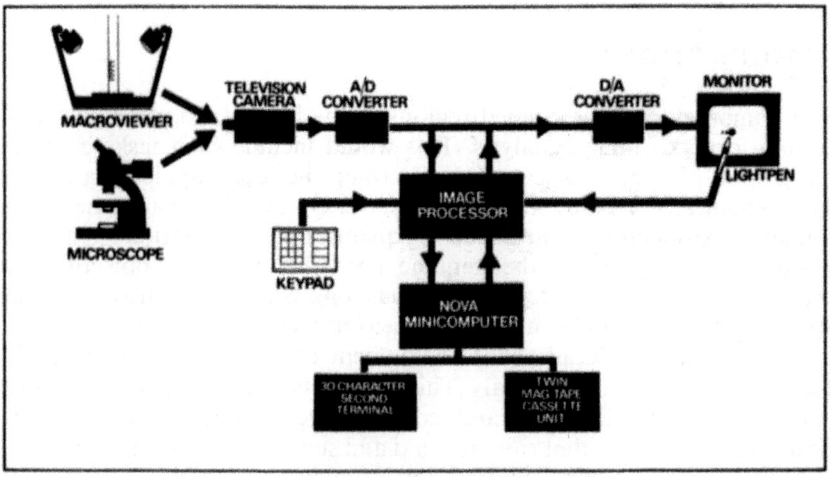

Figure 20.2 Schematic diagram showing the relationships of the components of the Magiscan image analysis system

INSTRUMENT DESIGN

The Magiscan system (Figures 20.1 and 20.2) uses a high-performance television camera to produce an electronic image of the specimen which is placed on either a microscope or macroviewer. This signal is digitized by a high-speed analogue/digital converter with a 6 bit resolution giving 64 levels of grey. Each scan line of the image is divided into small measuring elements by a system clock which controls the sampling rate and the scanning pattern of the TV camera to ensure precise registration of these picture points within a 512 × 512 matrix. This array of picture points is then introduced into a specially designed image processor which is interfaced to a minicomputer.

At this level, the operator can extract from the raw data features of interest and reject everything else by using standard programs. Generally, most problems can be solved by discriminating on the basis of grey levels and using routines written in MAGIC language. This is an easily learned, straightforward extension of BASIC computer language that contains special image analysis statements. For more complex problems, the operator can gain complete access to the array of picture points within the image processor by means of FORTRAN statements. This allows for more sophisticated types of analysis such as those based on textural differences or density gradients.

To ensure that the features of interest have been properly defined, the processed image is reconverted and displayed on a video monitor so that the operator can compare it with the original. Experience has shown that no matter how powerful the software routines may be, situations do occur where

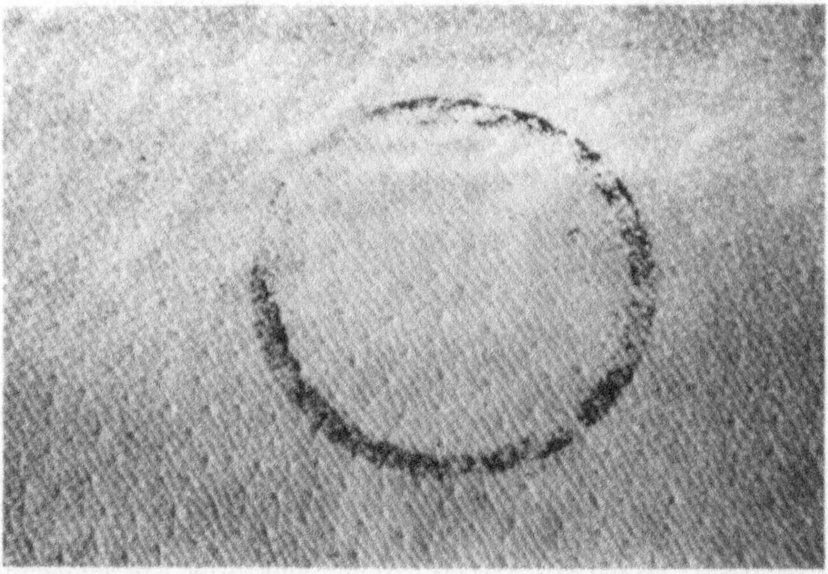

Figure 20.3 Macroscopic view of sweat gland impression taken from the volar forearm of a volunteer treated overnight with a 25% solution of aluminum chloride

it is necessary to interact directly with the image. By using an EDITOR lightpen, the operator can effectively interface himself into the system to change image details and select or reject designated regions. Examples of how this Editor feature can be used will be given in the application section.

Once the features of interest have been properly defined, Magiscan will perform measurements on either all features in a field together as a set or on each feature individually using the programmable geometric and densitometric algorithms described in Figure 20.3. The results and derived statistics are then displayed on the video monitor. Hard copy of histograms, plots, tabulations etc. can be produced on the terminal or photographed off the monitor. Since the instrument uses a standard Nova minicomputer, other peripherals, all of which can be operated off-line, can be used for data storage and retrieval.

To provide a practical understanding of the capabilities of the Magiscan system, a brief survey of possible applications will be given.

APPLICATIONS

Quantitative evaluation of antiperspirant efficacy

The magnitude of sweating can be non-intrusively evaluated using a modification of the silicone rubber impression technique introduced by Sarkany and Gaylarde[3] and later improved by Willis *et al*[4]. Our modification, devised by Dr Erhard Hoelzle, uses a 20:1 mixture of base and catalyst (Syringe Elasticon, Kerr, Romulus Michigan) which sets in a few minutes. To take an imprint the freshly prepared silicone rubber mix is applied as a thin film over the skin surface to be examined. Since this material is immiscible with water, each sweat droplet will produce a hole in the film. This film, which can serve as a permanent record of the distribution of active sweat glands, can then be easily peeled off with little or no discomfort to the subject.

A close-up of such a peel obtained from an area treated overnight with a highly effective antiperspirant (25% aluminum chloride) under a Duhring chamber 5 is shown in Figure 20.3. In the past we have assessed the sweating capacity by projecting the imprint on a screen and counting the number of holes per unit area by eye. However, we have found this approach to be not only time consuming and tedious but also subject to considerable error, especially when fatigue sets in.

In contrast, the Magiscan system allows highly reproducible assessments of the number of active glands to be made within seconds. In this manner even very subtle differences in sweat gland patterns as a result of age or sex in untreated areas may also be detected. Since the operator can store results from each frame analysed, it is possible to accumulate data from an entire group of subjects and run the statistical analysis of the results entirely within the system. Because we use standard sized groups of either 6 or 10 subjects with paired applications of candidate agents and proven agents, it is quite simple to use 'look-up' tables and provide probability values and efficacy judgments directly on the statistical summary.

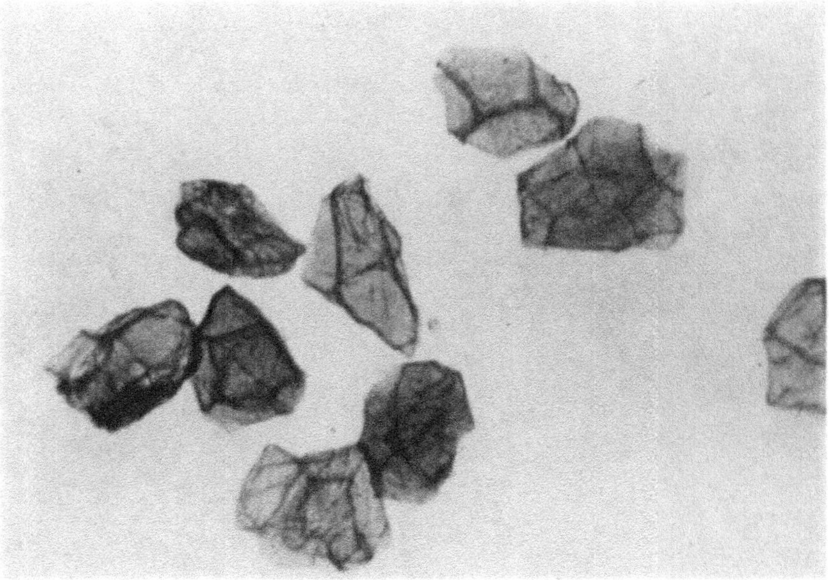

Figure 20.4 Corneocytes typically seen in scrub specimens obtained from healthy young donors (260 ×)

Automated classification of exfoliated corneocytes

Since cells are continually being shed from the skin surface, it is a simple matter to collect them for subsequent analysis. One method of examining the cytology of exfoliated cells is the detergent scrub technique devised by McGinley *et al*[6]. In this procedure a glass well covering a 3.8 cm^2 area is held firmly against the skin, buffered 0.1 % Triton X-100 (1 ml) is added and the skin surface gently rubbed with a Teflon scrubber for 1 minute. The wash fluid, which contains individual horny cells or corneocytes, can then be collected and processed in many ways[7].

This technique is especially valuable for making cytomorphological observations on these exfoliated skin cells. The array of cells shown in Figure 20.4 illustrates the typical features of this cell type. Note that these cells lack nuclei, one of the results of the terminal differentiation of epidermal cells. Normal corneocytes also tend to be polygonal in shape and to have relatively smooth borders. These cells often display a network of lines that corresponds to regions of overlap with neighbouring cells in the intact stratum corneum.

We strongly believe that much information about the physiological status of the underlying skin can be obtained in this manner without harm or inconvenience to the subject. For example, Figure 20.5 illustrates the typical features of cells obtained from an active lesion in a psoriatic patient. It is quite clear that these cells are abnormal and these cytomorphological alterations, especially retention of nuclei, are an expression of the complete differentiation of epidermal cells which is known to be a hallmark of this disease[5,8]. Other

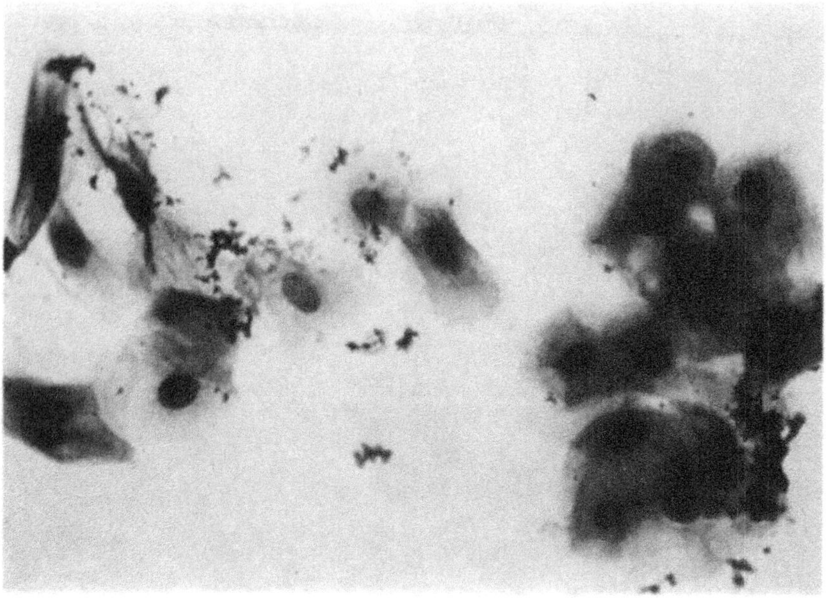

Figure 20.5 Corneocytes which have retained their nuclei can be seen in this scrub specimen obtained from an active lesion of a patient with psoriasis

cytomorphological changes have been reported in dandruff[9], chronic dermatitis[10], and ageing[7,11]. Anomalies include the retention of nuclei or nuclear remnants, loss of lines of overlap, the roughening of border edges and changes in tintorial properties.

Currently, we are exploring the possibility of using the Magiscan image analyser to automatically classify these cells by their geometric and staining characteristics. Such procedures already exist for several haematological and cytological applications[12,13]. This approach should make it possible to objectively describe cells and help identify abnormal and unusual cell types that may have diagnostic importance

Textural analysis of the desquamating skin surface

Another procedure for exfoliative cytological examination of human skin is the 'sticky slide' technique introduced by Goldschmidt and Kligman[14]. In this procedure, either thin uniform coats of 3M-EC791 adhesive or strips of Scotch brand double-coated cellophane tape are applied to glass microscope slides. When these 'sticky slides' are pressed firmly against the skin, specimens of the desquamating portion of the stratum corneum which retains the topographical relationships of the horny cells adhere to the slides. Figure 20.6 shows 'sticky' tape specimens obtained from an area used to evaluate the irritancy of soaps as described by Frosch and Kligman[15]. Although the extent

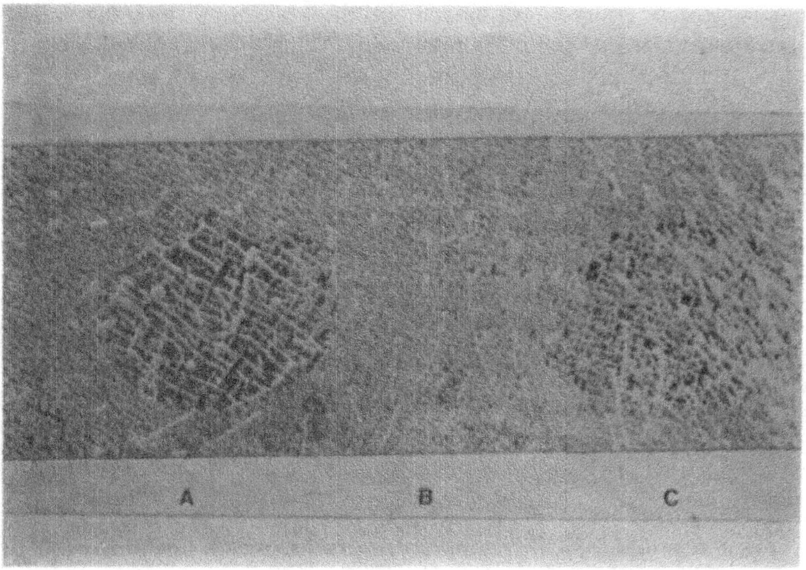

Figure 20.6 Macroscopic view of 'sticky slide' preparation taken from the volar forearm of a volunteer treated repeatedly for 3 days with 8% solution of commercially available soaps. Site A shows the effects of a fairly harsh soap; site B a very mild soap; site C a moderately harsh soap

of damage induced can be subjectively evaluated, this necessitates using a grading system (+1, +2, etc.) that is statistically weak. However, with the Magiscan system we can attain a highly objective method of evaluating the degree of harshness. Essentially, our program converts the various properties of the image into a single numerical value, similar to the procedure followed by NASA in selecting lunar landing sites from satellite photos.

In this case it was necessary to use the Image Editor lightpen to demark the centre area, which was treated with a mild soap that caused no appreciable damage. Currently we are using this method to not only determine the detrimental effects of various physical and chemical stresses on ageing skin but also to assess the benefits of various therapeutic and prophylactic regimens.

Histometric analysis of biopsy specimens

In agreement with the report of Picton *et al.*[16], we have also found that the image analysis systems can easily and rapidly perform measurements on stained and sectioned biopsy material. One parameter that is extremely important is that of the thickness of the viable epidermis. In the conventional approach[17], this value is determined as the average of 25–100 readings taken at random spots using an eyepiece micrometer. Such a procedure is not only time consuming and tedious but is also subject to considerable errors especially in specimens with a prominent rete-ridge pattern. With the

Magiscan system the area occupied by the viable epidermis within a standard sized field can be easily determined. Since this is essentially an integrated value of the area under the undulating dermoepidermal junction, we have a much better estimate of epidermal thickness than a simple average could provide.

It is also possible with the Magiscan system to evaluate the degree of undulation in the dermoepidermal junction by using the equation:

$$\text{undulation index} = \frac{\text{perimeter}}{2\,(\text{length})} \quad (1)$$

Once again, by using the Editor lightpen, this becomes a simple task and gives highly reproducible results with less than 1 % error within seconds.
Other histometric measurements which have been achieved by image analysis include determination of sebaceous gland volume and labelling indices of the epidermis by Picton's group[16]. In fact all presently known histometric and stereological methods of analysis can be handled in this fashion.

Dynamic analysis

With the Magiscan system the operator can store the detected features so that they can be compared with subsequent images. This is especially useful if one selects a standard field of reference against which all other results can be compared. Moreover, if the temporal aspects are known, it becomes a simple matter to produce a dynamic analysis from a series of frames of the same specimen taken at different times.

Although we have not yet made use of this feature, several applications of dynamic analysis have come to mind. One would be to obtain a dynamic analysis of the desquamation process by taking several photographs of dansyl chloride stain sites under Wood's Lamp lighting[18]. As a result of some preliminary studies along these lines we have gained the impression that people with dry skin show heterogeneous loss of stained material instead of the typical uniform fading. With the dynamic analysis feature of the Magiscan system, it would also be possible to evaluate the deformation of skin surface patterns at various stages of biomechanical tests. In fact any sequence of events which can be detected by the TV scanner could be handled in this fashion.

ACKNOWLEDGEMENTS

The author is indebted to Mr Robert Osgood and Mr John Fallon of Vickers Instruments Woburn, Massachusetts, who allowed the use of their facilities for this study. Portions of this investigation were supported by US Public Health Service Research Grant AG-00788 from the National Institute on Ageing.

References

1. Taylor, C. J., Brunt, J. N., Dixon, R. N. and Gregory, P. J. (1978). The Magiscan: a new generation software based automatic image analyser. *Pract. Metallogr.* **8**, 433

2. Fallon, J. P., Pycock, D. and Taylor, C. J. (1981). Quantitative cytophotometry and image analysis with the Magiscan. *J. Cytochem. Histochem.*, **27**, 1383
3. Sarkany, I. and Gaylarde, P. (1968). A method for demonstration of sweat gland activity. *Br. J. Dermatol.*, **80**, 744
4. Willis, I., Harris, D. R. and Moretz, W. (1973). Normal and abnormal variations in eccrine sweat gland distribution. *J. Invest. Dermatol.*, **60**, 98
5. Frosch, P. J. and Kligman, A. M. (1979). The Duhring chamber–an improved technique for epicutaneous testing of irritant and allergic reacions. *Contact Dermatitis*, **5**, 73.
6. McGinley, K. J., Marples, R. R. and Plewig, G. (1969). A method for visualizing and quantitating the desquamation portion of the human stratum corneum. *J. Invest. Dermatol.*, **53**, 107
7. Grove, G. L. (1979). Exfoliative cytological procedures as a nonintrusive method for dermatogerontological studies. *J. Invest. Dermatol.*, **73**, 67
8. Goldschmidt, H. W., Leyden, J. J. and Decherd, J. (1972). Exfoliative cytology in the diagnosis of psoriasis of the nail. *Cutis*, **10**, 701
9. Leyden, J. J., McGinley, K. J. and Kligman, A. M. (1975). Shorter methods for evaluating antidandruff agents. *J. Soc. Cosmetic Chem.*, **26**, 573
10. Hoelzle, E. and Plewig, G. (1976). Effects of dermatitis, stripping and steroids on the morphology of corneocytes, a new bioassay. *J. Invest. Dermatol.*, **66**, 259
11. Plewig, G. (1970). Regional differences of cell sizes in the human stratum corneum II: Effects of sex and age. *J. Invest. Dermatol.*, **54**, 19
12. Weid, G. and Bahr, G. (1971). Symposium by correspondance on automated cytology. *Acta Cytologica*, **15**, 1
13. Pycock, D. and Taylor, C. J. Progress towards automated uterine cancer cytology using the Magiscan image analyser. *Analyt. Quant. Cytol.* (In press)
14. Goldschmidt, H. and Kligman, A. M. (1967). Exfoliative cytology of human horny layer. *Arch. Dermatol.*, **96**, 576
15. Frosch, P. J. and Kligman, A. M. (1979). Soap chamber irritancy test. *J. Am. Acad. Dermatol.*, **1**, 35
16. Picton, W., Devitt, H. and Forgie, M. A. (1976). Practical applications of the quantimet 720, an image analysing computer, in the field of investigative dermatology. *Br. J. Dermatol.*, **95**, 341
17. Holt, P. J. A. and Marks, R. (1976). Epidermal architecture, growth and metabolism in acromegaly. *Br. Med. J.*, **1**, 496
18. Jansen, L. H., Hojyo-Tomoka, M. T. and Kligman, A. M. (1974). Improved fluorescence staining technique for estimating the turnover of the human stratum corneum. *Br. J. Dermatol.*, **90**, 9

21

A modified photometric technique for measuring sebum excretion rate

W. J. CUNLIFFE, J. N. KEARNEY and N. B. SIMPSON

INTRODUCTION

Patients with acne have seborrhoea, and there is a continued need to measure sebum excretion to more fully understand the already reasonably well investigated field of sebaceous gland physiology. In addition, sebum excretion rate (SER) measurements are necessary to evaluate the possible anti-seborrhoeic effect of oral and topically applied drugs, such as antiandrogens or 13-*cis* retinoids, as any drug which will appreciably reduce SER will considerably improve acne.

There are two widely used techniques of measuring SER: the gravimetric technique of Strauss and Pochi and Cunliffe and Shuster; and the photometric technique as devised by Schaefer and Kuhn-Bussius. Both techniques are accirate and yield reproducible results but the gravimetric technique is time consuming to the patient and staff and the latter technique is less accurate at high levels of sebum production.

We have helped in the design of a lipometre originally devised by L'Oreal, Paris. The machine produced is light and robust and is the equipment of choice for measuring SER or the 'casual level' of sebum.

The equipment consists of a light emitter, which is a high-brightness light-emitting diode (LED), and a light sensor which is a NPN phototransistor. Both of these are located coaxially on the frosted side of a sapphire plate. The non-frosted side of the plate is metal-coated under a vacuum and consequently acts as a mirror, reflecting the light delivered from the LED to the phototransistor. An automatic system allows the reading of the clean plate to be adjusted to zero, and the memorizing of this value. After the plate has been charged with fat, the machine gives a direct reading in $\mu g/cm^2$ which is the exact difference between the fatty plate value and the memorized zero value.

USE OF LIPOMETRE

The patient's forehead is prepared by the successive application of absorbent papers, held in position by gauze and an elasticated broad band fixed at the back of the head with nylon mesh. This preliminary part of the test is identical with the gravimetric technique and the aim, by changing the papers in contact with the skin every 10 minutes, is to produce the so-called 'follicular pattern'. This takes 3–4 changes, i.e. about 30–40 minutes, by which time the emerging sebum is derived from the pilosebaceous units, rather than collecting lipid from the pilosebaceous units and the epidermis. The time when this pattern occurs is noted and the patient's forehead is then left uncovered.

Three hours later a sapphire plate in a light-weight holder is applied to the skin for 20 seconds, the sampler being so designed to give a pressure of 0.55 kg/cm^2. Before and after sampling the sapphire plate is placed in the lipometre which gives a linear read-out of surface lipid in $\mu g/cm^2$. The sapphire plate on each occasion is rotated through 90° into four fixed positions in the lipometre so that four areas of the plate are recorded. The sampler is reapplied to the skin area sampled on three further occasions immediately after taking the first sample.

REPRODUCIBILITY

Experiments have shown that four samples taken at one site give a SER not significantly different from the SER as determined by the gravimetric technique ($n = 30$; $t = 1.19$). The reproducibility and accuracy of the modified photometric technique has been determined by comparing the SER on both sides of the forehead ($r = 0.89$) and by studying 22 males at weekly intervals for 5 weeks. The latter study showed that there was no significant variation over these 5 weeks. Means $\pm$SE over these 5 weeks were: 0.79 ± 0.04; 0.83 ± 0.04; 0.80 ± 0.05; 0.83 ± 0.06; and 0.83 ± 0.05.

We have previously investigated the time course of SER and have shown that there is less variance over a 3 hour collection time than for collections of 1 or 2 hours. For this reason we believe that it is wisest to collect sebum for 3 hours when determining SER. We were, however, concerned that during this 3 hours the patient could touch the forehead skin and significantly reduce the amount of lipid collected. We therefore asked a group of patients to wipe one side of their forehead 1 hour before the end of the study. This did not significantly reduce the SER on the wiped side, suggesting that the surface lipid has a greater affinity for the forehead skin than for hand skin. We have shown a correlation of SER and acne grade in 70 subjects that was almost identical with that as measured by the gravimetric technique.

'CASUAL LEVELS' OF SEBUM

The term 'casual level' refers to the skin-surface lipid collected at any time without any prior preparation. Although this is less accurate than a SER assessment, there is some correlation between SER and casual level. The modified photometric technique permits measurement of casual levels at sites where it is almost impossible to measure SER, such as the nose and the chin, and also permits assessment of SER much more easily at sites which are difficult with the gravimetric technique, such as on the back and on the chest.

A further advantage of the modified photometric technique is that it is time saving for the operator. In the study of 20 patients we have shown that the gravimetric technique takes 8 hours longer than the modified photometric technique. We believe that the modified photometric technique is probably the method of choice for measuring SER and casual levels of sebum.

Reference

Cunliffe, W. I., Kearney, J. N. and Simpson, N. B. (1980). A modified photometric technique for measuring sebum excretion rate. *J. Invest. Dermatol.*, **75**, 394

22

Determination *in vivo* of water concentration profile in human stratum corneum by a photoacoustic method

I. SIMON, A. G. EMSLIE, C. M. APT, I. H. BLANK and R. R. ANDERSON

The purpose of this study was to examine the use of photoacoustic spectroscopy for measuring the concentration profile of water in human stratum corneum *in vivo*. Stratum corneum is normally 'dry' at its surface and 'wet' at its base where it is in contact with the moist part of the epidermis. The concentration of water at the surface of the skin, the depth profile of water within the stratum corneum, and the overall water content of stratum corneum vary with the condition of the skin and the relative humidity of the ambient air. Changes in hydration of the stratum corneum are known to affect plasticity, conductivity, and diffusion of substances into the skin.

The concentration of water in the stratum corneum as a function of depth is unknown, and only recently have attempts been made at its measurement with the photoacoustic spectroscopy method[1,2]. However, we may surmise that the water is distributed as shown in Figure 22.1.

Generally, photoacoustic spectroscopy measures the acoustic signals arising when modulated optical radiation is absorbed near a sample's surface. In our method, absorption of optical radiation by water within the stratum corneum results in localized heating in proportion to the amount of water present at any given depth. Because the incident radiation is modulated, the heat generated in any volume element propagates as a periodic heat wave travelling both inwards and outwards through the stratum corneum. The superposition of these thermal waves gives rise to periodic temperature variations at the surface of the skin. These periodic temperature variations at the skin surface cause periodic pressure fluctuations which are then detected as sound by the microphone within the closed photoacoustic cell.

Our method, in its present form, enables us to determine only a quantity *a*, the depth of the water-depleted zone, in the simple model profile represented in Figure 22.1 as a rectangular step. The relation of the step model to the actual

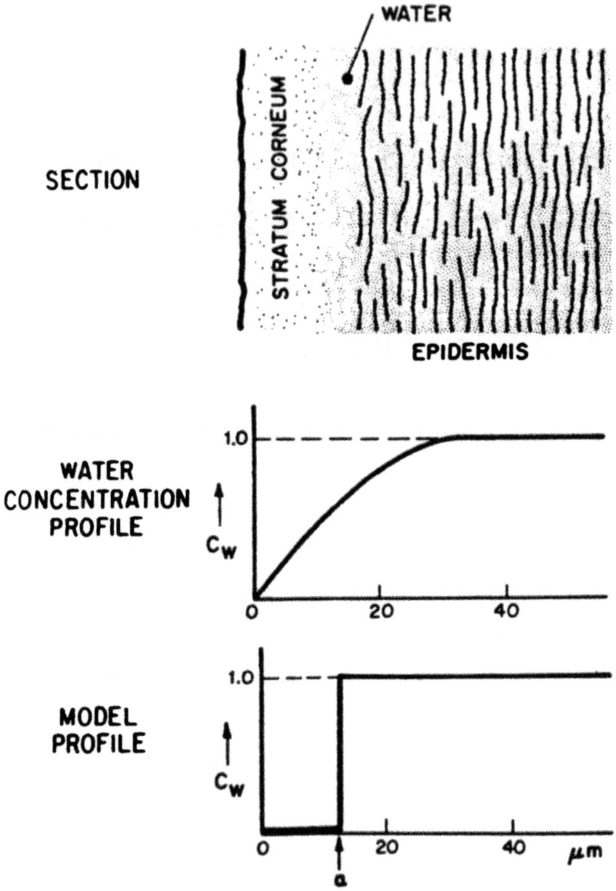

Figure 22.1 Water concentration in the outer layers of skin; experimental and model profiles.. For key see text

concentration profile is governed by the requirement that the total water deficiency of the depleted zone should be the same in the two cases. The theoretical treatment used in our method also allows us to consider model profiles other than the rectangular step, in particular the 'boxcar' profile; at present, however, the experimental accuracy is not sufficient to distinguish between such models.

The method uses monochromatic 1.95 μm wavelength infrared radiation, corresponding to one of the characteristic vibrational absorption bands of the water molecule, the $v_1 + v_2$ stretching–bending vibration. This particular absorption band was selected rather than any of the much stronger fundamental bands because it occurs at a wavelength at which the dry stratum corneum is effectively transparent. At longer wavelengths the absorption

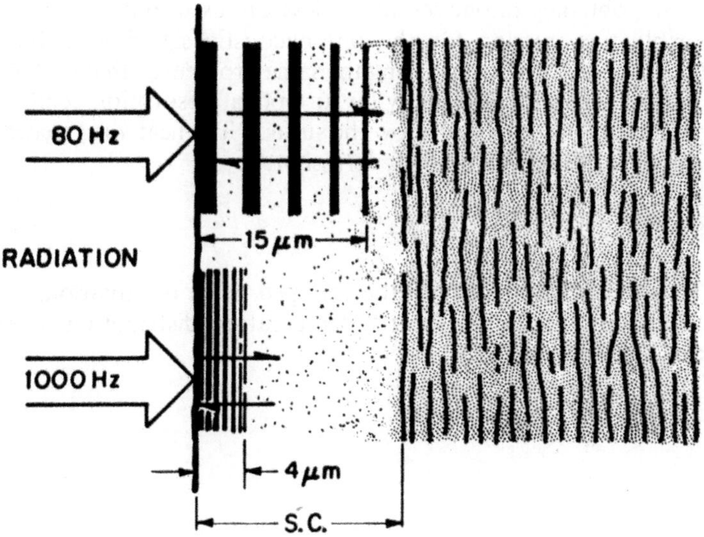

Figure 22.2 Effect of frequency of modulation on the depth of penetration of the periodic heat wave. SC = stratum corneum

bands of water overlap those due to the protein constituents of the skin and hence are not useful for our purpose.

The basis of our method, which we have called acoustic domain photoacoustic spectroscopy (ADPAS), is the notion that the penetration depth of a periodic heat wave into a solid depends on its frequency. This is shown in Figure 22.2. With a typical value of thermal diffusivity (κ) of about 5.5×10^{-4} cm^2/s, the penetration depth for frequencies in the audible range is of the same order as the thickness of stratum corneum. By varying the frequency from about 40–2000 Hz optical absorption of the skin for depths 3–20 μm below the surface can, in principle, be probed.

When the amplitude and the phase of the resulting photoacoustic signal over a wide range of frequencies have been recorded, the acoustic spectrum (amplitude of signal against frequency; $S\,v\,f$) may be inverted into the water concentration profile (concentration of water against depth below surface; $c_w\,v\,x$). The inversion transform cannot be obtained in an explicit form but it can be handled by a computational algorithm, provided a set of accurate data are obtained over a wide range of frequencies. The latter proved to be a difficult task because changes in the distribution of water in the stratum corneum can occur when the skin is enclosed in a photoacoustic cell.

The theoretical axis of the present form of the ADPAS method considers a model concentration profile for which the inversion transform can be obtained explicitly. The need for large-scale computation is thus eliminated and the necessary number of independent data points drastically reduced, from about

40 to 2 or 3. A complete set of data at three different frequencies (e.g., 250, 480 and 780 Hz), obtained in one minute or less, are used to determine a and c_w.

The explicit expression for the photoacoustic signal as a function of frequency is obtained by considering the same problem as that of Rosencwaig and Gersho[3], generalized to allow for an optical absorption coefficient $\alpha(x)$ that is a function of the depth below the surface. The heat, $q(x)$, generated per unit volume is then given by the expression:

$$q(x) = \tfrac{1}{2} I_0 \alpha(x) \mathrm{e}^{-\int_0^x \alpha(x)\,\mathrm{d}x} \tag{1}$$

where I_0 is the peak intensity of the chopped beam of radiation.

The differential equation for the temperature distribution $T(x)$ in the sample is given by:

$$\frac{\mathrm{d}^2 T}{\mathrm{d}x^2} - \gamma^2 T = -\frac{1}{K} q(x) \tag{2}$$

where

$$\gamma = (i\omega C\rho/K)^{1/2} \tag{3}$$

and ω is the angular chopping frequency, and C, ρ, K are the specific heat, density, and thermal conductivity of the sample respectively, which at first are assumed to be independent of x.

Since the thermal conductivity of the air in the acoustic cell is very small compared with that of the sample, the boundary conditions at the air–sample interface are such that T is continuous and $\mathrm{d}T/\mathrm{d}x$ is zero. Under these conditions, the solution of (2) for the amplitude, T_0, of the oscillatory surface temperature is:

$$T_0 = \frac{1}{K\gamma} \int_0^\infty q(x) \mathrm{e}^{-\gamma x}\, \mathrm{d}x \tag{4}$$

On inserting the expression (1) for $q(x)$ and integrating by parts, we obtain the following relation between the surface temperature, amplitude and the distribution of the absorption coefficient:

$$T_0 = \frac{I_0}{2K} \left(\frac{1}{\gamma} - \int_0^\infty \mathrm{e}^{-\gamma x - \int_0^x \alpha(x)\,\mathrm{d}x}\, \mathrm{d}x \right) \tag{5}$$

The inversion problem is to determine $\alpha(x)$, given a set of experimental values of T_0 at several chopping frequencies. We can obtain a useful first approximation to the solution by assuming that $\alpha(x)$ has a form of a step function, being zero in a water-depletion layer of thickness, a, adjacent to the surface of the sample, and attaining a constant value, α_0 for large values of x. Then equation (5) becomes:

$$T_0 = \frac{I_0}{2K} \left(\frac{1}{\gamma} - \frac{1}{\alpha_0 + \gamma} \right) \mathrm{e}^{-\gamma a} \tag{6}$$

The signal amplitude, S, and the phase, ϕ, can therefore be expressed by the equation:

$$S\,e^{i\phi} = \left(\frac{1}{\gamma} - \frac{1}{\alpha_0 + \gamma}\right) e^{-\gamma a} \tag{7}$$

which is the explicit solution used for interpreting experimental data in terms of a.

Note that K, the thermal conductivity, is treated above as a constant, which simplifies the model. If we now allow K to vary with water concentration, but hold $(K/C\rho)^{1/2}$ constant, we find that equation (7) becomes:

$$S\,e^{i\phi} = \frac{\dfrac{1}{\gamma} - \dfrac{1}{\alpha_0 + \gamma}}{\dfrac{1}{2}\left[\left(1 + \dfrac{K_1}{K_2}\right) e^{\gamma a} + \left(1 - \dfrac{K_1}{K_2}\right) e^{-\gamma a}\right]} \tag{8}$$

where $K = K_1$ for $x \leqslant a$ and $K = K_2$ for $x > a$. For the case $\alpha_0 = 71.8\,\text{cm}^{-1}$ (liquid water at 1.95 μm), $(K/C\rho)^{1/2} = 5.5 \times 10^{-4}\,\text{cm}^2/\text{s}$ (stratum corneum), $a = 5\,\mu\text{m}$, $f = 300$ Hz, and amplitude S has the values 15.7 and 17.8 for $K_1/K_2 = 1.0$ and 0.5, respectively. Thus the signal is not very sensitive to variation of K with water concentration. Therefore, in practical use, the thermal conductivity, K, is assumed to be constant.

In a preliminary investigation of the thermal properties of excised (plantar) stratum corneum, we found that K increased from 1.55×10^{-3} to $2.32 \times 10^{-3}\,\text{W cm}^{-1}\,\text{deg}^{-1}$, i.e. by less than 50%, when water concentration was increased from 0 to 0.69 g H_2O per g of dry stratum corneum. The increase in $K/C\rho$ was found to be even smaller, probably because of the increase in C concurrent with K. Therefore, we consider the use of equation (7) with K and γ constants justified for our purpose.

The evaluation of the experimental data is performed by using a set of theoretical curves of amplitude as a function of frequency computed for several values of a, the depth of the water-depleted zone (Figure 22.3). For $a = 0$, a uniformly water-saturated stratum corneum, the amplitude against frequency dependence reduces to a proportionality between S and f^{-1} which, on a log–log scale, is represented by a straight line with -1 slope, as shown in the diagram. As the thickness of the water-depleted layer increases, the S against f curves become steeper and curved more downwards. Typically, a set of three experimental data points (shown as black circles in Figure 22.3) is slid along the vertical axis until a 'best fit' to one of the curves, and hence the value of a, is obtained.

The same single-layer model also permits estimation of the mean value of water concentration in the stratum corneum corresponding to any particular value of a. For this purpose we prepared a thin slab of freshly cast gelatin of an exactly known water content ($c_w = 0.75$–0.80 g H_2O/g of wet gelatin) and determined the amplitudes of the photoacoustic signal, S_{gel}, at the same frequencies used with the skin. The values of the ratio S_{skin}/S_{gel}, representing c_w in skin normalized to the particular value of water concentration used in

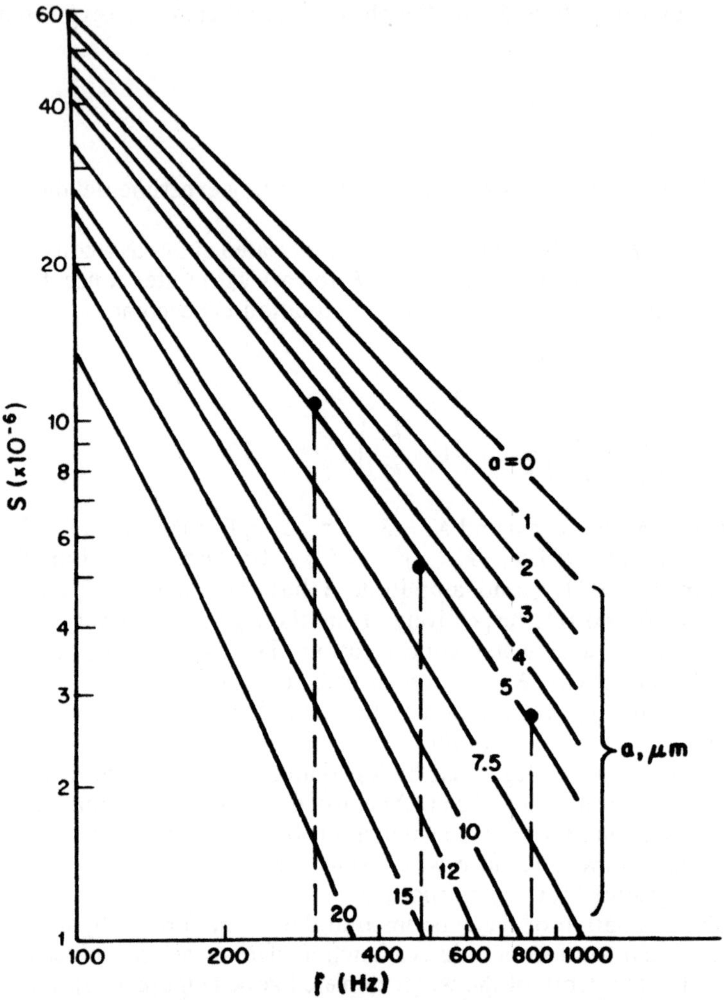

Figure 22.3 A set of computed curves of signal against frequency (f) for determining the depth of the water-depleted zone (a) by finding the best fit of experimental points. For rest of key see text

the gelatin, are then taken and averaged at each of the three acoustic frequencies. The numbers obtained in this manner are taken to be the 'water concentration' (liquid H_2O per g wet tissue) in the skin under test.

The experimental apparatus is shown in Figure 22.4. At present the apparatus is adapted for measurements on skin of the proximal surface of the left forearm, although other parts of the body could be tested as well. The subject is seated next to the photoacoustic cell with the arm resting on padded supports, lightly pressed against the orifice of the cell. The skin is used in its prevailing condition, without any preparation (such as shaving, scrubbing, stripping, etc) and without any sealants or adhesives at the contact with the cell.

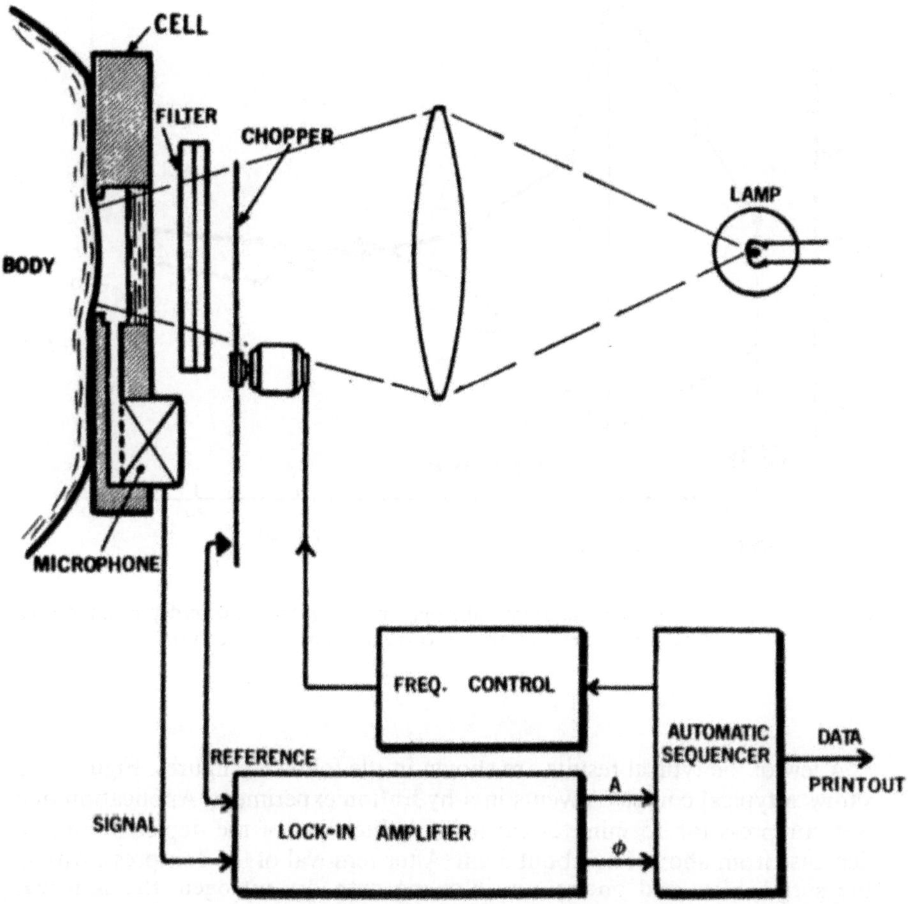

Figure 22.4 Diagram of the photoacoustic cell, illuminator, chopper, and electronics

Minute, involuntary movements of the forearm are a source of noise which degrades the quality of the data, particularly at low frequencies. Statistical studies of the signal against noise levels showed that the standard deviation of S may be as large as 25–35% of the mean (at low frequencies) and, at best, no less than about 10% of the mean.

Many preliminary experiments were performed to explore the capability of the ADPAS method to measure changes in the (a) and c_w in skin subjected to alternate conditions of hydration and dehydration. Hydration was conveniently induced by a wet compress applied for 15–60 minutes to the test area, measuring about 5 × 5 cm. Dehydration was accomplished by passing a stream of dry nitrogen over the skin using a plastic, open-bottom container cupped over the test area for the desired time (15–30 minutes).

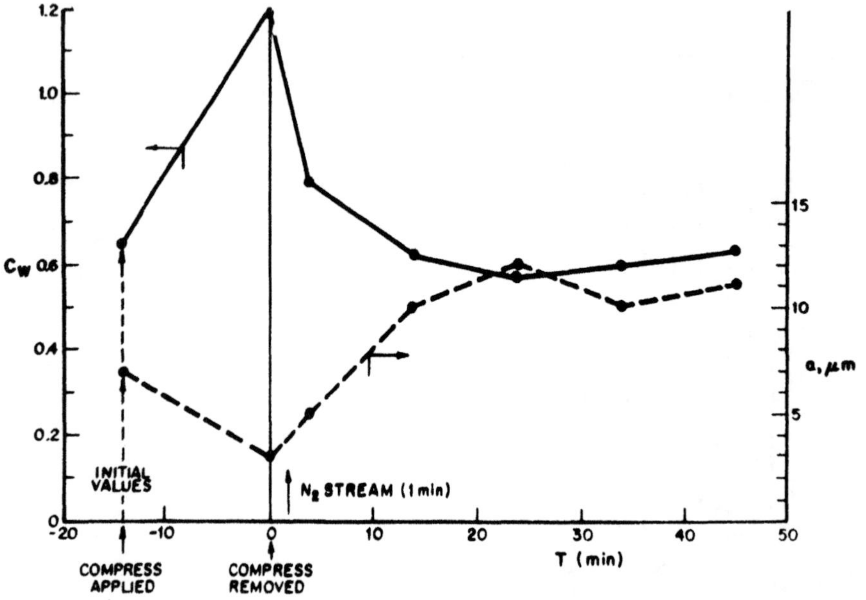

Figure 22.5 Effect of application and removal of wet compress on the depletion depth (*a*) and water concentration (c_w) in stratum corneum (proximal forearm)

A few of the typical results are shown in the following figures. Figure 22.5 shows a typical course of events in a hydration experiment. Application of a wet compress for 15 minutes caused the thickness of the depleted zone to decrease from about 6 to about 3 μm. After removal of the compress, wiping the surface dry, and one-minute blowing with dry nitrogen, the skin was allowed to come to equilibrium with ambient air at 58% relative humidity (RH). In this instance the skin eventually ended up with a depleted zone deeper (about 11 μm) than the initial value (about 6 μm). However, the average water content seemed to return almost exactly to the initial value ($c_w = 0.70$). The excessive value (1.18) of c_w at the time of removal of the compress was possibly due to a residue of liquid water left at the surface after wiping (and subsequently removed by evaporation).

Figure 22.6 shows the results of an experiment in dehydration of the skin. As could be expected, the course of events is somewhat a mirror image of those observed in the hydration experiment. The depth of the water-depleted zone increased from 4 to 10 μm after 15 minutes of dehydration and the water content decreased from 0.57–0.48. During the next 30 minutes of recovery (at 58% ambient RH), *a* and c_w tended to each final values slightly higher than the initial values (7 μm and 0.66 respectively).

We are not prepared at this time to interpret these observations; they are presented here merely as an illustration of typical experiments that are still in progress.

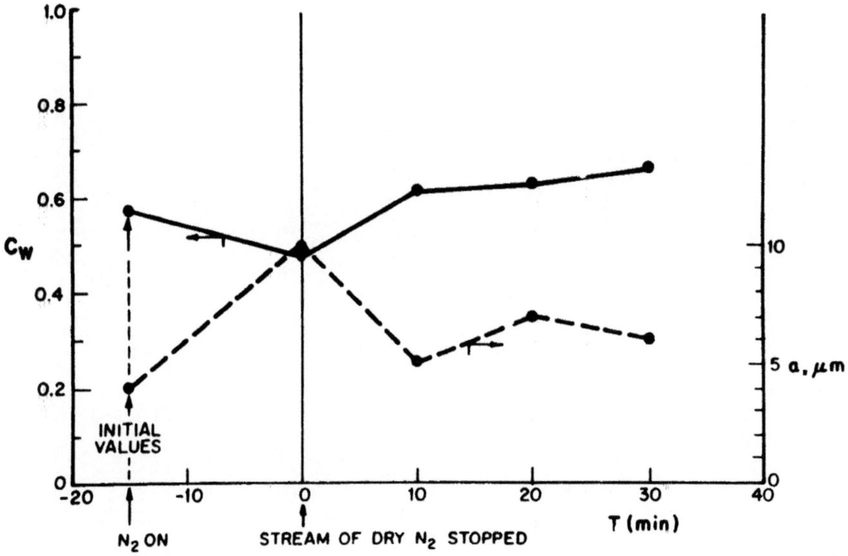

Figure 22.6 Changes in depletion depth (a) and water concentration (c_w) of the stratum corneum of proximal forearm during and after exposure to a stream of dry nitrogen

References

1. Rosencwaig, A. and Pines, E. (1977). A photoacoustic study of newborn rat stratum corneum. *Biochim. Biophys. Acta*, **493**, 10
2. Campbell, S. D., Yee, S. S. and Afromowitz, M. A. (1979). Applications of photoacoustic spectroscopy to problems in dermatology research. *IEEE Trans. Biomed. Eng.*, **BME-26**, 220
3. Rosencwaig, A. and Gersho, A. (1976). Theory of the photoacoustic effect with solids. *J. Appl. Phys.*, **47**, 64

23
Epidermal water and electrolyte content and the thermal, electrical and mechanical properties of skin

M. A. STEINMETZ and T. ADAMS

SUMMARY

The epidermis seems to be freely permeable to both water and electrolytes. Gradients which are established within the epidermis together with the epidermal permeability coefficients determine the three-dimensional movement and storage of water and electrolytes in the skin. Data are presented which support the hypothesis that the thermal, electrical and mechanical energy transfer properties of the skin are influenced by these epidermal water and salt concentrations. Consequently, the skin's hydration state may affect the perception of environmental energies impacting the skin surface, as well as being an important variable in studies of the skin's physical properties.

INTRODUCTION

The water storage function of the stratum corneum has been recognized for some time[1]. It has been shown that water enters this avascular, epidermal layer by absorbtion from the environment, by diffusion from the dermis and by lateral movement from the helically coiled, distal tubules of secreting atrichial (eccrine) sweat glands. Water absorbtion by these mechanisms is important in establishing hydration gradients within the stratum corneum. The passive absorbtion of water from the environment is directly related to ambient relative humidity[2,3], and the permeability is similar for exposure to water in either a liquid or a saturated vapor state[4,5]. Previous studies in our laboratory[6,7] in which sweat gland secretory rates have been experimentally controlled indicate that both sweat water and electrolytes diffuse from the distal, atrichial sweat gland tubule into the relatively drier stratum corneum. The amounts of water stored in the stratum corneum and evaporated from its surface are directly related to the rate and duration of sweat gland secretion. These experiments examined the relationships between atrichial sweat gland activity and the hydration of the stratum corneum, as well as rates of diffusion of water from the dermis to the skin surface, for the glabrous, tactile skin of both the domestic cat and the North American raccoon.

Other investigators have shown that adult human skin contains a large quantity (about 9 liters) of water[8] and that percutaneous water transfer rates within the stratum corneum are directly related to the amount of water stored in the skin[9]. Observations that the stratum corneum is freely permeable to water[10] and that it can absorb up to six times its weight when fully hydrated[11] further support the concept that hydration gradients and water density of the superficial layers of the epidermis, especially those of the stratum corneum, are quite labile.

Water in the skin has been described as being in one of two states: either as 'free' water (available to dissolve electrolytes), which accounts for about 20% of the water in skin; or as 'bound' water (not available to dissolve electrolytes), which accounts for about 80% of the total water stored[12].

Both the transient and steady-state water contents of the stratum corneum are important as storage phenomena. They also modify the transcutaneous transduction of environmental stimuli delivered to the skin surface. The thermal, electrical and mechanical properties of the skin have been shown to vary as functions of the hydration and salinization of the stratum corneum. Both the 'free' and 'bound' water pools have been implicated in the hydration effect on these energy transfer processes. A basic concept is that the energy transfer characteristics of the epidermis are not the same for skin in different hydration states. One implication is that the activation of thermal receptors, which are located deep in the skin, depends not only on the intensity and duration of environmental, thermal energy impacting the skin surface, but also on the effective thermal conductivity of overlying skin layers, which varies as a function of its hydration state.

Some of the experiments which show stratum corneal hydration as a variable in transcutaneous energy transfer have been performed on humans, but most of our data come from studies using the glabrous, central footpad of the anaesthetized, domestic cat. This animal model offers important advantages for studying the functional relationships among atrichial sweat gland activity, skin hydration and cutaneous energy transfer. The central footpad of the cat has a thick, relatively uniform stratum corneum which contains a rich supply of only one type of exocrine gland, the atrichial sweat gland. Electrical stimulation either within the central nervous system[13] or at the peripheral nerve supplying the footpad[14] allows the secretory rates of these cholinergic, sympathetically innervated glands to be precisely regulated independently of the animal's intrinsically active central nervous system. Peripheral vasomotor changes consequent to neural stimulation are simply and effectively eliminated using an alpha-adrenergic blocking agent such as phenoxybenzamine. The use of a sensitive detector for measuring rates of evaporative water loss from the skin surface[15] while controlling the secretion rate of the atrichial sweat glands provides a valuable model for deducing the interactions among sweat gland activity, skin hydration and the concomitant effects on energy transfer. An important bonus in these experiments is that these measurements can be made in intact skin, which retains its normal relationship with the underlying dermis and subdermis and has an undisturbed circulation.

THERMAL PROPERTIES OF THE SKIN

The skin is an important mass and energy boundary layer between the body's internal and external environments. Thermal characteristics of the skin are important in defining net heat transfer pertinent to whole body thermoregulation. They also influence the perception of thermal stimuli by the skin. Most, if not all, of the skin's thermal receptors are bare nerve endings which lie adjacent to or below its dermoepidermal junction, and which respond most acutely to changes in their temperature. The rate at which these receptors change temperature and the patterns of encoded, afferent action potentials they produce and deliver to the central nervous system depend on the rate of heat transfer from or to fluids (liquids or gases) or solid objects at the skin surface. The rate of heat flux across the skin varies as a function of thickness, its area, the difference in temperature between the skin and the environment, and its intrinsic quality for heat transfer, expressed as its effective thermal conductivity coefficient. Since skin is a non-homogeneous, anisotropic mass, its intrinsic quality for heat transfer is accurately expressed as an effective thermal conductivity coefficient (k'), in contrast with materials with uniform thermal properties, for which the thermal conductivity coefficient (k) defines this characteristic.

These interrelationships for heat transfer through the skin from the level of its thermal receptors to its surface are expressed in a form of the Fourier heat flow equation as:

$$Q_h = -k'A\frac{T_0 - T_j}{x} \tag{1}$$

where Q_h equals the total transcutaneous heat flow (in watts), k' equals the effective thermal conductivity of the skin between its thermal receptors and its surface (watts $m^{-1}\,°C^{-1}$), T_0 equals skin surface temperature (°C), T_j equals thermal receptor temperature (°C), A equals skin area exposed to the environment (m^2), and x equals the distance between thermal receptors and the skin surface (m). In practice, heat flow between a warm-water thermode and the skin surface is measured with a calibrated heat-flow disc and the temperature difference is measured with a thermopile, with one junction on the skin surface and the other inserted to the depth of the dermoepidermal junction (Figure 23.1)[16].

Because blood vessels in the skin lie below the level of its thermal receptors, skin blood flow does not directly affect the k' of the tissue mass between them and the skin surface, but the k' of the epidermis is strongly influenced by its hydration state (density of water). Skin blood flow, of course, can change the temperature at any site within the skin and it is an important determinant of T_j. This is a separate consideration to the relationship expressed in equation (1). It

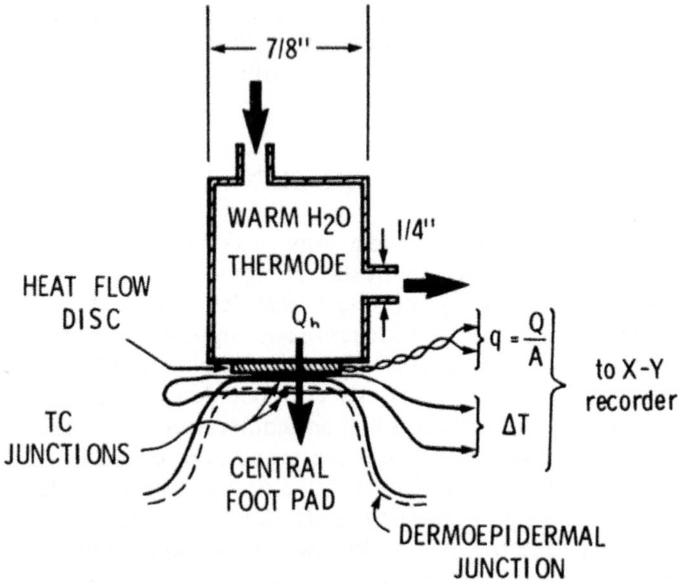

Figure 23.1 Apparatus for determining epidermal thermal conductivity by measuring heat flow between a warm water thermode and the skin surface

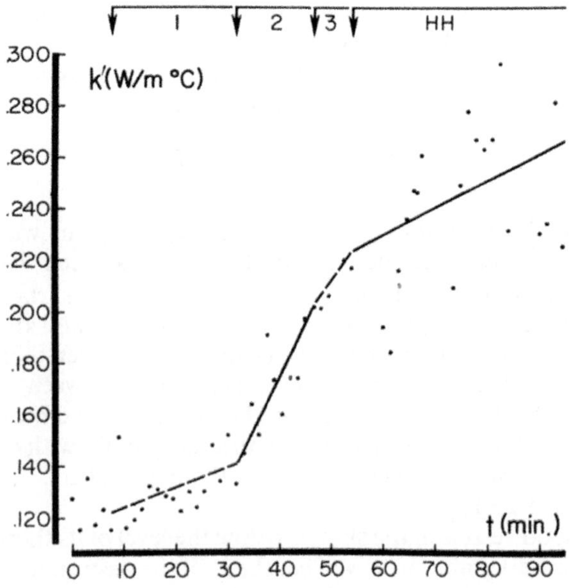

Figure 23.2 Effective thermal conductivity (k') of the cat footpad epidermis as a function of time, during progressive stratum corneal hydration by either electrical stimulation (pulse trains/s) of sweat gland nerves or exposure to high ambient relative humidity (HH)

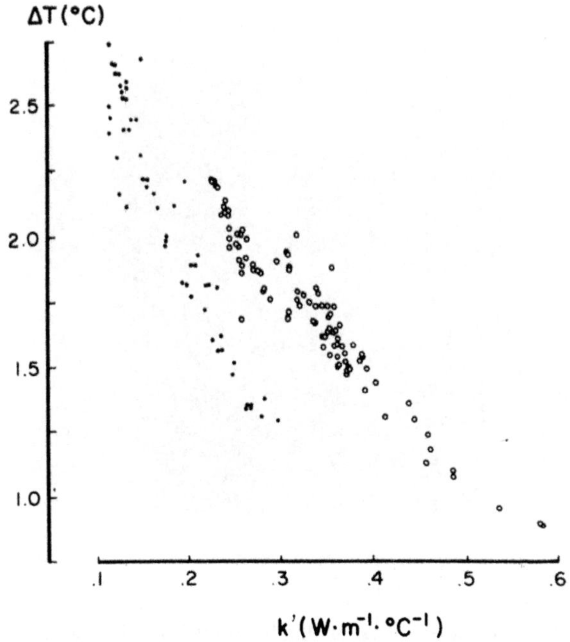

Figure 23.3 Transepidermal temperature difference (ΔT) and epidermal effective thermal conductivity (k') during increasing skin hydration

has been shown that for the same conditions of skin area and transcutaneous temperature gradient $(T_0 - T_1)/x$ (Figure 23.1), heat transfer rates through the skin are different for different states of epidermal hydration. Skin which is fully hydrated by absorbtion of water from the environment, from the dermis or from sweat gland activity has a higher k' (by a factor of 2 or more) than it does when its water density (and its k') is lower (Figure 23.2)[16].

An important conclusion from this study is that not only are heating and cooling rates higher for hydrated skin, but also (for skin in a thermal and hydration steady-state) that cutaneous thermal receptors rest at a temperature closer to that of the environment when the epidermis is hydrated than when it is drier (Figures 23.3 and 23.4)[16].

An important implication of these conclusions is that thermal perception and judgement of thermal comfort are affected by physiological and environmental conditions which modify epidermal hydration, namely sweat gland activity and ambient relative humidity. This phenomenon is perhaps obvious to anyone who reports that he or she 'feels colder' on a wet–cold day than on a dry–cold one, or who 'feels hotter' on a wet–hot day than on a dry–hot one. Certainly, other conditions – such as water-vapor pressure gradients, temperature, air flow, evaporation, heat-transfer coefficients at the skin surface, cutaneous blood flow rate, as well as whole body and regional (muscle) metabolic heat production – also modify thermal perception. The change in skin heat-transfer properties (especially its k') owing to epidermal

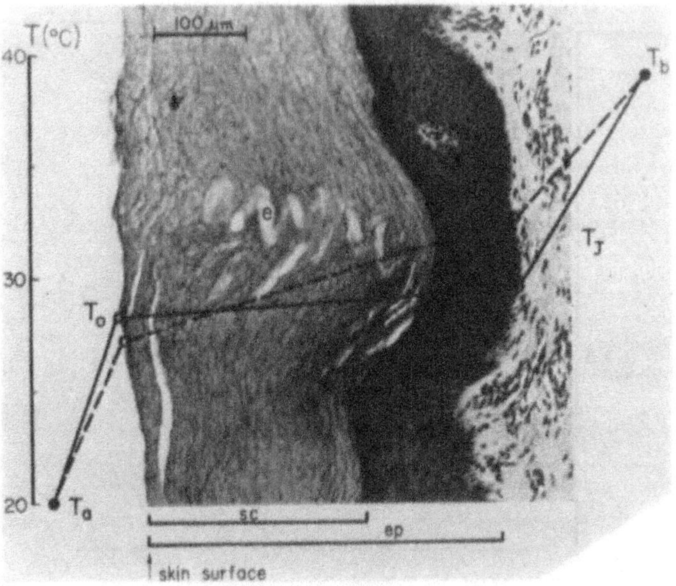

Figure 23.4 Histological section of cat hindlimb central footpad showing temperature profiles across epidermis (ep) and stratum corneum (sc) at different skin surface (T_0) and dermoepidermal (T_i) temperatures, and constant ambient (T_a) and deep body (T_b) temperatures. ○---○ = At low k'. ○——○ = At high k'. Note helical penetration of distal atrichial sweat gland duct (e) through stratum corneum

hydration has been unrecognized before as an influential factor in cutaneous thermal perception and in the evaluation of thermal comfort levels. Perhaps this phenomenon would be valuable when considering thermal injury to the skin[17], and in studies in which cutaneous heat transfer rates are used as a basis for indirectly estimating skin thickness[18].

ELECTRICAL PROPERTIES OF THE SKIN

Electrical properties of the skin are indexed either as its electrical potential, or as some measurement of its resistance to electrical current flow. Skin electrical potential is usually measured using electrodes attached at two different skin-surface sites. Because no electrical energy other than that arising from the skin itself is used for this measurement, skin electrical potential has been called an 'endosomatic' property. Measurements of the skin's resistance to electrical current flow are usually made by applying an external voltage across two skin-surface sites, and determining the transient and/or steady-state patterns of transcutaneous electrical current flow between them. Data for these so-called 'exosomatic' measurements are expressed in resistance (ohms), conductance (mhos) or impedence (ohms at a fixed frequency), the last of which also provides information about the skin's electrical capacitance by using alternating rather than direct currents.

The precise relationships among the many physical and physiological factors that affect skin electrical phenomena are not completely known [19,20]. Most of the epidermis seems to be freely permeable to water and salts, as well as other molecules[19], and the movement and storage of these substances are clearly important determinants of the skin's electrical properties, regardless of the type of measurement.

The electrical potential and resistive measures of the skin have both been used as indices of the intrinsic activity of the central nervous system, especially its autonomic divisions. The basis for this deduction is that peripheral vascular smooth muscle receives both sympathetic and parasympathetic nerve branches, and the atrichial sweat glands are sympathetically innervated. Also, sweat glands are activated simultaneously with either voluntary or reflex contractions of striated, somatic muscles in an unidentified manner. Changes in peripheral blood flow and atrichial sweat gland activity individually and together modify patterns of intracutaneous water and electrolyte gradients and density, and thereby indirectly affect both the endosomatic and exosomatic electrical phenomena recorded in, on or across the skin.

A strong correlation exists between skin electrical resistance and skin surface evaporative water loss rates when these measurements are made simultaneously on palmar skin of resting, human subjects (Figure 23.5)[21]. There have been many attempts to sort out the complex and interdependent relationships between the skin's electrical features, peripheral vasomotion and atrichial sweat gland activity. These studies are exceptionally difficult in humans, because of uncontrolled and unrecorded spontaneous variations in voluntary and autonomic central nervous system activities. There is a large advantage in using an animal model, in which all efferent neural activity and the resulting peripheral vasomotor and neurosecretory responses can be controlled when attempting to quantify the interactions among skin electrical potential, resistance, water and electrolyte densities. The transfer function of the atrichial sweat glands has been described in the cat footpad model by measuring precisely their output (in terms of the volume flow rate of secreted water) to a rigidly controlled input (the frequency of action potentials experimentally triggered on the peripheral nerve fibers supplying the sweat glands)[6]. With this model, atrichial sweat gland activity may be precisely regulated, peripheral vasomotion controlled, and skin surface evaporation measured while the skin's electrical phenomena are simultaneously recorded.

The results of studies of the electrical properties of human skin[21] suggested experiments on the cat footpad model[22,23] which have provided quantitative information about the interactions between atrichial sweat gland activity and skin electrical phenomena. Measurements were made with the sweat glands and the skin in steady-state hydration states of different magnitudes and during controlled variations in hydration, dehydration, salinization and desalinization. These studies using the anaesthetized cat provided data which reinforce the impression gained in tests in human subjects that both skin electrical potential and resistance characteristics are closely related to sweat gland activity. Figure 23.6 shows the results of such a study[22]. The skin potential

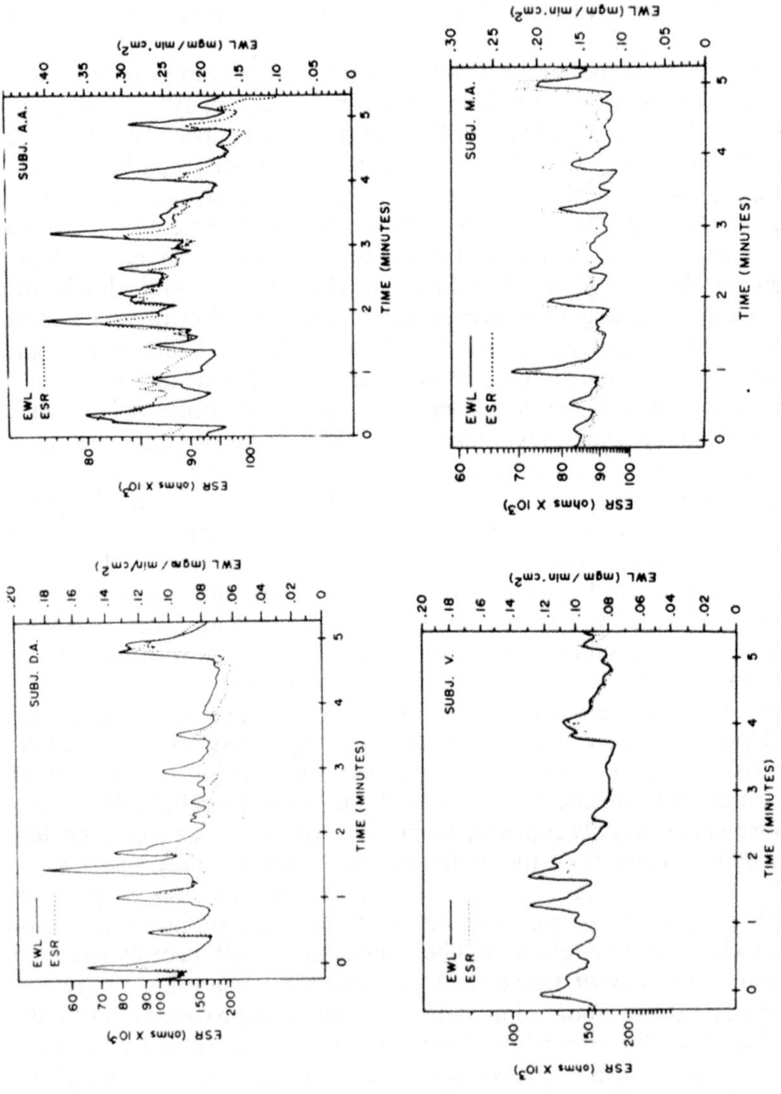

Figure 23.5 Superimposed tracings of electrical skin resistance (ESR) and skin surface evaporation rate (EWL) in four subjects who were either resting quietly (low EWL, high ESR) or were engaged in provocative conversation (high EWL, low ESR). Note the similarity in time course for these two variables regardless of basal EWL and ESR values

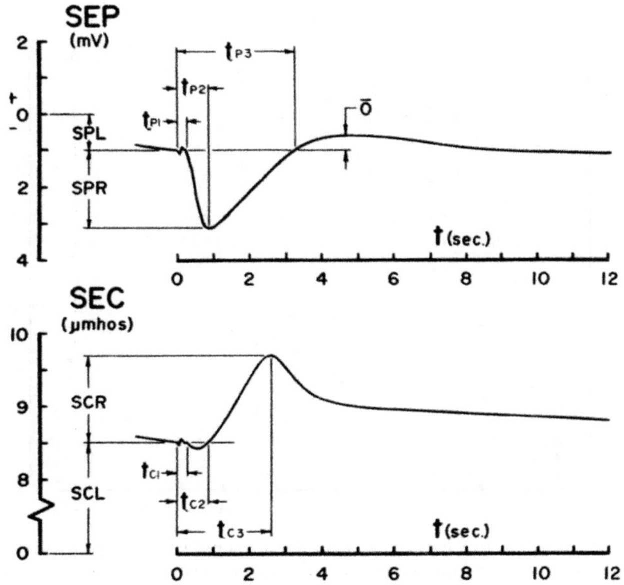

Figure 23.6 Representative records of skin electrical potential (SEP) and skin electrical conductance (SEC) from cat footpad after the delivery of an electrical shock to the peripheral nerves innervating the sweat glands ($t = 0$). SPL = Skin potential level. SCL = Skin conductance level. SPR and SCR = Maximum deviation from rest of skin potential and conductance responses. t_{p1}, t_{c1}, etc. = Time-course characteristics. $\overline{O}$ = Overshoot

and conductance levels, maximum deviation from rest of skin potential and conductance responses, and the time-course characteristics and overshoot are shown during one evoked response.

For both skin electrical potential (SEP) and skin electrical conductance (SEC), the time course and steady-state values of these measures are predictable on the basis of controlled sweating rates and the concomitant skin hydration (Figures 23.7 and 23.8)[22]. These data show that the water gradient and density in the epidermis are two important variables in establishing both the electrical potential and resistive features of the intact skin.

Recently available data make it clear that many of the skin's electrical properties depend on not only its water density, but also its electrolyte content. Data shown in Figures 23.9 and 23.10 were collected in experiments in which skin salt and water contents were independently varied[23]. In the first set of experiments (Figure 23.9), dried skin from cat central footpad was progressively hydrated by electrical stimulation (8 V DC; 2 ms pulses; 3 Hz) of sweat gland nerves; concentrations of epidermal electrolytes were altered beforehand by soaking in either deionized water or hypertonic or isotonic saline followed by air drying. In the second set of experiments (Figure 23.10), dried skin from cat central footpad underwent progressive epidermal

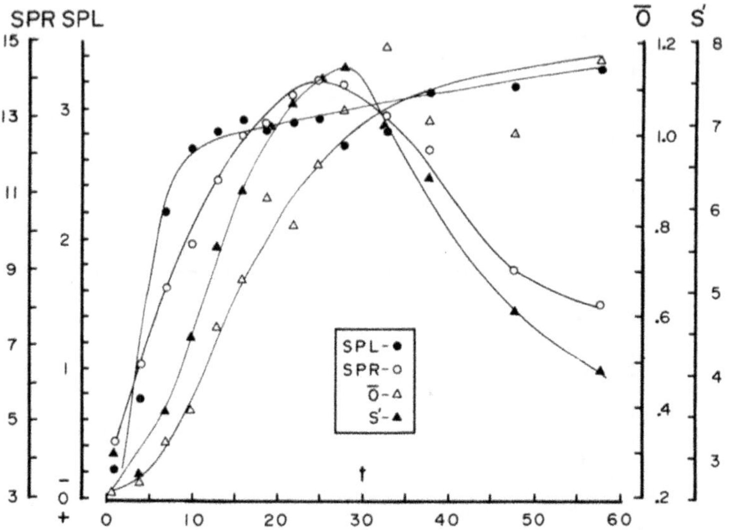

Figure 23.7 Mean values in 8 cats of the skin electrical potential baseline (SPL; mV), response (SPR; mV), overshoot (O) and maximum rate of return to baseline (*S'*; mV/s) shown as functions of dehydration time; dried air was blown at 15 l/min across the footpad previously hydrated by 3 minutes of electrical stimulation (8 V DC; 2 ms pulses; 8 Hz) of sweat gland nerves

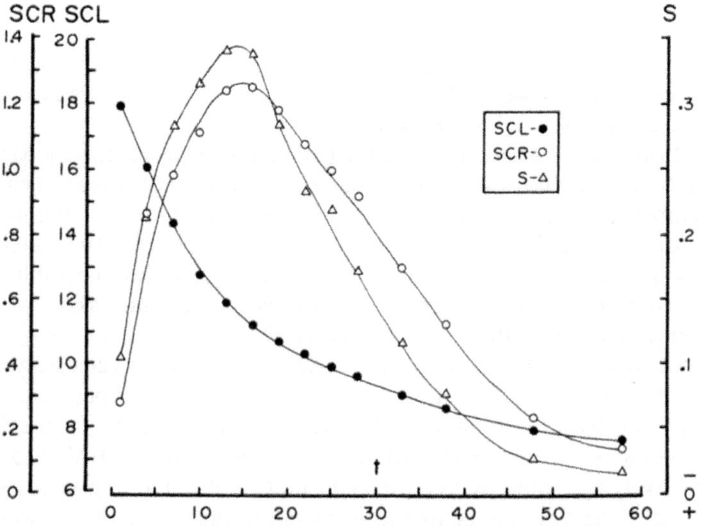

Figure 23.8 Mean values in 8 cats of the skin electrical conductance baseline (SCL; μmhos), response (SCR; μmhos) and maximum rate of return of SCR to baseline (*S*; μmhos/s) shown as functions of dehydration time; method as described in Figure 23.7

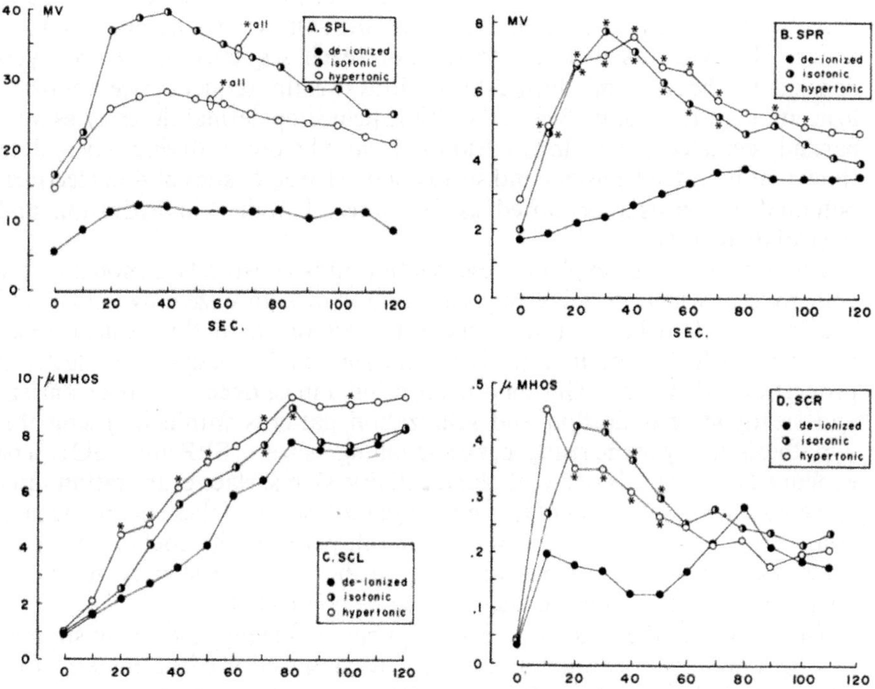

Figure 23.9 (a) Skin potential level (SPL); (b) skin potential response (SPR); (c) skin conductance level (SCL); and (d) skin conductance response (SCR) from cat central footpad shown as functions of time starting from dried skin and during progressive skin hydration. Asterisks mark data points significantly different from deionized skin ($p < 0.05$)

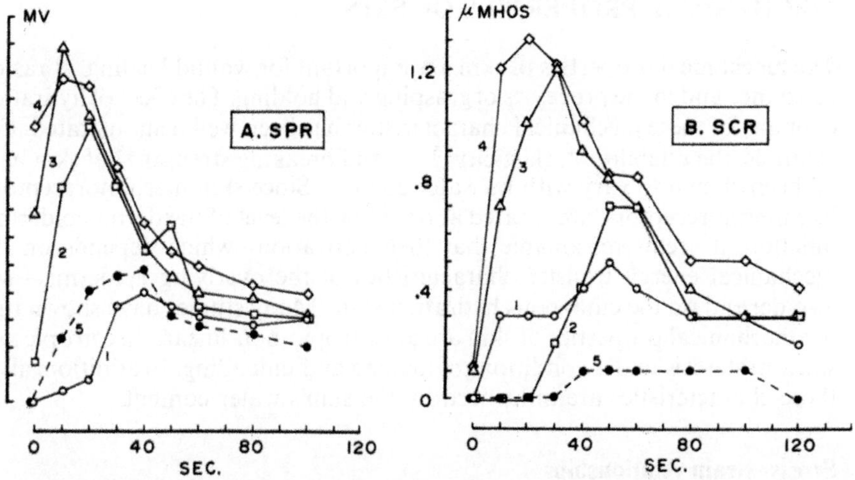

Figure 23.10 (a) Skin potential response (SPR); and (b) skin conductance response (SCR) from cat central footpad shown as functions of time starting from a dried skin and during progressive epidermal hydration. Numbers in graph refer to sweating sequence.

hydration by electrical stimulation (8 V DC; 2 ms pulses; 3 Hz) of sweat gland nerves. Before the first sweating sequence, the skin was soaked in deionized water. The skin was then air dried before each sequence, but electrolytes deposited in the epidermis during the previous sweating sequences were allowed to accumulate in the skin. Before the fifth sequence, epidermal electrolytes were partially depleted by soaking in deionized water before air drying. These data show that both the transient and steady-state characteristics of skin electrical potential and resistance varied as functions of both skin hydration and electrolyte density.

One hypothesis to explain these relationships constructs a model of the epidermis in which its thick, superficial layer of keratinized, avascular and anucleated cells (the stratum corneum) is viewed as a three-dimensional electrical volume conductor. The transient and steady-state electrical properties of this electrically conducting cellular mass depend on the transient and steady-state hydration and salinization patterns within it. Using this hypothesis, many of the rising, peak and falling values of SEP and SEC can be explained on the basis of sweat gland activity, skin surface evaporation rates and epidermal water and salt exchange. Quantification of these phenomena in relationship to intracutaneous water and salt permeability coefficients, mass transfer and electrical volume conductor properties of the skin as well as the anatomy of its boundary layers, is yet to be completed.

The data from these experiments on human and feline tactile skin support the suggestion that the water and electrolyte composition of the epidermis are important in defining its electrical properties. Imagining the epidermis to be a variable, electrical volume conductor, and considering sweating as a mechanism for altering its hydration and salinization, and thereby its electrical features, is an attractive working hypothesis.

MECHANICAL PROPERTIES OF SKIN

The mechanical properties of skin are important for wound healing, abrasion resistance and in the processes of grasping and holding. The effect of hydration on many of these mechanical characteristics has been well demonstrated. For example, the pliability[24], elasticity[25–27] and breaking strength[28] of skin have all been shown to vary with its water content. Since skin mechanoreceptors, like thermoreceptors, are located at or below the level of the dermoepidermal junction, it seems reasonable that their activation – which depends on the mechanical energy transfer characteristics of the overlying epidermis – will also depend on the cutaneous hydration state. Many studies have shown that the mechanical properties of skin are anisotropic, non-linear, thixotropic and show hysteresis under conditions of loading and unloading. In addition, all of these characteristics are influenced by the skin's water content.

Stress–strain relationships

The mechanical properties of skin have been quantified by various testing methods. *In vitro* tests in which the skin is loaded either uniaxially or biaxially have provided important information about the directional anisotropicity and the hydration effect[29–31]. Studies of intact human skin, in which loading was either uniaxial[32], torsional[33], or by vertical extension[34], suction[35] or ballistometry[36] have also made important contributions to our understanding of the mechanical nature of skin. Although these studies on human skin offer the advantage of intact skin, which retains its normal relationship with the underlying tissues and an undisturbed circulation, reproducibility in these experiments depends on the loading and hydration conditions, which are difficult to control. Nevertheless, these studies have documented the effects of processes such as ageing, radiation exposure, and delipidization, as well as the plasticizing effect of hydration on the mechanical properties of intact skin.

Although it has been proposed that much of the hydration effect on the mechanical properties of skin results from the interaction between water molecules and dermal collagen fibres, tests in both excised[26–28] and intact[37] skin samples indicate that the hydration state of the stratum corneum is also an important determinant of these characteristics. Owing to the extremely labile three-dimensional hydration profiles that exist in the stratum corneum, and which depend on ambient relative humidity, the dermal hydration state and amount of sweat gland activity, the mechanical properties of this avascular layer are likely to be extremely variable, even in a normal skin. Changes in hydration of the stratum corneum because of physiological, pharmacological, pathological or environmental conditions will probably affect the mechanical properties of skin as well as the transduction and perception of environmental stimuli.

Frictional properties

The properties of static and dynamic friction between different objects and the skin surface have also been examined[38–41]. Several of these experiments have focused on the effect of hydration on frictional forces generated at the skin surface. Figure 23.11 shows the results of such an experiment[41]. The horizontal force needed to move a plastic plate across the cat footpad surface at a constant velocity, and the amount of water accumulated in the skin after 3 minutes of sweat gland nerve stimulation were measured at different frequencies of peripheral nerve stimulation. Both variables were measured in sequential tests in the same preparation. Frictional forces increased with a small addition of water to the epidermis then decreased with further hydration; liquid water never appeared on the skin surface at these low stimulation frequencies. Although the precise mechanism for this effect is not completely understood, it has been alternatively proposed to be related to changes in skin surface area, surface tension, deformability and cohesion between corneocytes.

These data further show the importance of epidermal hydration as a determinant of the mechanical characteristics of skin. Although the molecular interactions within the dermis are certainly important in the effect of hydration on the mechanical properties of skin, the contribution of the epidermis should not be overlooked.

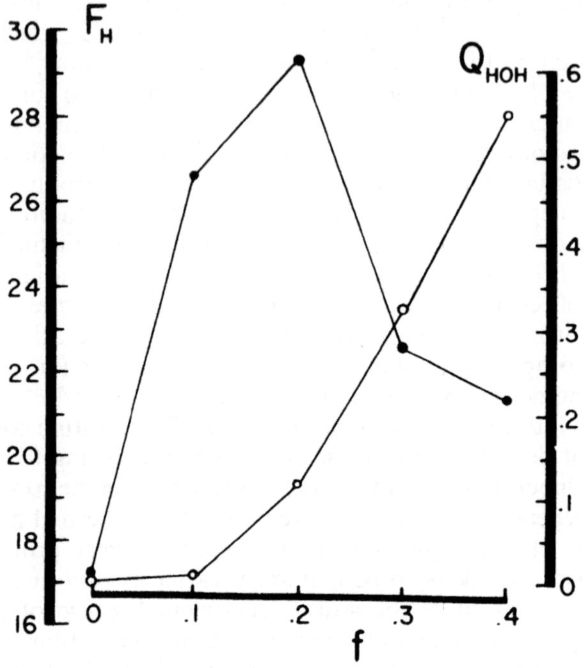

Figure 23.11 Horizontal force (F_H; ●——●) required to move a plastic plate across cat footpad surface at a constant velocity, and the amount of water (Q_{HOH}; O——O) accumulated in the skin after sweat gland nerve stimulation shown as functions of different frequencies of peripheral nerve stimulation (f)

CONCLUSIONS

The results of our experiments and those of others which have examined the relationships among epidermal water content and the skin's energy and mass transfer characteristics suggest an hypothesis for further tests. (Figure 23.12.) This model proposes that the density of water and electrolytes at any point in the epidermis at any time is a combined function of diffusion of water and electrolytes from the vascularized dermis to the skin surface and the diffusion from the liquid filled ducts of the atrichial sweat glands. In the model a hydration gradient for a skin sample of given area is established across the thickness of the epidermis at a low ambient relative humidity and water-vapor partial pressure, such that the water content at the skin surface is less than that at the dermoepidermal junction. Without sweat gland activity water passively diffuses along this gradient from the body, depending on the rate of diffusion from the dermis and on the water diffusion coefficient in that direction, and establishes the 'insensible water loss' rate. With electrical stimulation of sweat gland nerves, secreted sweat either diffuses into the epidermis or escapes at the skin surface. Sweat that moves laterally from the helically coiled, distal sweat gland tubule–which depends on the rate of secretion, the hydration gradient, and the diffusion coefficient in that direction – diffuses across the skin surface,

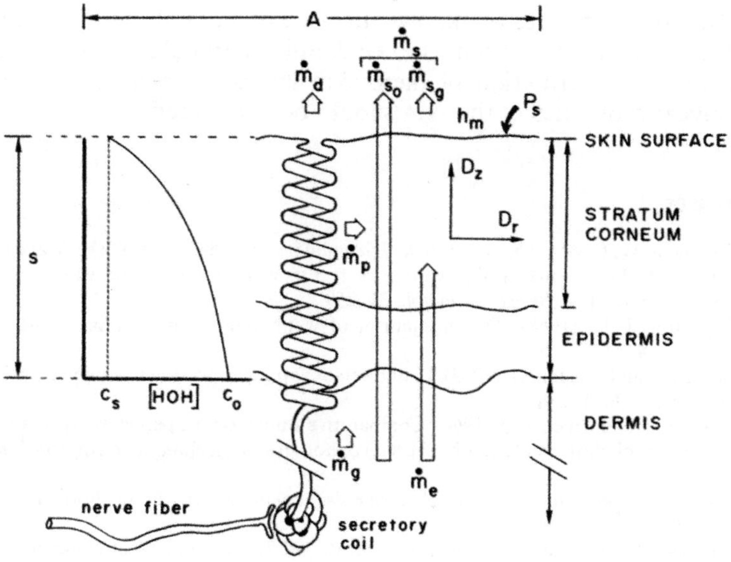

Figure 23.12 Model of sweat gland–skin interaction. HOH = Hydration gradient. A = Skin-sample area. s = Thickness of epidermis. h_m = Low ambient relative humidity. P_s = Low water-vapor partial pressure. c_s = Water content at skin surface. c_o = Water content at dermoepidermal junction. $\dot{m}_e$ = Rate of diffusion from the dermis. D_z = Water diffusion coefficient from dermis. $\dot{m}_{so}$ = 'Insensible water loss' rate (basal water diffusion rate). $\dot{m}_g$ = Secreted sweat. $\dot{m}_p$ = Secreted sweat diffusing into epidermis. $\dot{m}_d$ = Secreted sweat escaping at surface. D_r = diffusion coefficient from sweat gland tubule. $\dot{m}_{sg}$ = Sweat diffusing across skin surface. $\dot{m}_s$ = Total rate of transcutaneous water loss

which adds to the basal water diffusion rate to form the total rate of transcutaneous water loss. Both the transient and steady-state storage and movements of water and electrolytes in the epidermis affect the thermal, electrical and mechanical energy transfer properties of this skin layer.

The rates of these diffusion processes depend on the water and electrolyte permeability coefficients within the epidermis – both normal and parallel to the skin surface – especially within its outermost layer, the stratum corneum. These permeability coefficients are probably different from one another, considering the anatomical arrangement of cells which form the stratum corneum. Either one of them also probably varies as a function of the distance from the skin surface at which it is measured. The rate at which water and electrolytes move through the skin predictably depends on both the salinization and hydration states of the epidermis. The concentrations of water and electrolytes are important in establishing the direction and steepness of the associated gradients for passive diffusion.

Although the thermal, electrical and mechanical properties of the skin – especially the epidermis – have been described as varying with its water and salt contents, the underlying physical and chemical principles which determine not only rates and patterns of water and salt distribution but also the effect of these on the energy transfer properties are either poorly understood or completely unknown. Because of the labile nature of water and electrolyte gradients within the epidermis, particularly in the stratum corneum, the contribution of these skin layers to thermal, electrical and mechanical properties of the skin should be considered.

References

1. Montagna, W. (1965). *The structure and function of skin.* 454. (New York: Academic Press)
2. Blank, I. H. (1953). Further observations on factors which influence the water content of the stratum corneum. *J. Invest. Dermatol.*, **21**, 259
3. Middleton, J. D. (1968). The mechanism of water binding in stratum corneum. *Br. J. Dermatol.*, **80**, 437
4. Stannett, J. and Yasuda, H. (1963). Liquid versus vapor permeation through polymer films. *J. Polymer Sci., B.*, **1**, 289
5. Sivadjian, J. and Ribeiro, D. (1964). Comparative studies on the permeability to water and to water vapor of plastic materials by the hygrophotographic technique. *J. Appl. Polymer Sci.*, **8**, 1403
6. Adams, T. (1966). Characteristics of eccrine sweat gland activity in the footpad of the cat. *J. Appl. Physiol.*, **21**, 1004
7. Steinmetz, M. A., Adams, T. and Heisey, S. R. (1977). Atrichial sweat gland activity in the raccoon (*Procyon lotor*). *The Physiologist*, **20**, 91
8. Skelton, H. (1927). The storage of water by various tissues of the body. *Arch. Intern. Med.*, **40**, 140
9. Grice, K., Sattar, H. and Baker, H. (1972). The effect of ambient humidity on transepidermal water loss. *Dermatology*, **58**, 343
10. Scheuplein, R. J. and Blank, I. H. (1971). Permeability of the skin. *Physiol. Rev.*, **51**, 702
11. Scheuplein, R. J. and Morgan, L. J. (1967). 'Bound water' in keratin membranes measured by a microbalance technique. *Nature*, **214**, 456
12. Yates, J. R. (1971). Mechanism of water uptake by skin. In Elden, H. R. (ed.) *A Treatise of Skin. Vol. I. Biophysical Properties of the Skin*, pp. 485–512. (New York: Wiley-Interscience)
13. Celesia, G. G. and Wang, G. H. (1964). Sudomotor activity induced by single shock stimulation of the hypothalamus in anesthetized cats. *Arch. Ital. Biol.*, **102**, 599
14. Langley, J. N. (1891). On the course and connections of the secretory fibres supplying the sweat glands of the feet of the cat. *J. Physiol.* [*Lond.*], **12**, 347
15. Adams, T., Funkhouser, G. E. and Kendall, W. W. (1963). The measurement of evaporative water loss by a thermal conductivity cell. *J. Appl. Physiol.*, **18**, 1291
16. Holmes, K. R. and Adams, T. (1975). Epidermal thermal conductivity and stratum corneum hydration in cat footpad. *Am. J. Physiol.*, **228**, 1903
17. Stoll, A. M., Chianta, M. A. and Piergallini, J. R. (1979). Thermal conduction effects in human skin. *Aviation Space Environ. Med.*, **50**, 778.
18. Stoll, A. M. (1977). Thermal properties of human skin related to nondestructive measurement of epidermal thickness. *J. Invest. Dermatol.*, **69**, 328
19. Edelberg, R. (1971). Electrical properties of skin. In Elden, H. R. (ed.) *A Treatise of Skin. Vol. I. Biophysical Properties of the Skin*, pp. 513–550. (New York: Wiley-Interscience)
20. Venables, P. H. and Christie, M. J. (1973). Mechanisms, instrumentation, recording techniques, and quantification of responses. In Prokasy, W. F. and Raskin, D. C. (eds.) *Electrodermal Activity in Psychological Research*, pp. 1–124. (New York: Academic Press)
21. Adams, T. and Vaughan, J. A. (1965). Human eccrine sweat gland activity and palmar electrical skin resistance. *J. Appl. Physiol.*, **20**, 980
22. Stombaugh, D. P. and Adams, T. (1971). Skin electrical phenomena, sweat gland activity, and epidermal hydration of the cat footpad. *Am. J. Physiol.*, **221**, 1014
23. Adams, T., Steinmetz, M. A. and Heisey, S. R. (1980). Effect of epidermal hydration and osmolarity on skin phenomena. *Bioengineering and the Skin*, **2**, 52
24. Blank, I. H. (1952). Factors which influence water content of the stratum corneum. *J. Invest. Dermatol.*, **18**, 433

25. Laden, K. and Morrow, R. (1970). Torsional measurements on skin. *J. Soc. Cosmetic Chem.*, **21**, 417
26. Park, A. C. and Baddiel, C. B. (1972). Rheology of stratum corneum – I: a molecular interpretation of the stress–strain curve. *J. Soc. Cosmetic Chem.*, **23**, 3
27. Park, A. C. and Baddiel, C. B. (1972). Rheology of stratum corneum – II: a physico-chemical investigation of factors influencing the water content of the corneum. *J. Soc. Cosmetic Chem.*, **23**, 13
28. Wildnauer, R. H., Bothwell, J. W. and Douglass, A. B. (1971). Stratum corneum biomechanical properties I. Influence of relative humidity on normal and extracted human stratum corneum. *J. Invest. Dermatol.*, **56**, 72
29. Daley, C. H. and Odland, G. F. (1979). Age-related changes in the mechanical properties of human skin. *J. Invest. Dermatol.*, **73**, 84
30. Lanir, Y. and Fung, Y. C. (1974). Two-dimensional mechanical proprties of rabbit skin. I. Experimental system. *J. Biomechanics*, **7**, 29
31. Lanir, Y. and Fung, Y. C. (1974). Two-dimensional mechanical properties of rabbit skin. II. Experimental results. *J. Biomechanics*, **7**, 171
32. Burlin, T. E., Hutton, W. C. and Ranu, H. S. (1977). A method of *in vivo* measurement of the elastic properties of skin in radiotherapy patients. *J. Invest. Dermatol.*, **69**, 321
33. Barbenel, J. C. and Evans, J. H. (1977). The time-dependent mechanical properties of skin. *J. Invest. Dermatol.*, **69**, 318
34. Pierard,, G. E. and Lapiere, Ch. M. (1977). Physiopathological variations in the mechanical properties of skin. *Arch. Dermatol. Res.*, **260**, 231
35. Cook,, T., Alexander, H. and Cohen, M. (1977). Experimental method for determining the 2-dimensional mechanical properties of living human skin. *Med. Biol. Eng. Comput.*, **15**, 381
36. Tosti, A., Compagno, G., Fazzini, M. L. and Villardita, S. (1977). A ballistometer for the study of the plasto-elastic properties of skin. *J. Invest. Dermatol.*, **69**, 315
37. Christensen, M. S., Hargens, C. W., Nacht, S. and Gans, E. H. (1977). Viscoelastic properties of intact human skin: instrumentation, hydration effects, and the contribution of the stratum corneum. *J. Invest. Dermatol.*, **69**, 282
38. Naylor, P. F. D. (1955). The skin surface and friction. *Br. J. Dermatol.*, **67**, 239
39. Comaish, S. and Bottoms, E. (1971). The skin and friction: Deviations from Amonton's laws, and the effects of hydration and lubrication. *Br. J. Dermatol.*, **84**, 37
40. Highley, D. R., Coomey, M., DenBeste, M. and Wolfram, L. J. (1977). Frictional properties of skin. *J. Invest Dermatol.*, **69**, 303
41. Adams, T. and Hunter, W. S. (1969). Modification of skin mechanical properties by eccrine sweat gland activity. *J. Appl. Physiol.*, **26**, 417

24
Xeroradiographic and ultrasound techniques in the assessment of skin disorder

C. Y. TAN, R. MARKS, E. ROBERTS and E. GUIBARRA

SUMMARY

Patients with localized benign and malignant skin lesions have been investigated with xeroradiographic imaging and an ultrasound scanning (A mode) method. Xeroradiography provided useful information, especially in benign lesions, while ultrasound scanning in the A mode gave no additional advantage. These two methods were also used to detect dermal thinning after applying topical corticosteroid preparations. A good correlation was obtained between the two methods for normal as well as thinned skin. The ultrasound technique has the advantages of using a relatively safe type of radiation and permitting multiple observations.

INTRODUCTION

The diagnosis of skin tumours and swellings is based traditionally on clinical examination and biopsy. The management of patients with these conditions often necessitates two operations, one for diagnosis and another for treatment. For this reason we believe that the development of non-invasive investigative techniques which could aid in diagnosis is desirable.

Xeroradiography is now an accepted method of evaluating soft tissue masses, especially in breasts. The technique provides more information and has a higher resolution than the conventional X-ray film[1]. Although ultrasound techniques are widely used in medical practice, they have rarely been used in dermatological practice. However, an ultrasound method has recently been used to assess the depth of burns[2] and has been evaluated as an aid to diagnosis[3]. Because skin lesions are usually small, any ultrasound scan device has to be used with a high frequency probe to obtain the necessary resolution. High frequency ultrasound scans of A and B modes have been successfully used in ophthalmology for detecting ophthalmological lesions for some years[4,5].

We have used a xeroradiographic method for accurately measuring dermal thickness in the detection of dermal atrophy due to corticosteroids. For a review see Dykes and Marks[6]. A more recently introduced method of measuring skin thickness is based on the use of high frequency (15 MHz) pulsed ultrasound, using an A-scan system[7]. We have also adopted this technique and have shown that the results of xeroradiographic and ultrasound measurements in the same subjects have a strong correlation[8].

MATERIALS AND METHODS

Skin lesion study

Forty-one patients with a variety of malignant and benign skin lesions were studied (Table 24.1). For xeroradiographic imaging tangential views of the lesions were taken at a voltage of 120 kV.

The ultrasound A-scan system is similar to that described by Alexander and Miller[7]. A transducer designed to operate at 15 MHz is driven by a very fast rise-time pulse from the transmitter using transistors operating in the avalanche mode. The receiver input stage comprises a dual-gate field effect transistor and the overall range resolution of the system is such that interfaces separated by about 0.25 mm can be distinguished on the oscilloscope display. The transducer is offset from the skin by a water path to avoid confusion between the transmitter pulse and the first skin echo. However, it has not proved necessary to incorporate a vacuum device to display the dermis–subcutaneous fat interface.

Table 24.1 Skin lesions studied by xeroradiography and ultrasound scanning

Lesions	*No of patients*
Squamous cell carcinoma	5
Basal cell carcinoma	3
Myxoid cyst	4
Epidermoid cyst	5
Pilar cyst	3
Lipoma	2
Ganglion	2
Lymphoedema	2
Histiocytoma	3
Keloid	1
Cellular naevus	4
Skin lymphoma	1
Rheumatoid nodule	1
Pretibial myxoedema	1
Vascular malformation of fingers	1
Leiomyoma	1
Bowen's disease	1
Morphoea	1

When examining the skin lesions with the ultrasound scan, the ultrasound probe was manoeuvred so that the probe with its open-ended column of water was situated on the lesion itself. The display obtained was compared with that of the perilesional skin.

The lesions were then photographed and biopsied for histological examination.

Skin thickness study

The detection of dermal thinning by ultrasound and xeroradiography was performed using two randomized double-blind clinical trials. The first trial compared the dermal thinning capacity of the following preparations:

(A) cream base
(B) hydrocortisone 1.0% cream (Hydrocortisyl)
(C) hydrocortisone 17 butyrate 0.1% cream (Locoid)
(D) clobetasone butyrate 0.05% cream (Eumovate)
(E) betamethasone valerate 0.1% cream (Betnovate)
(F) clobetasol propionate 0.05% cream (Dermovate)

Forty-eight volunteers applied two of the six preparations in the manner AB, BC, CD, DE, EF or AF, twice daily without occlusion on the flexor aspect of the forearm. The pairs of test materials were allocated at random. They were rubbed in for one minute and one centimetre lengths of cream from the nozzle of the tubes were squeezed onto the skin surface at each application.

Skin thickness measurements by xeroradiography and ultrasound were made on day 0 and immediately after six weeks of treatment. To measure dermal thickness with ultrasound the probe was placed on the desired site and the transit time (T) between the water–stratum corneum interface and the dermis–subcutaneous fat interface was read directly from the oscilloscope display. The dermal thickness (L) was then determined by the formula $L=\frac{T}{2}\times V$, where V is the acoustic velocity as determined by Daly and Wheeler[9] (1518 m/s).

In the second trial the dermal thinning effect of a new steroid compound Org 6216, was compared with the base hydrophilic gel. Nine men and nine women applied the two preparations three times daily for 8 weeks on their right and left forearms according to a randomized code. Skin thickness measurements were determined with xeroradiography and ultrasound at week 0, week 4 and week 8. Skin thickness measurements with ultrasound alone were also determined at week 1, week 2 and week 6.

RESULTS

Imaging of skin lesions

Xeroradiography was helpful in confirming the presence of some benign intracutaneous and subcutaneous lesions, such as cysts. Such lesions were

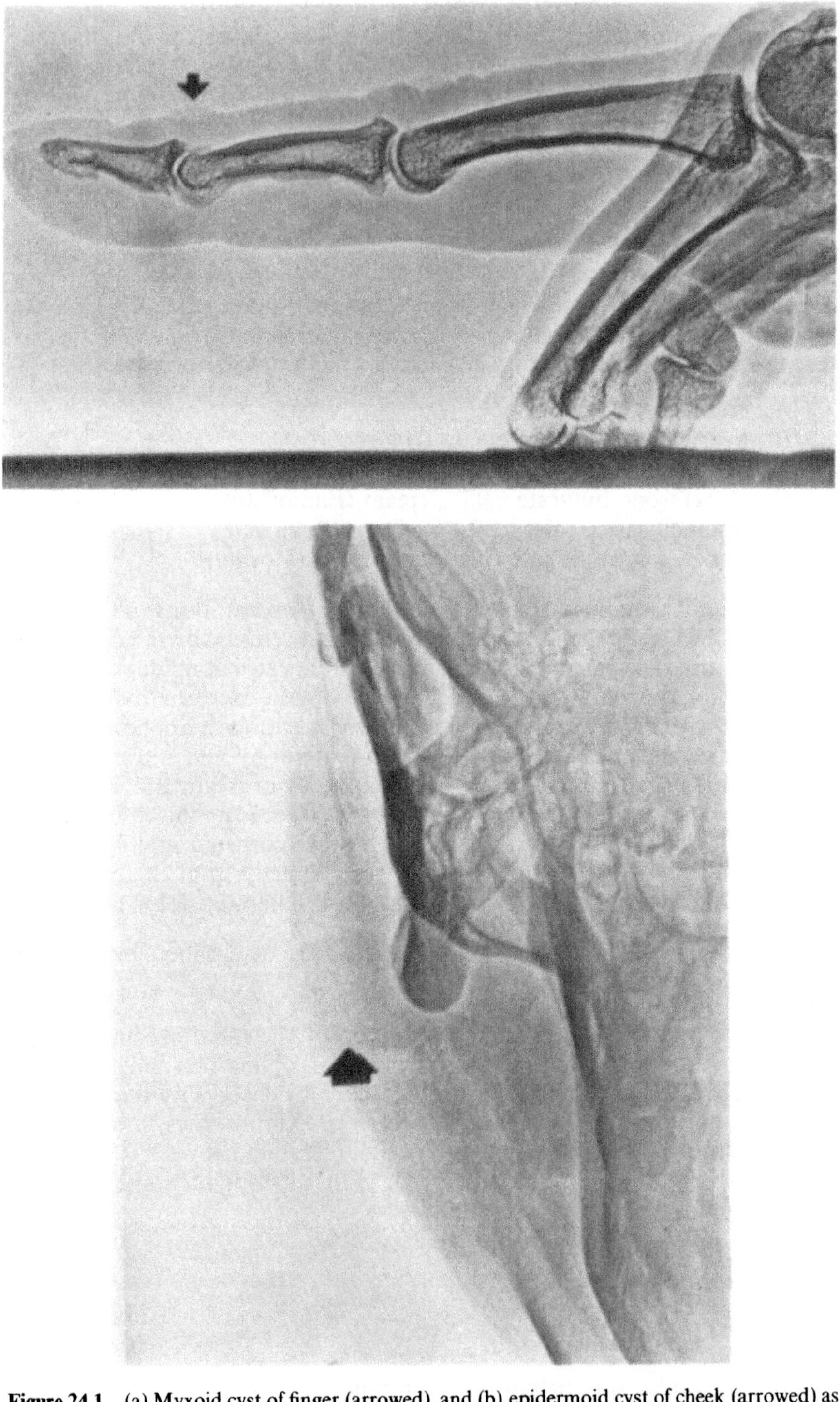

Figure 24.1 (a) Myxoid cyst of finger (arrowed), and (b) epidermoid cyst of cheek (arrowed) as detected with xeroradiography

circumscribed and homogenous (Figure 24.1). Lipoma was seen as a well-defined lesion of the same X-ray 'density' as the surrounding fat (Figure 24.2). In one patient with vascular malformation of the fingers, severe soft-tissue hypertrophy could be seen (Figure 24.3). Pretibial myxoedema was depicted as an undulating, homogenous dermal band (Figure 24.4). Other small benign lesions such as histiocytoma, cellular naevus and keloid did not show any helpful features, although the depth and extent of the lesions could be ascertained.

Of the malignant lesions examined, squamous cell and basal cell carcinomas did not show any distinguishing features, either from each other or from benign lesions. A plaque-like lesion of a lymphocytic lymphoma in one patient affecting the skin showed features suggesting invasion into the subcutis from the dermis, despite the apparent superficiality of the lesions clinically (Figure 24.5).

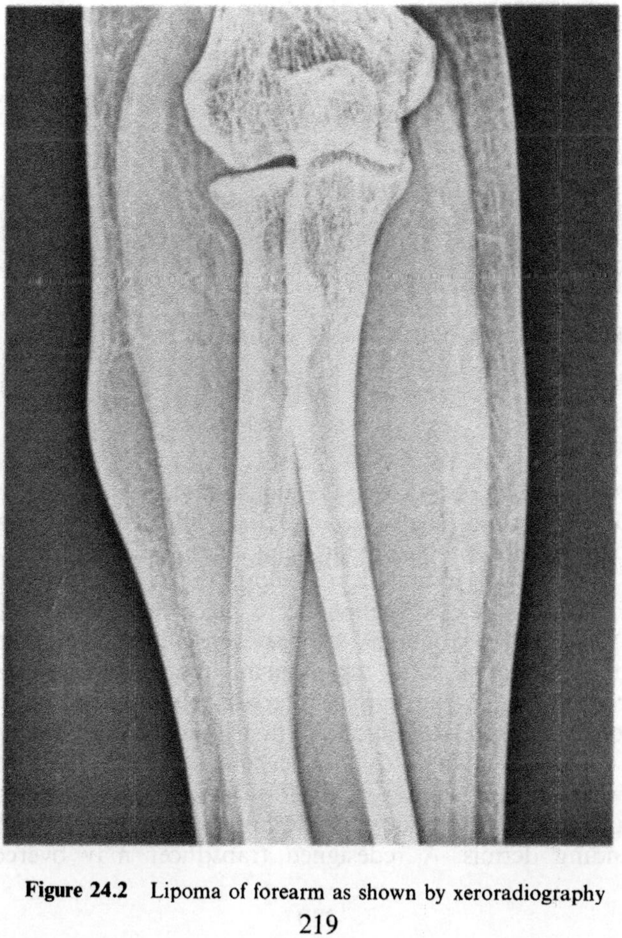

Figure 24.2 Lipoma of forearm as shown by xeroradiography

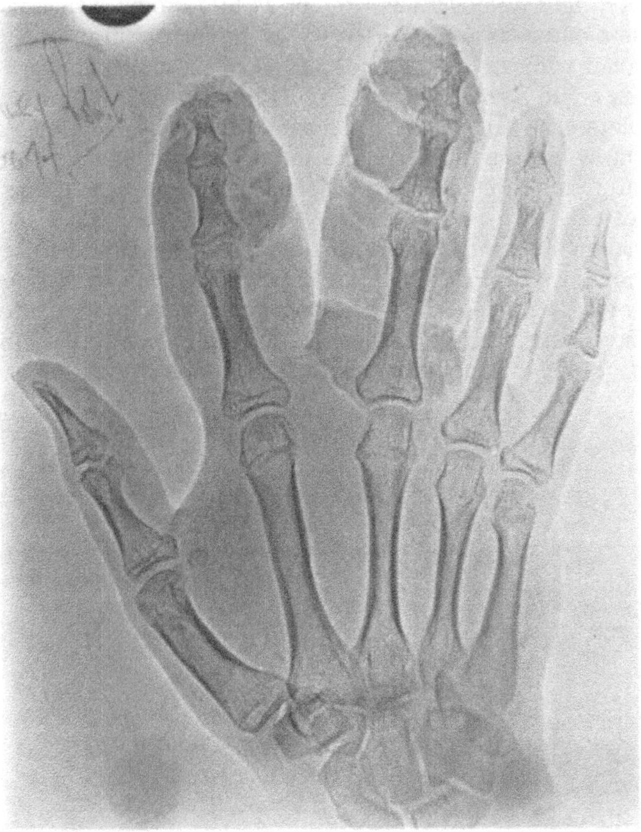

Figure 24.3 Vascular malformation of fingers as shown by xeroradiography

Ultrasound scanning of skin lesions yielded inconsistent results which were often difficult to interpret. There was a practical difficulty in placing the probe on the lesions without bubbles of air being introduced into the water bath because of the surface irregularity of both the lesions and the various anatomical sites. Where it was possible to place the probe on the lesions without introducing air bubbles, echoes were often encountered – either inconsistently or not at all – after the initial major echo which represented the reflection from the water–skin surface interface. Most ultrasound waves were probably reflected and scattered at the first interface and scattering at the subsequent interface, owing to its unevenness, meant that little of the reflections were detected. However, in a few cases it was possible to detect an increase in the depth of the tissue plane occupied by the lesions compared with the surrounding dermis. A redesigned transducer may overcome these problems.

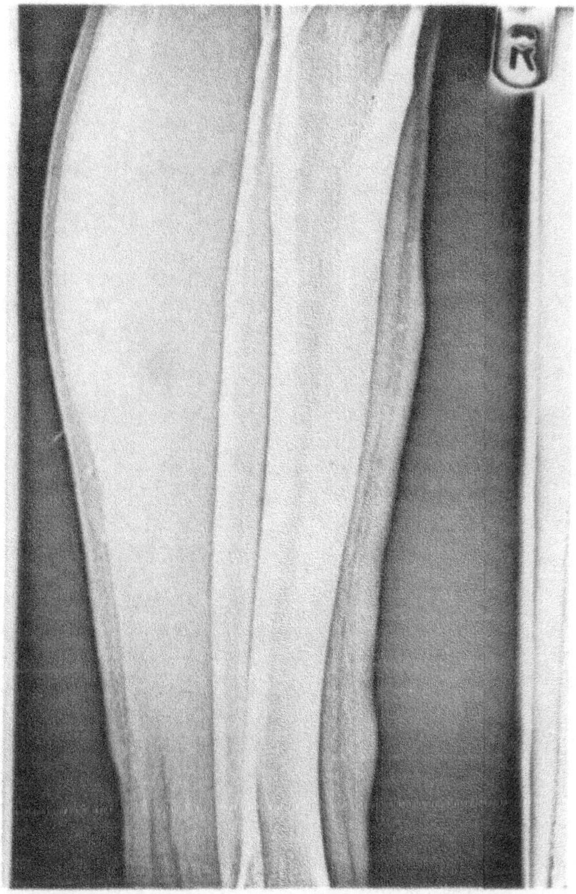

Figure 24.4 Pretibial myxoedema of leg as shown by xeroradiography

Detection of dermal thinning

The ability of xeroradiography to measure dermal thickness accurately and detect dermal thinning was once again shown in these clinical studies. From the data derived from the second study a strong linear correlation between the X-ray measurements and the ultrasound measurements was obtained, as illustrated in Figures 24.6 and 24.7. A similar, strong correlation was obtained from the first study (pretreatment $r = 0.89$, $p < 0.001$; post-treatment $r = 0.96$, $p < 0.001$). However, there was a consistent difference of about 0.15 mm in the values of the dermal thickness determined by X-ray and ultrasound, the X-ray measurements being greater. Table 24.2 shows the mean pre- and post-treatment values of dermal thickness in subjects in the first study. Significant dermal thinning (paired t-test, $p < 0.001$) was detected after

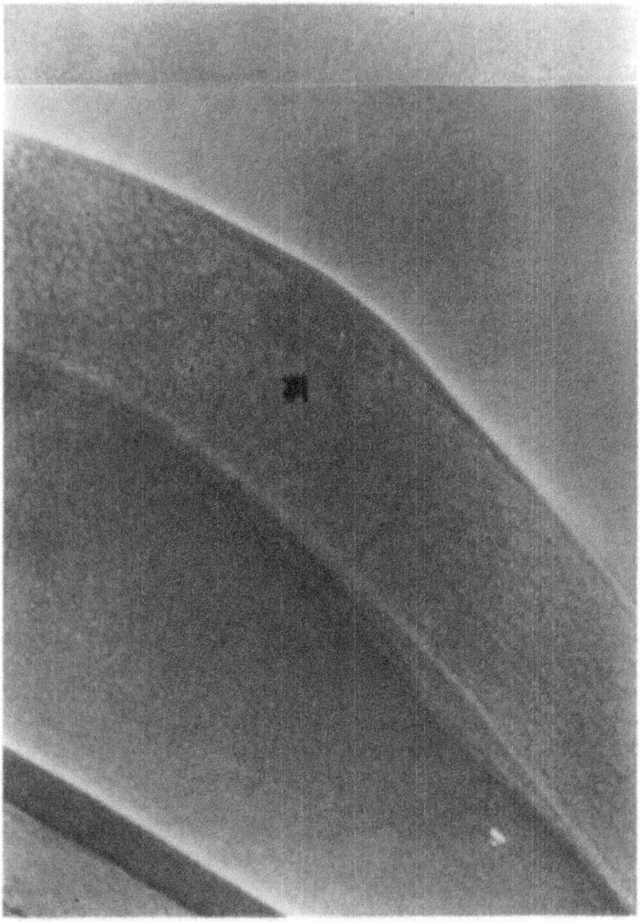

Figure 24.5 Skin lymphoma (arrowed) invading subcutis of upper arm as detected with xeroradiography

6 weeks of treatment with all the preparations, including cream base. Table 24.3 shows the time course of the mean dermal thickness of subjects treated with Org 6216 and base over the eight week period. Significant dermal thinning (paired *t*-test, $p < 0.05$) was detected after four and eight weeks of treatment with Org 6216 when compared with base.

DISCUSSION

Xeroradiography of skin lesions was useful in detecting and differentiating some benign lesions. It would thus aid in planning treatment for these lesions. Unfortunately, it does not seem from our results that xeroradiography would

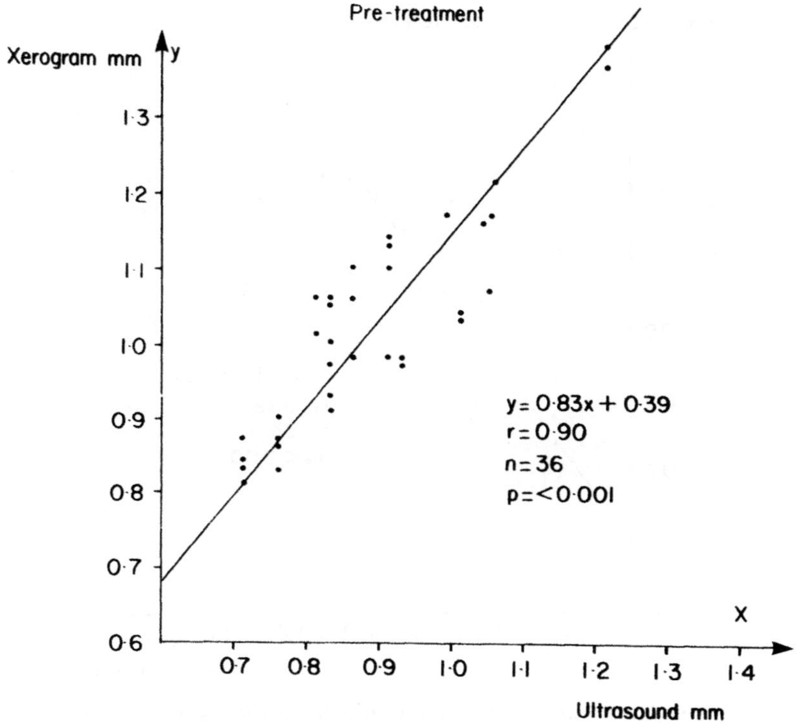

Figure 24.6 Correlation between measurements of skin thickness with xeroradiography and ultrasound scanning before treatment in second study

Table 24.2 Mean (± SD) dermal thickness (mm), in 48 subjects before and after treatment with various preparations, determined with ultrasound scanning and xeroradiography

Treatment	*Ultrasound scanning*		*Xeroradiography*	
	*Before**	*After†*	*Before**	*After†*
Clobetasol propionate 0.05%	1.00 ± 0.10	0.78 ± 0.16	1.22 ± 0.14	1.00 ± 0.18
Betamethasone butyrate 0.1%	0.98 ± 0.15	0.83 ± 0.19	1.18 ± 0.21	1.09 ± 0.20
Hydrocortisone 17 butyrate 0.1%	0.98 ± 0.12	0.82 ± 0.20	1.19 ± 0.17	1.04 ± 0.24
Clobetasone butyrate 0.05%	0.99 ± 0.13	0.90 ± 0.17	1.22 ± 0.20	1.14 ± 0.20
Hydrocortisone 1.0%	0.91 ± 0.12	0.79 ± 0.16	1.10 ± 0.18	1.01 ± 0.20
Cream base	0.95 ± 0.12	0.86 ± 0.12	1.16 ± 0.15	1.11 ± 0.15

* At day 0.
† At day 42.

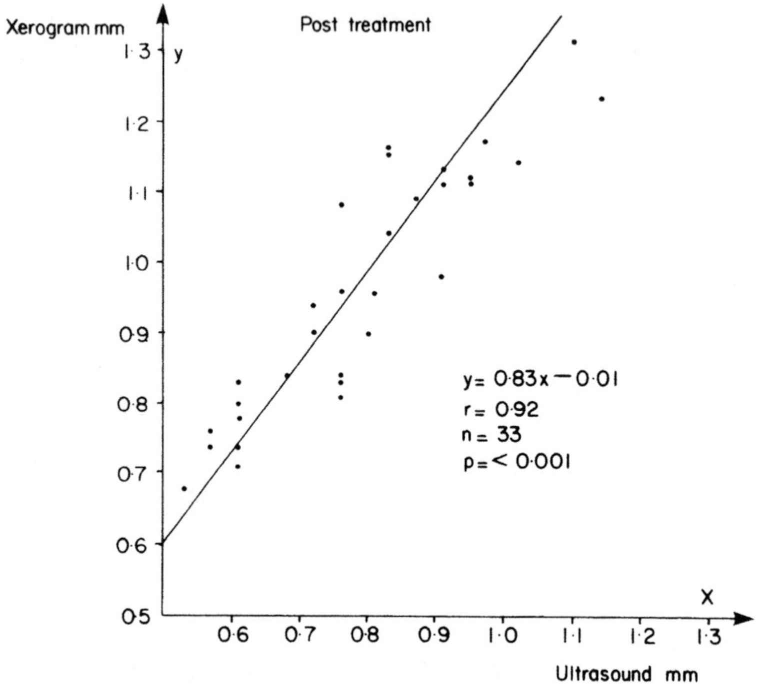

Figure 24.7 Correlation between measurements of skin thickness with xeroradiography and ultrasound scanning after 8 weeks' treatment in second study

provide much assistance in planning treatment for most malignant skin lesions, although it may help in detecting deep invasion. We believe that this simple manoeuvre can be regarded as a useful adjunct in diagnosing localized skin lesions.

The results of ultrasound scanning (A mode) of skin lesions were disappointing, as they did not provide any additional information on the nature of the skin lesions examined. In contrast, ultrasound has proved to be a

Table 24.3 Mean (±SD) dermal thickness (mm) in 18 subjects treated with org 6216 and base, determined with ultrasound scanning and xeroradiography

Treatment	*Weeks of treatment* 0	1	2	4	6	8
			Ultrasound scanning			
Org 6216	0.89 ± 0.14	0.85 ± 0.12	0.83 ± 0.13	0.80 ± 0.11	0.76 ± 0.13	0.75 ± 0.16
Base	0.89 ± 0.13	0.89 ± 0.12	0.88 ± 0.13	0.86 ± 0.11	0.89 ± 0.14	0.85 ± 0.15
			Xeroradiography			
Org 6216	1.02 ± 0.15			0.92 ± 0.16		0.91 ± 0.18
Base	1.03 ± 0.14			1.00 ± 0.19		1.00 ± 0.16

useful and accurate method for measuring skin thickness and detecting dermal thinning. It has the added advantages of using a relatively safe type of radiation, having instantly available results, being simple to operate and having the capacity for multiple observations. It should be noted that the skin thickness values as determined with xeroradiography are consistently greater than those determined with ultrasound scanning. One possibility which accounts for this discrepancy is that the two methods do not detect the same 'dermal–subcutaneous fat interface' as defined by the histological dermal collagen and subcutaneous fat junction. Alternatively, the value of acoustic velocity through skin (1518 m/s) used to calculate the dermal thickness may be lower than its true value. Earlier studies, however, suggested that the acoustic velocity through soft tissue approached 1500 m/s[10,11], and more recently values nearer 1550 m/s have been found[12].

References

1. Wolfe, J. N. (1969). Xeroradiography of bones, joints and soft tissues. *Radiology*, **93**, 583
2. Kalus, A. M., Aindow, J. and Caulfield, M. R. (1979). Application of ultrasound in assessing burn depth. *Lancet*, **1**, 188
3. Rukavina, B. and Mohar, N. (1979). An approach of ultrasound diagnostic techniques of the skin and subcutaneous tissue. *Dermatologica*, **158**, 81
4. Coleman, D. J., Abramson, D. H., Jack, R. L. and Franzen, L. A. (1974). Ultrasonic diagnosis of tumours of the choroid. *Arch. Opthalmol.*, **91**, 344
5. Coleman, D. J., Konig, W. F. and Katz, L. (1969). A hand operated ultrasound scan system for ophthalmic evaluation. *Am. J. Ophthalmol.*, **68**, 256
6. Dykes, P. J. and Marks, R. (1979). An appraisal of the methods used in the assessment of atrophy from topical corticosteroids. *Br. J. Dermatol.*, **101**, 599
7. Alexander, H. and Miller, D. L. (1979). Determining skin thickness with pulsed ultrasound. *J. Invest. Dermatol.*, **72**, 17
8. Tan, C. Y., Marks, R. and Payne, P. (1981). Comparison of xeroradiographic and ultrasound detection of corticosteriod induced dermal thinning. *J. Invest. Dermatol.*, (In press)
9. Daly, C. H. and Wheeler, J. B. (1971). The use of ultrasonic thickness measurements in the clinical evaluation of oral soft tissue. *Int. Dent. J.*, **21**, 418
10. Goldman, D. E. and Hueter, T. F. (1956). Tabular data of the velocity and absorption of high frequency sound in mammalian tissues. *J. Acoustic Soc. Am.*, **28**, 35
11. Lees, S. (1971). Ultrasonics in hard tissues. *Int. Dent. J.*, **21**, 403
12. Wells, P. N. T. (1977). *Biomedical Ultrasonics*. (London: Academic Press)

25
Is there a relationship between corneocyte size and stratum corneum function *in vivo*?

S. NICHOLLS, C. S. KING and R. MARKS

SUMMARY

The effects of differences in corneocyte dimensions on the function of stratum corneum were investigated. Intracorneal cohesion and transepidermal water loss were correlated with mean corneocyte area in the horny layers in different subjects. When mean corneocyte area and the measures of function chosen were compared graphically a hyperbole was obtained.

INTRODUCTION

Several techniques have been used for collecting the desquamating portion of the horny layer. Goldschmidt and Kligman[1] removed corneocytes with an adhesive slide technique; McGinley *et al.*[2] scrubbed the skin surface manually in the presence of detergent phosphate buffer to detect individual cells and to make cell counts. Nicholls and Marks[3] modified the technique and applied a standardised stimulus to the skin to obtain an index of 'forced desquamation'. Recently, Roberts and Marks[4] compared this type of measurement with a 'passive collection' of corneocytes in small capsules and found a strong correlation. Christensen *et al.*[5] used a self-hardening gel to remove superficial corneocytes and performed protein estimations on the collected material.

In 1970 Plewig and Marples[6], and Plewig[7] used the detergent scrub method to collect corneocytes from various body sites and measured corneocyte dimensions. They discovered site, sex and age differences in corneocyte size. Small cells occurred at the body periphery, and men possessed smaller cells than women. Corneocyte size increased with age. In a study of the changes with age in human epidermis, Marks[8] studied both keratinocyte and corneocyte dimensions. Keratinocytes become smaller with age, but mean corneocyte area increased. This was attributed to a longer period of maturation and increased corneocyte 'flattening' in the older age groups.

Hölzle and Plewig[9] discovered abnormally small corneocytes from the skin of patients with allergic contact dermatitis and in stripped skin of normal subjects. Treatment with topical steroids increased corneocyte size. Smaller corneocytes have also been found in patients with ichthyosis[10]. However, the question remains as to whether the size of the corneocytes is related to the function or properties of the stratum corneum *in vivo*. To answer this question we have investigated corneocyte size in relation to transepidermal water loss (TEWL) and intracorneal cohesion (ICC).

METHODS USED

Corneocyte collection

Corneocytes were collected in 0.1% Triton X-100 buffered phosphate by rubbing the skin with a blunted glass rod for about 10 seconds. Collections were made from the forehead, forearm, wrist and palm in 15 normal subjects (mean age 42.7; 8 women, 7 men). Cells were also collected from the forearms of individuals with scaling disorders (Table 25.2). Samples of each suspension were taken up into Pasteur pipettes and 'smears' made onto glass microscope slides. Specimens were left to dry in air and subsequently examined with a Nikon (Type 102) binocular microscope with a projection side arm. The images of at least 30 cells in each sample were projected onto graph paper and mean cell area (μm^2) calculated. Figure 25.1 shows typical cells obtained from a smear. Figure 25.2 is an example of a projected working drawing.

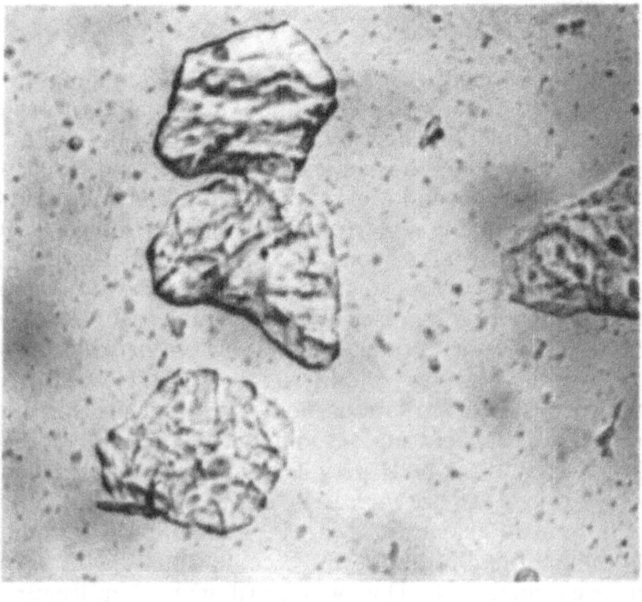

Figure 25.1 Corneocytes from a dry smear preparation

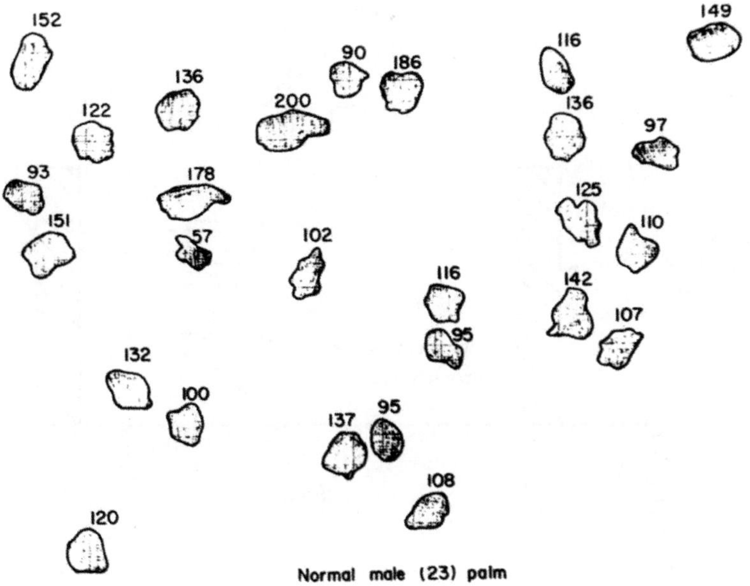

Figure 25.2 Working drawing of corneocytes from palmar skin in normal male subject projected onto graph paper and their projected area measurements (μm^2); subject was aged 23

Rate of transepidermal water loss (TEWL)

TEWL was measured *in vivo* from each site investigated with an evaporimeter constructed by the Bioengineering Unit UHW. Its function is based on the principle described by Nilsson, Sedin and Oberg[11] and is discussed fully elsewhere (Chapter 19). Basically, the rate of water loss from a surface is calculated by measuring the vapour pressure gradient just above the skin surface. Thus:

$$\frac{1}{A}\frac{dm}{dt} = -D\frac{dp}{dx} \tag{1}$$

where A is surface area, D is a constant, $\frac{dm}{dt}$ is the rate of evaporation and $\frac{dp}{dx}$ is the vapour pressure gradient. The apparatus is simple and rapid to use. The readings are reproducible in normal ambient conditions, and it has a minimal effect on the skin surface microclimate. Figure 25.3 shows the measuring capsule.

Intracorneal cohesion (ICC)

An index of binding forces within the stratum corneum (intracorneal cohesion; ICC) was obtained *in vivo* using a 'cohesograph'[3]. A reading is obtained by sticking a metal cap to the skin surface with a cyanoacrylate

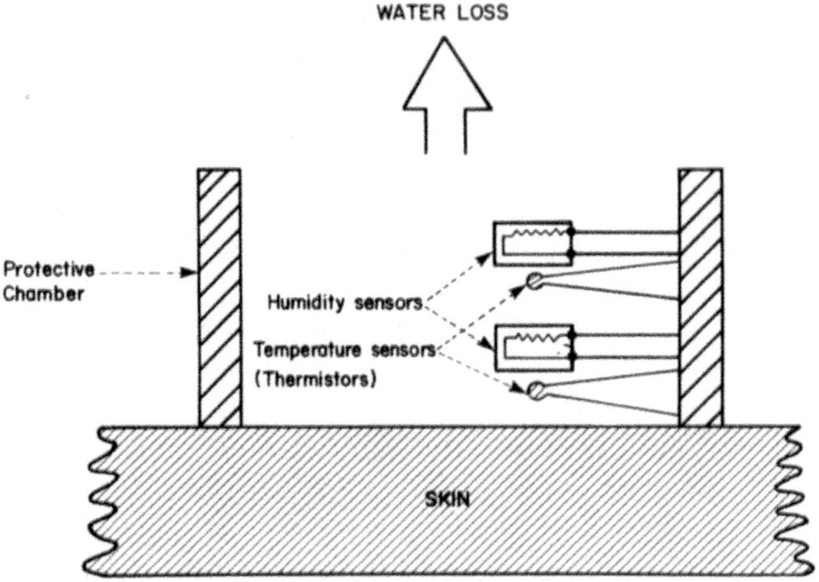

Figure 25.3 Evaporimeter-measuring capsule

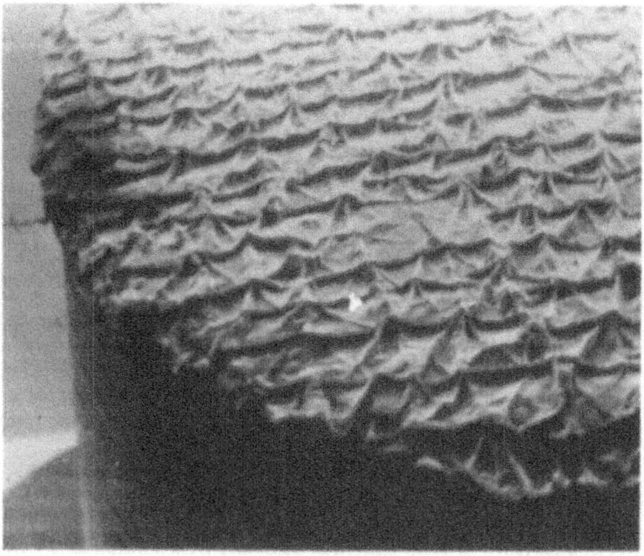

Figure 25.4 Scanning electron micrograph of a cohesograph metal cap plus stratum corneum subsequent to a reading (×15)

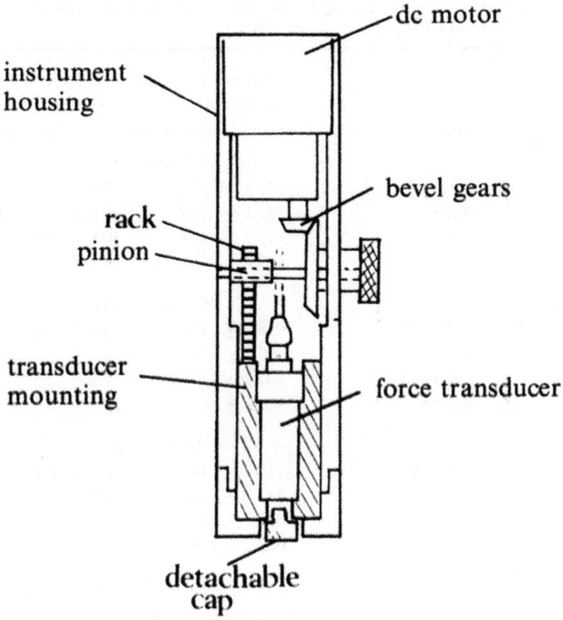

Figure 25.5 The cohesograph

adhesive (Powabond). The cap is distracted vertically from the skin surface removing partial thickness stratum corneum attached to the cap (Figure 25.4). A force transducer measures the force, which is then displayed on a chart recorder in gram force. The instrument used for this study has been modified to incorporate a DC motor so that the cap is removed automatically at 1 cm/s. Thus, vertical breaking strength of stratum corneum *in vivo* is determined mechanically. Figure 25.5 shows the cohesograph in cross section.

RESULTS

Table 25.1 summarises the mean corneocyte area, and the TEWL and ICC from each body site. There seems to be a consistent decrease in cell size with age from forearm, wrist, forehead, with the smallest cells on the palm. This occurs in each age group and the results agree with those of Plewig[7]. Corneocytes also increase in area in older individuals. Of interest is that body sites possessing smaller corneocytes are regions of increased water loss through the skin in normal subjects. Figure 25.6 shows the relationship between mean corneocyte area and TEWL. The relationship is not simple but the experimental data fits the theoretical curve of a rectangular hyperbole. Figure 25.7 shows the relationship between mean corneocyte area and ICC. There is a similar relationship as with TEWL, although it is not as well defined as with TEWL. There is certainly a trend for sites with 'small cells' to have increased stratum corneum cohesion (notably the wrist and palm).

Table 25.1 Mean (±SD) corneocyte area, TEWL and ICC at four different sites in normal subjects (relative humidity 54–60%; temperature 20–22°C)

Age group (years)	*Site*	*Corneocyte Area* (μm^2)	*TEWL* ($g/m^2/h$)	*ICC* (g force)
		Men ($n = 4$)		
20–40	Forearm	933.8 ± 52.2	5.8 ± 2.1	96.5 ± 17.0
	Wrist	726.1 ± 77.4	10.5 ± 4.3	123.0 ± 38.2
	Forehead	675.7 ± 36.6	10.6 ± 0.9	98.0 ± 19.6
	Palm	609.0 ± 66.0	48.7 ± 10.5	310.0 ± 36.0
		Women ($n = 4$)		
	Forearm	1038.5 ± 149.0	2.8 ± 0.6	103.0 ± 5.2
	Wrist	799.7 ± 126.6	3.7 ± 1.3	114.0 ± 30.6
	Forehead	707.2 ± 114.0	7.8 ± 2.5	104.5 ± 14.3
	Palm	706.6 ± 110.0	22.5 ± 9.1	295.0 ± 63.6
		Men ($n = 3$)		
45–75	Forearm	1173.9 ± 49.2	4.2 ± 0.7	91.5 ± 5.0
	Wrist	889.7 ± 102.1	6.3 ± 1.6	92.0 ± 23.0
	Forehead	878.3 ± 37.7	7.7 ± 3.1	94
	Palm	662.2 ± 104.4	34.5 ± 7.7	180
		Women ($n = 4$)		
	Forearm	1060.8 ± 109.3	3.6 ± 0.6	100.0 ± 48.1
	Wrist	864.3 ± 37.7	4.4 ± 1.3	115.0 ± 64.4
	Forehead	767.9 ± 20.4	6.7 ± 2.2	90
	Palm	627.7 ± 96.9	30.0 ± 3.0	360

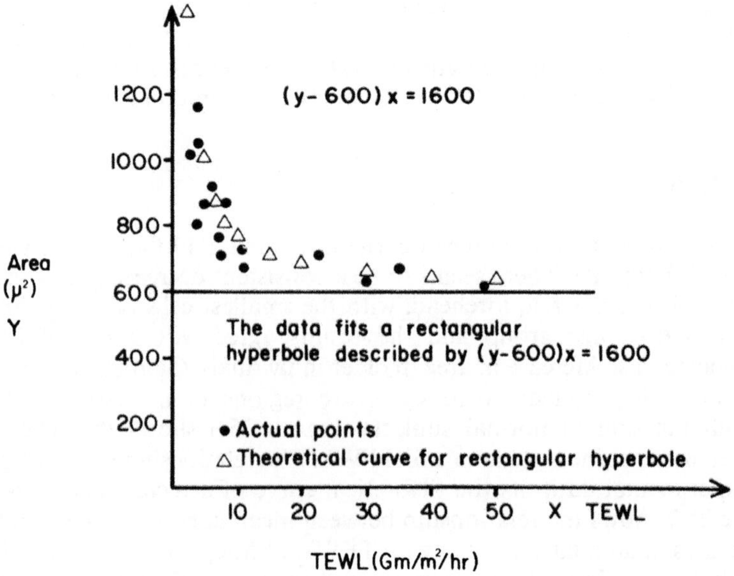

Figure 25.6 Mean corneocyte area and local TEWL at different sites in normal subjects

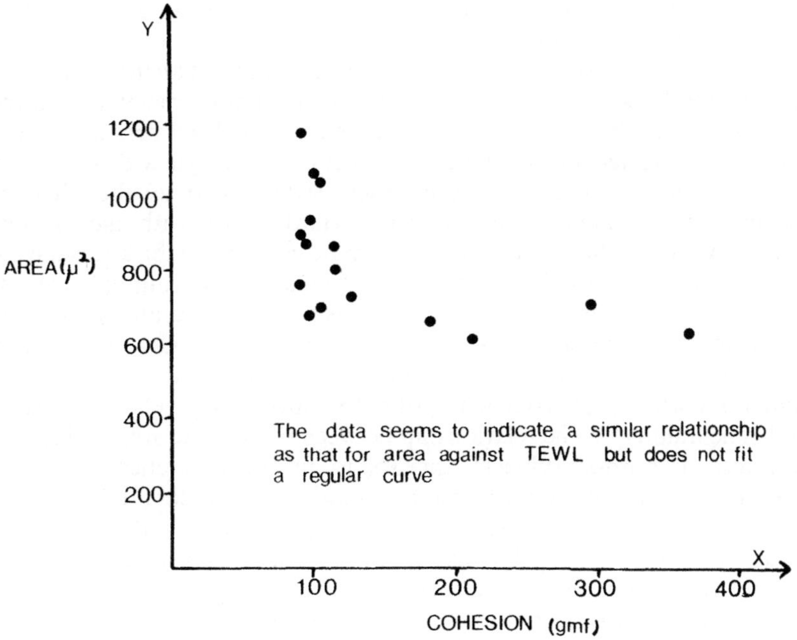

Figure 25.7 Mean corneocyte area and ICC at different sites in normal subjects

Table 25.2 shows details of the results of cell size, water loss and ICC from arms of patients with scaling disorders. Once again sites with small corneocytes show increased TEWL and/or ICC.

DISCUSSION

The aim of this study was to determine whether change in corneocyte size has any implication for stratum corneum function as measured by transepidermal water loss (TEWL) and intracorneal cohesion (ICC). The area of a corneocyte will reflect the number of junctional sites within the whole stratum corneum

Table 25.2 Preliminary measurements of mean (±SD) corneocyte area, TEWL and ICC in scaling dermatoses on the arm in subjects aged under 45

Dermatoses	*Corneocyte area* (μm^2)	*Corneocyte diameter* (μm^2)	*TEWL* ($g/m^2/h$)	*ICC* (g force)
Psoriasis lesion ($n = 2$)	685 ± 30	33.3 ± 2.0	10.0 ± 3.0	145
Atopic dermatitis ($n = 2$)		33.8 ± 1.5	5.1 ± 0.6	151.2 ± 16.6
Ichthyotic disorders ($n = 4$)	891 ± 189.4		6.6 ± 1.1	169.5 ± 64.4

membrane per unit volume. Thus, there is a greater junctional space for sites with small corneocytes than for sites with large corneocytes.

The increase in corneocyte area with age could reflect an increased transit time and the longer the cell remains in the stratum corneum the flatter it becomes. Marks[8] has shown that the viable epidermal cell volume decreases with age. If in old age this smaller cell produces a corneocyte with a larger area than in younger age groups (because of increased transit time), then there may be a disproportionate decrease in corneocyte thickness with age. However, studies in which we have been able to estimate corneocyte thickness have not indicated that there is a disproportionate decrease[12]. We cannot explain this anomaly. The decreased cell area in psoriasis lesions and atopic dermatitis may reflect the decreased transit time in these disorders. However, turnover time in dominant ichthyosis is considered normal[13], and the smaller area found for corneocytes from ichthyotic conditions may reflect changes in membrane fluidity[10]. The relationship between corneocyte area and regional water loss in normal subjects appears to fit the theoretical curve of a rectangular hyperbole. Baker and Kligman[14] also noted the variation in TEWL with different body sites, even after correcting for thickness of the stratum corneum. It may be that when the surface corneocyte reaches a critically small area the intercorneocyte area is no longer effectively 'rate limiting' for water loss from the skin. The same consideration applies to ICC. Since ICC drops near the stratum corneum surface and permits desquamation, the intercorneocyte space may be the critical determining site within the stratum corneum subserving desquamation.

References

1. Goldschmidt, H. and Kligman, A. M. (1967). Exfoliative cytology of human horny layer. *Arch. Dermatol.*, **96**, 572
2. McGinley,, K. J., Marples, R. R. and Plewig, G. (1969). A method for visualising and quantitating the desquamating portion of the human stratum corneum. *J. Invest. Dermatol.*, **53**, 107
3. Nicholls, S. and Marks, R. (1977). Novel techniques for the estimation of intracorneal cohesion *in vivo*. *Br. J. Dermatol.*, **96**, 595
4. Roberts, D. and Marks, R. (1980). The determination of regional and age variations in the rate of desquamation: A comparison of four techniques. *J. Invest. Dermatol.*, **74**, 13
5. Christensen, M. S.,, Nacht, S., Kantor, S. L. and Gans, E. H. (1978). A method for measuring desquamation and its use for assessing the effects of some common exfoliants. *J. Invest. Dermatol.*, **71**, 289
6. Plewig, G. and Marples, R. R. (1970). Regional differences of cell sizes in the human stratum corneum I. *J. Invest. Dermatol.*, **54**, 13
7. Plewig, G. (1970). Regional differences of cell sizes in the human stratum corneum II. *J. Invest. Dermatol.*, **54**, 19
8. Marks,, R. (1980). Measurement of biological ageing in human epidermis. In preparation.
9. Hölzle,, E. and Plewig, G. (1977). Effects of dermatitis, stripping and steroids on the morphology of corneocytes. A new bioassay. *J. Invest. Dermatol.*, **68**, 350
10. Marks, R. (1980). Quantitative aspects of keratinisation. In Marks, R. and Christophers, E. (eds.) *The Epidermis in Disease*. (Lancaster: MTP)

11. Nilsson, G. E., Sedin, G. and Oberg, P. A. (1975). A transducer for measurement of evaporation from the skin. In *International Conference on Biomedical Transducers. Paris B5.3*, pp. 71–76.
12. Marks, R., Nicholls, S. and King, C. S. (1980). Corneocytes and stratum corneum function. *Int. J. Cosmetic Sci.* (In press)
13. Frost, P. (1973). Ichthyosiform dermatoses. *J. Invest. Dermatol.*, **60**, 541
14. Baker, H. and Kligman, A. M. (1967). Measurement of transepidermal water loss by electrical hygrometry. *Arch. Dermatol.*, **96**, 441

26
Relationship of intracorneal cohesion to rates of desquamation in the scaling disorders

C. S. KING, S. NICHOLLS and R. MARKS

SUMMARY

The relationship between intracorneal cohesion (ICC) and numbers of single corneocytes released after a scrub stimulus has been investigated in normal skin, the skin of patients with a variety of scaling disorders and in tape-stripped normal skin. A weak negative correlation has been found between ICC and the numbers of corneocytes released. The importance of this relationship in the pathophysiology of scaling is discussed.

INTRODUCTION

The stratum corneum (SC) maintains a constant ordered structure appropriate to the various body sites despite its perpetual dissolution and reformation. The constancy is a prerequisite for normal function and implies some regularity in the process of desquamation. Desquamation normally takes place by the loss of single corneocytes at the surface. There is very little information available on the method by which the intercorneocyte bond ruptures. It seemed that more information could be obtained about the process of desquamation by determining whether the loss of intercorneocyte cohesion takes place 'suddenly' near the surface or more gradually as a function of the distance from the surface. We therefore measured intracorneal cohesion throughout the depth of the SC[1]. In this experiment we used an instrument termed the cohesograph[2,3] in which the breaking strength of the SC is measured vertically. The original instrument has been improved by adding an automated constant speed withdrawal system and is illustrated in Figure 25.5 (p. 231). When this technique is used a rapidly bonding cyanoacrylate adhesive is placed on the surface of a detachable cap which is screwed into a piston. The cap is stuck to the surface of the skin at the site being investigated and the force

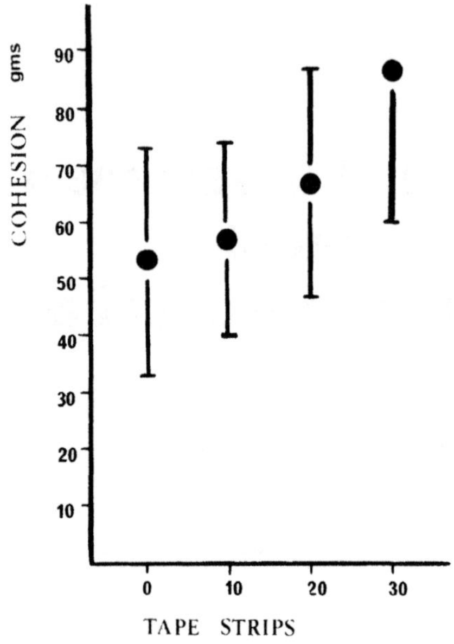

Figure 26.1 Mean (±SD) ICC measurement taken at various depths within the stratum corneum after tape stripping

required to distract the piston and cap vertically is measured electronically. The cap always pulls several (3–5) layers of SC with it and thus the rupture always occurs within the SC itself. The intracorneal cohesion (ICC) was measured repeatedly at the same site while the SC was progressively removed by adhesive tape stripping. The results are illustrated in Figure 26.1 and it is obvious that, at least at the forearm site in normal subjects, there is a progressive increase in ICC with increased depth of SC.

Many common dermatoses are characterized by scaling at the skin surface. We detect scaling either visually or by touch largely because corneocytes fail to separate as individual cells, but detach in clumps known as lamellae (or scales). The partially detached lamellae disrupt the normal smooth contour of the skin surface and feel 'prickly' to the sensing finger or reflect light abnormally, giving the appearance of 'scaliness'. The failure of corneocytes to separate implies retention of intercorneocyte bonds and maintenance of an intracorneal cohesion more appropriate to the deeper parts of the SC.

In this study we have attempted to relate ICC to the rates of desquamation in disordered scaling skin. To measure desquamation we have used an improved and hand-held 'scrub' apparatus[4] which is based on the same standardized rotational stimulus principle as the original 'scrub' apparatus described by Nicholls, *et al.*[2] The improved version is illustrated in Figure 26.2. The torque applied is variable over a wide range and is preset by an electronic module as is the duration of the scrub. The number of corneocytes released after a 'standardized scrub' is taken as an index of the usual rate of corneocyte loss, as Roberts and Marks[5] have shown that there is a strong correlation

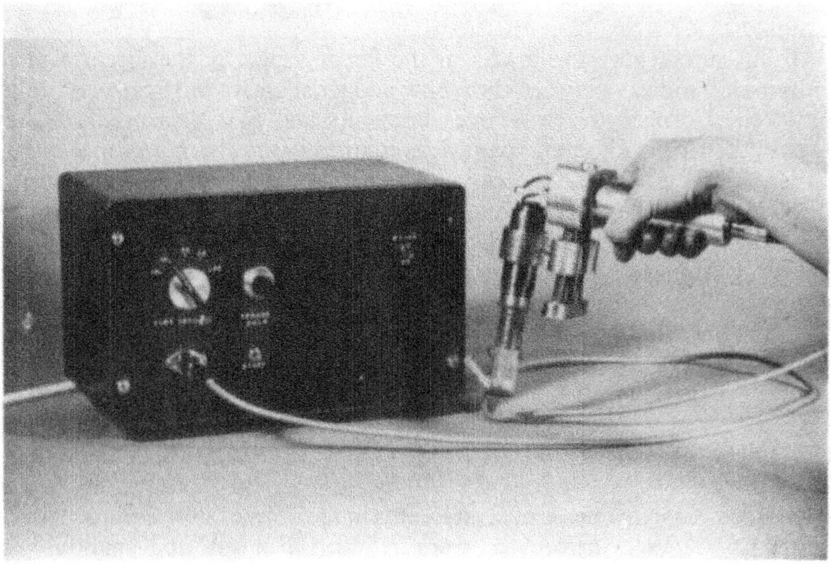

Figure 26.2 The automatic hand-held 'scrub' apparatus with control box

between this figure and the number that are 'passively' released into an attached chamber.

Technical details

To measure ICC the automatic cohesograph was used with a piston withdrawal rate of 1 cm/s. The detachable metal caps had a diameter of 8 mm. The hand-held scrub instrument was preset for 10 seconds of scrubbing and a torque of 2 g/cm. 1 ml phosphate buffer with 0.05% Triton X-100 was introduced into the scrub chamber into which the cells were released. The

Table 26.1 Patients with scaly skin disorders studied. All investigations were performed on flexor aspect of forearm and palm

Case No.	*Sex*	*Age (years)*	*Disorder*
1	F	42	Psoriasis
2	F	60	Eczema
3	M	33	Ichthyosis
4	M	48	Scaly skin associated with coelic disease
5	M	38	Psoriasis
6	M	16	Ichthyosis
7	M	43	Ichthyosis
8	M	31	Ichthyosis
9	F	18	Eczema
10	M	23	Ichthyosis
11	M	43	Psoriasis

width of the perspex blade (hence the diameter of the area scrubbed) was 18 mm.

All measurements were made on the flexor aspect of the forearm at the centre point midway between the elbow and wrist crease in 16 normal healthy volunteers. Corneocyte release measurements and ICC measurements were always taken from adjoining sites. Measurements were also made on the palms of these subjects. To induce scaling artificially, between 30 and 35 strips of adhesive tape (Sellotape) were applied and removed consecutively at a site 8 cm from the elbow crease on the flexor aspect of the forearm in eight normal healthy volunteers. Measurements of ICC and corneocyte release after scrubbing were made from these sites and from the contralateral normal forearm (control values). Eleven subjects with various scaling disorders (Table 26.1) were also examined.

RESULTS

ICC and corneocyte release measurements in the normal subjects are set out in Figure 26.3. Few corneocytes were released from normal palmar skin (mean = 7 per haemocytometer field). ICC measurements were difficult to take from the palms, as in some cases the value obtained was too high to be recorded on the chart recorder; in others the rupture took place at the glue–skin interface rather than within the SC (which indicated a very high

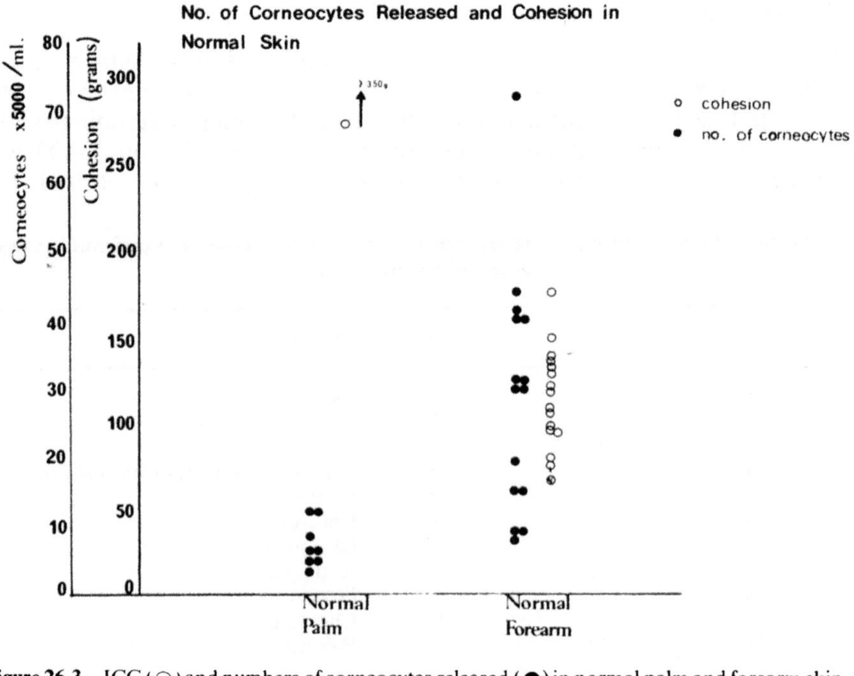

Figure 26.3 ICC (○) and numbers of corneocytes released (●) in normal palm and forearm skin

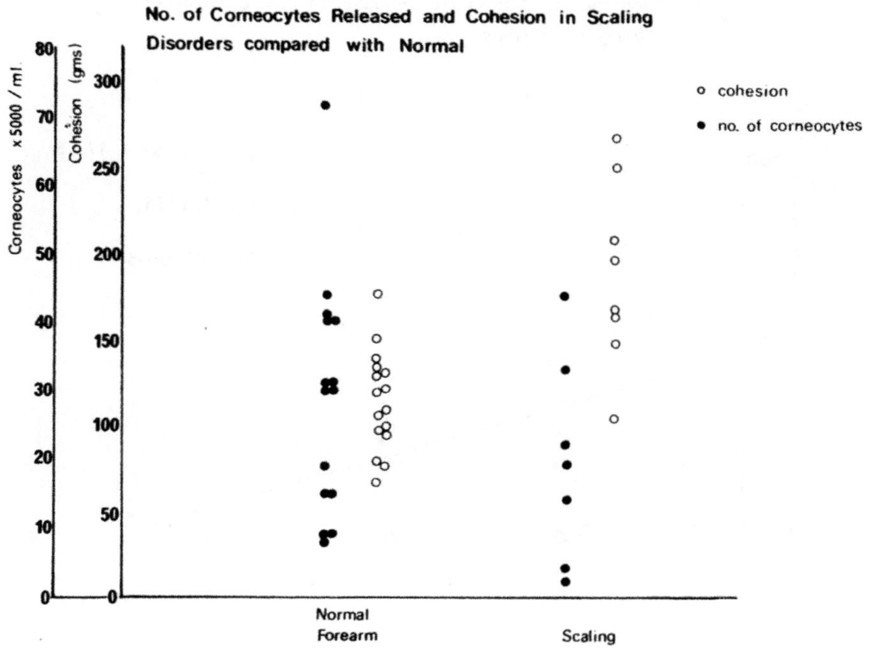

Figure 26.4 ICC (○) and numbers of corneocytes released (●) in forearm skin of normal subjects and those with scaling disorders

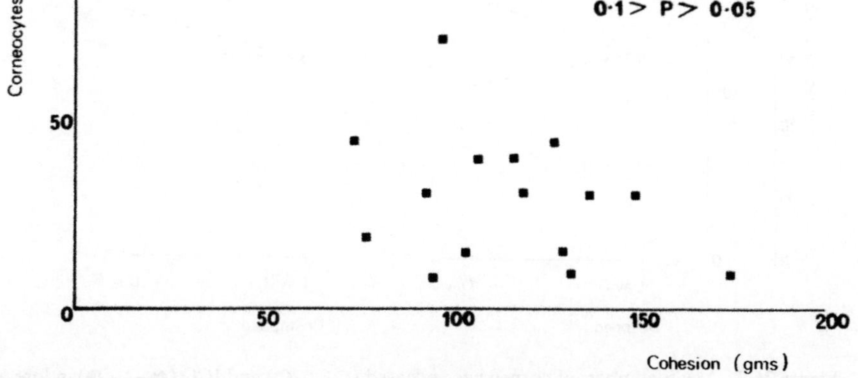

Figure 26.5 Scatter diagram of ICC against numbers of corneocytes released in normal skin (16 subjects)

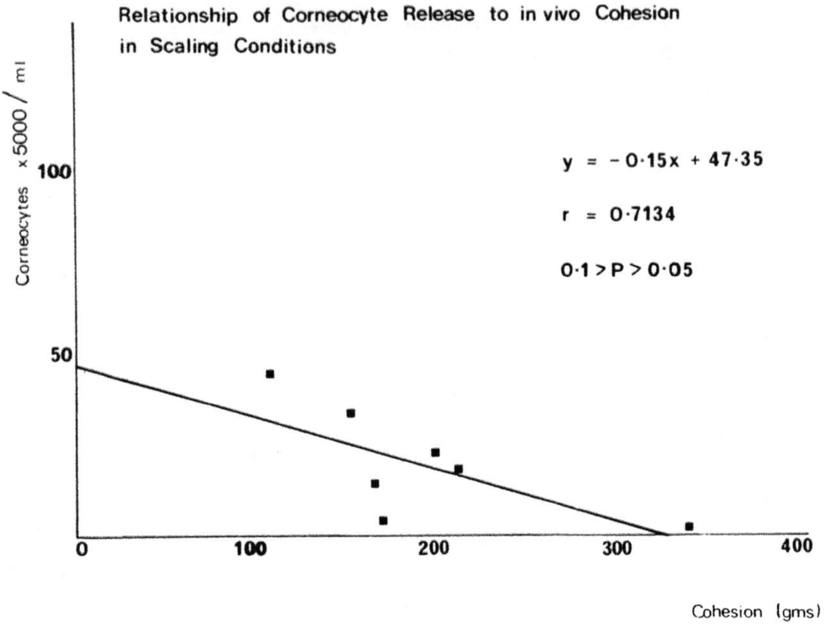

Figure 26.6 Scatter diagram of ICC against numbers of corneocytes released in skin from subjects with scaling disorders

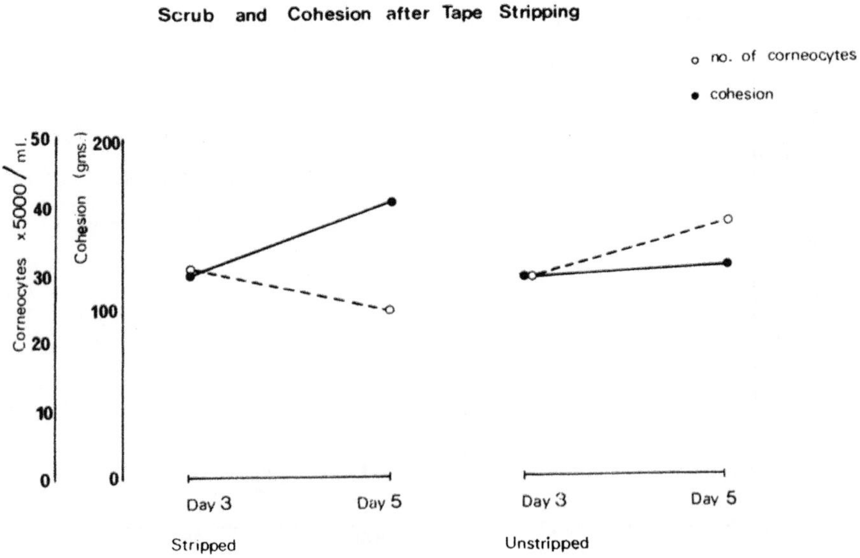

Figure 26.7 Mean numbers of corneocytes released (○---○) and ICC (●——●) in forearm skin subjected to tape stripping ($n = 8$) and in unstripped control skin ($n = 8$)

ICC). The lowest registrable force for normal palmar skin was 272 g (Figure 26.3). The mean number of corneocytes released from normal forearm skin was 33 (significantly higher than from palm $p < 0.001$). Mean ICC from forearm skin was 113 g.

Figure 26.4 shows the differences in ICC and numbers of corneocytes released on forearm skin between patients with the scaling disorders and the normal volunteers. The number of corneocytes released is lower in scaling skin (mean $\pm$SD = 19 $\pm$ 14) but not significantly different, and the ICC was considerably higher (mean = 206 g, $p < 0.001$).

Figure 26.5 shows the scatter obtained for measurements in normal skin when the numbers of corneocytes released are plotted against the ICC; a weak negative correlation was found. A stronger negative correlation was seen (Figure 26.6) for the skin of patients with scaling disorders.

The tape-stripping experiment was designed to produce a model of scaliness but did not achieve the clinical result intended. It is, none the less, possible to detect a significant increase in cohesion ($p < 0.02$) from day 3 to day 5 after stripping which did not occur in unstripped control arm (Figure 26.7). The number of corneocytes released from the stripped arm dropped from day 3 to day 5 but this difference was not significant ($p > 0.3$).

COMMENT

There was a negative correlation between ICC and the number of cells released by scrubbing from adjoining sites in normal skin, diseased scaling skin and in experimentally stripped skin. However, in none of these cases was this relationship strong. It was strongest with diseased skin and least strong in stripped skin after 5 days.

Several points should be borne in mind when interpreting the significance of this data. As release of intercorneocyte bonds and drop in ICC seem to occur near the surface of the SC[1] and ICC is increased in all the scaling disorders investigated, including psoriasis[6], ichthyosis[7] and the dry skin of atopic dermatitis[8], we suggest that failure of the normal loss of ICC results in abnormal desquamation and scaling. An inverse relationship between ICC and numbers of corneocytes released (only single corneocytes being estimated) is therefore not unexpected. The relatively weak negative correlations may be explained by: (a) the use of a detergent buffer to collect the corneocytes, which may break up some of the corneocyte clumps; (b) use of adjoining sites for the two measurements, which may not have exactly the same physiological properties; (c) the application of a rotational stress to collect the corneocytes rather than a stress perpendicular to the skin to measure ICC; and (d) the presence of other anatomical biochemical and/or biophysical considerations in the scaling disorders – e.g. the invasion of the SC in psoriasis with polymorphonuclear leukocytes.

The inconsistent results obtained after stripping may have been a result of the readings being taken at inappropriate time intervals; or possibly a result of the stripping stimulus being irregular, and sometimes excessive and at others inadequate.

CONCLUSION

There is an inverse relationship between ICC and the numbers of single corneocytes released by a scrub stimulus in both normal and abnormal skin. We believe that this is helpful in understanding the pathophysiology of the scaling dermatoses.

References

1. King, C. S., Barton, S. P., Nicholls, S. and Marks, R. (1979). The change in properties of the stratum corneum as a function of depth. *Br. J. Dermatol.*, **100**, 165
2. Nicholls, S. and Marks, R. (1977). Techniques for the estimation of intracorneal cohesion *in vivo*. *Br. J. Dermatol.*, **96**, 595
3. Marks, R., Nicholls, S. and Fitzgeorge, D. (1977). Measurement of intracorneal cohesion in man using *in vivo* techniques. *J. Invest. Dermatol.*, **61**, 299
4. Guibarra, E. J. (1979). An instrument for the stimulated release of corneocytes. *Bioengineering and the Skin*, **1**, 17
5. Roberts, D. and Marks, R. (1980). The determination of regional and age variations in the rate of desquamation: a comparison of four techniques. *J. Invest. Dermatol.*, **74**, 13
6. King, C. S., Nicholls, S., Barton, S. and Marks, R. (1979). Is the stratum corneum of uninvolved psoriasic skin abnormal? *Acta Dermatovener*, **59**, 95
7. Nicholls, S., King, C. S. and Marks, R. (1978). Morphological and quantitative assessment of physical changes in the horny layer in ichthyosis. In Marks, R. and Dykes, P. J. (eds.) *The Ichthyoses*. (Lancaster: MTP)
8. Finlay, A. Y., Nicholls, S., King, C. S. and Marks, R. (1980). The 'dry' non-eczematous skin associated with atopic eczema. *Br. J. Dermatol.* (In press)

27
Propagation of waves in hair

J. C. VAN DER LEUN

INTRODUCTION

Waves in hair have attracted much attention. Hair-styling is one of the oldest achievements of man. Much effort is spent in removing existing curls and in making artificial ones, and there is more knowledge available about the mechanisms of artificial waving than about natural waves.

In permanent waving, there is first a relaxation stage in which the disulphide bridges between the keratin main-chains are broken and reduced to –SH side-chains. Once the desired curvature is reached, an oxidizing agent recombines the keratin –SH groups to form new disulphide bridges, now between the extended chains. The new cross-linkages hold the hair in the set state[1].

It may be imagined that natural waves have a similar basis. The bends would be caused not by external chemical or physical manipulation, but by some influence in the follicle during the growth of the hair. A twist in the follicle[2] and asymmetric keratinization[3] have been suggested as such influences. Such mechanisms may well play a role; they are not sufficient, however, to explain a coherent pattern of waves in a head of hair. Order has to originate somewhere just as in permanent waving, where hair is arranged by the hairdresser .

Let us suppose for a while that the bends, which together form a wave, result from past action in the respective follicles; if this were correct, the organization of a wave would require the carefully concerted action of thousands of follicles, each effecting a bend at the proper time. The bends would have a fixed location on the individual hairs; consequently the wave could move over the head only at a velocity determined by the growth of the hairs, and over a distance limited by their length. The results presented here do not confirm such expectations.

MEASUREMENTS

The positions of hair waves were measured with time in three Caucasian volunteers with clearly defined natural wave patterns. The measurements were performed with calipers, as shown in Figure 27.1. The position of a wave crest was measured relative to the hairline on the forehead.

The measurements were performed almost daily. Figure 27.2 shows the measurements in subject 1, a dark-haired male Caucasian aged 40 years.

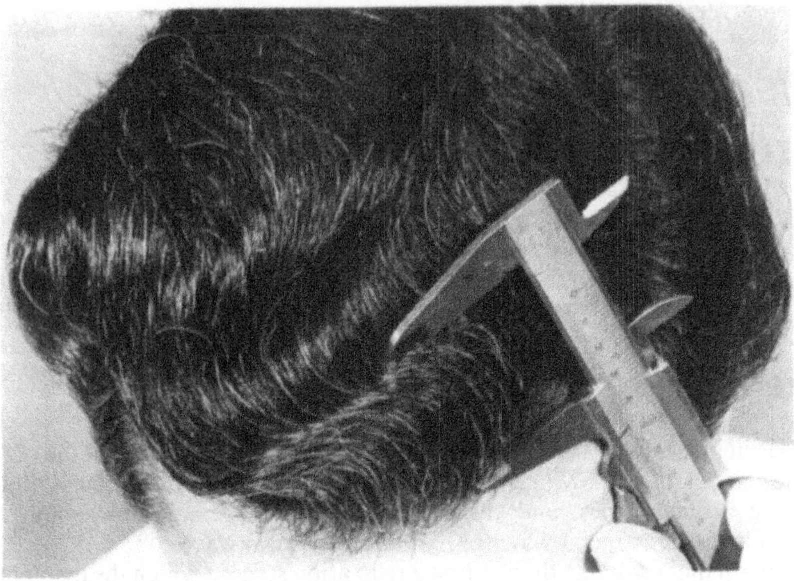

Figure 27.1 Subject 1. Measuring the position of a wave crest

As may be seen from Figure 27.2, the wavelength in this case was about 120 mm. The waves propagated over the head in about 14 days. This correponds to an average velocity of 9 mm/day, which greatly exceeds the velocity of hair growth. The lines in Figure 27.2 are slightly curved, indicating

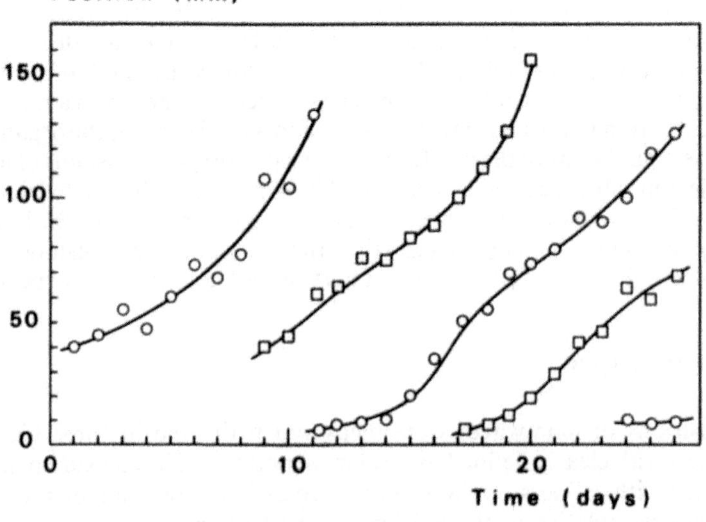

Figure 27.2 Subject 1. Positions of successive wave crests. Circles and squares denote wave crests with the bends in opposite directions. A full wavelength is the distance between two equivalent symbols

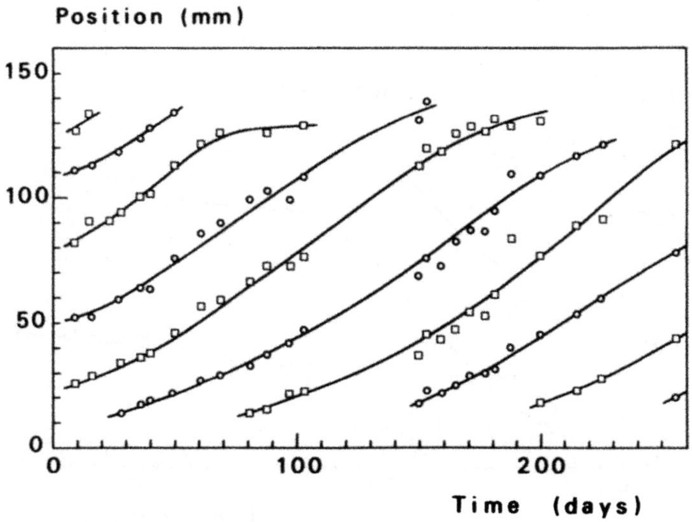

Figure 27.3 Subject 2. Propagation of waves. For key see Figure 27.2

that the velocity of propagation tended to increase as the wave progressed over the head. The waves could be followed over distances well exceeding the average length of the hairs.

Figure 27.3 shows similar measurements in subject 2, a male Caucasian volunteer aged 38 years. He was fair haired and had a more rigid wave pattern than subject 1. The waves in this case propagated at an average velocity of 0.54 mm/day, much slower than in subject 1. The curvature of the lines in Figure 27.3 reflects an initial acceleration of the waves and a deceleration towards the end. This was in line with a relatively small wavelength near the forehead and on the back of the head (45 mm on average) and a larger wavelength – up to 70 mm – near the middle of the scalp. The waves again moved over distances well exceeding the length of the hairs. This subject had several haircuts in the period of study; any influence of these on the propagation of the waves could not be detected.

Subject 3 was a blonde Caucasian woman aged 25 years. The average wavelength was 24 mm. The average velocity of the waves was 0.35 mm/day; the velocity increased as the waves moved from the forehead backwards.

To examine a possible influence of external circumstances on the propagation of the waves, the effect of water was measured. Water was selected because it is known to greatly influence keratin structures[4]. The measurements were performed in subject 1, who in the previous measurements had the highest velocity of propagation.

For 2 weeks, the subject moistened his hair once a day and for another 2 weeks he kept water away from his hair as much as possible. Throughout these two phases, the positions of the hair waves were measured. The results are shown in Figure 27.4.

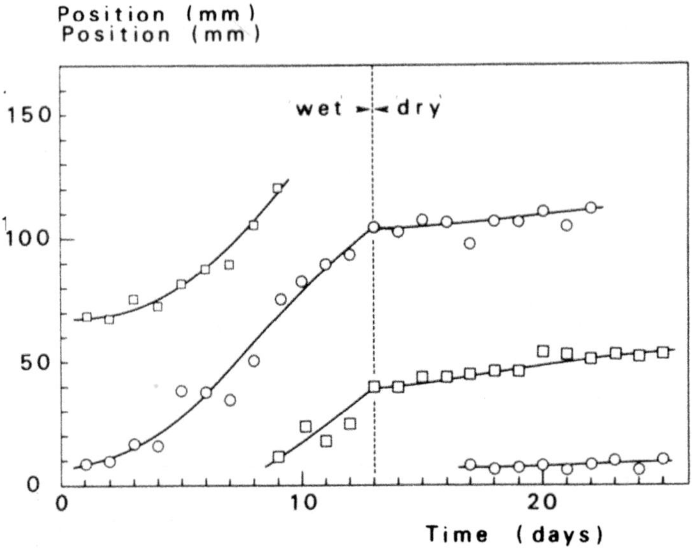

Figure 27.4 Subject 1. Influence of water on the propagation of waves. For key see Figure 27.2

The availability of water greatly influenced the propagation of the waves. In the 'wet' phase, the average velocity of propagation was 7 mm/day. This was less than the velocity measured in the same subject without controlling the water supply to the hair (Figure 27.2), but of the same order of magnitude. In the 'dry' phase, however, the average velocity was about 1 mm/day – a large reduction. Prolonging the dry phase for 3 more weeks gave a further reduction of the average velocity to 0.4 mm/day.

DISCUSSION

Several observations indicated that natural waves may propagate through a head of hair. This could have been concluded, without making any measurements, from the observation that the wavelength often increases from the forehead backwards. This cannot be explained by waves being formed by bends which have a fixed location on the hairs.

The same conclusion is supported by the measurements showing that the waves may propagate over the head at a velocity exceeding the velocity expected on the basis of hair growth. Whether the slower-moving waves, such as those in Figure 27.3, also had a greater velocity than that caused by hair growth requires more careful consideration. The average velocity of hair growth on the human scalp is 0.32 mm/day[5]. Any point on a hair would be displaced over the scalp at a lower velocity, however, because the hair is waved and the displacement is measured along a straight line. Measurements (Figure 27.1) showed that a distance measured along a hair exceeded the corresponding distance measured along a straight line by about 40 %. A point on such a hair, moved along by hair growth, would have a velocity relative to the scalp of (1:1.40) × 0.32 = 0.23 mm/day. All wave velocities measured in the present study were greater. All waves examined, therefore, propagated with respect to the hairs.

In two of the three subjects examined, the waves propagated over distances exceeding the length of the individual hairs. This means that the waves are passed on from one hair to the next. This in itself could be explaimed by a concerted action of the hair roots over the entire scalp. A more likely explanation in view of the other findings, however, is that the waves result from interactions between the hairs; passing of the wave from one hair to the next is then to be expected.

The idea of a wave formed by interactions between the hairs is also supported by a direct observation. In measuring the positions of the waves an attempt was made to measure a new wave arising on the forehead as soon as possible. The waves became sharp enough to be measured, however, only at a distance of 5–10 mm from the hairline. This is a direct indication that the waves are not preformed in the follicles, but arise afterwards.

The large influence of water on waves already formed (Figure 27.4) also suggests that the waves depend on interactions between the hairs.

The follicles may well play a role in providing the necessary conditions for waves to occur. There have been several reports stating that the waviness of hair depends on physical characteristics of the individual hairs, such as the shape of the cross-sections[6]. Such properties, as well as the asymmetric keratinization mentioned earlier, may be necessary conditions for waviness. The waves, however, seem to result from interactions between the hairs.

Acknowledgement

I thank Mr A. H. van Veen and Mrs B. J. R. Blom-Veldhuyzen, MD, for their helpful co-operation in this investigation.

References

1. Astbury, W. T. (1962). The molecular structure and elastic properties of hair. In Savill, A. and Warren, C. (eds.). *The Hair and Scalp*, 5th Ed., pp. 66–79. (London: Edward Arnold)
2. Savill, A. and Warren, C. (1962). *The Hair and Scalp*. 5th Ed., p. 57. (London: Edward Arnold)
3. Ryder, M. L. (1973). *Hair*. p. 12. (London: Edward Arnold)
4. Blank, I. H. (1952). Factors which influence the water content of the stratum corneum. *J. Invest. Dermatol.*, **18**, 433
5. Pecoraro, V. and Astore, J. P. L. (1979). Messungen des Haarwachstums unter physiologischen und pathologischen Bedingungen. In Orfanos, C. E. (ed.) *Haar und Haarkrankheiten*, p. 211. (Stuttgart: Gustav Fischer Verlag)
6. Hayashi, S., Okumura, T. and Ishida, A. (1976). Preliminary study on racial difference in scalp hair. In Kobori, T. and Montagna, W. (eds.) *Biology and Disease of the Hair*, pp. 555–561. (Baltimore: University Park Press)

Thermal, Acoustic, Optical and Electrical Properties

28
Optical radiation transfer in the human skin and applications in *in vivo* remittance spectroscopy

R. R. ANDERSON, J. HU and J. A. PARRISH

SUMMARY

Transmittance spectra (from ultraviolet (UV) to near infrared) and remittance (reflectance) spectra of human dermis and epidermis *in vitro* were analyzed using simple radiation transfer models. Fresh skin from operations and autopsy specimens was separated into layers by various techniques. Absorption spectra for melanins (synthetic, natural, and *in situ*) were determined, and the importance of the pigmented epidermal basal layer in Caucasian skin as a barrier to penetration of ultraviolet radiation into the dermis was documented. Previous studies of epidermal transmittance in Caucasian skin have failed to examine the basal layer, thus overestimating the fraction of incident UV radiation reaching the dermis. The transfer of 320–1000 nm radiation within washed dermis *in vitro* was shown to be primarily determined by optical scattering, not by absorption. This is because normal dermis lacks large quantities of pigments other than blood which absorb these wavelengths. Dermal scattering coefficients varied inversely with wavelength. These data and models can therefore be used to quantitate changes in both the amounts and depth profiles of pigments within the skin *in vivo* by interpreting *in vivo* remittance spectra. *In vivo* remittance spectra of erythematous skin induced by UV exposures were determined and interpreted in this fashion.

INTRODUCTION

Diffuse reflectance, or remittance, is defined as that fraction of incident optical radiation which is returned from a turbid sample. As shown by Edwards and Duntley[1] in their classic study of skin color, it is relatively easy to identify minima in the 400–700 nm diffuse reflectance spectra of human skin as absorption bands of the cutaneous pigments melanin, oxy- and reduced haemoglobin, and carotenoids (mainly β-carotene). Further studies of reflectance spectra of human skin over the 250–3000 nm spectral region[2–5] have shown that water (near infrared absorption bands), bilirubin[6,7] ($\lambda_{max} \simeq 460$ nm), and epidermal aromatic amino acids[8,9] ($\lambda_{max} = 280$ nm) may be added to the list of chromophore molecules exerting an important effect on diffuse reflectance spectra of human skin. Absorption spectra of some of these pigments are shown in Figure 28.1. In normal skin melanin is localized in the epidermis and stratum corneum, haemoglobin is confined to vessels in the dermis, and bilirubin is largely bound to serum albumin and may be increasingly extravascular in the presence of oedema. β-carotene is lipophilic and tends to concentrate in subcutaneous fat and, to some extent, stratum corneum[5], while water is found in high concentration throughout the epidermis and dermis.

It was recognized long ago that measurements of remittance might be used to quantify cutaneous pigments. These studies used reflectance spectra, or diffuse reflectance measurements, at carefully chosen wavelengths corresponding to absorption maxima, minima, or isosbestic points of pigments of interest. With varying degrees of success, investigators were able to quantify, on relative scales, melanin pigmentation[10–12], erythema[10–13], oxygen saturation of cutaneous blood[14,15], bilirubin[6,7], and β-carotene[5,16]. Since the structure and optics of skin is complex, most of these workers approached such measurements empirically, choosing wavelengths by inspection and establishing relative scales for pigments measured by reflectance spectroscopy.

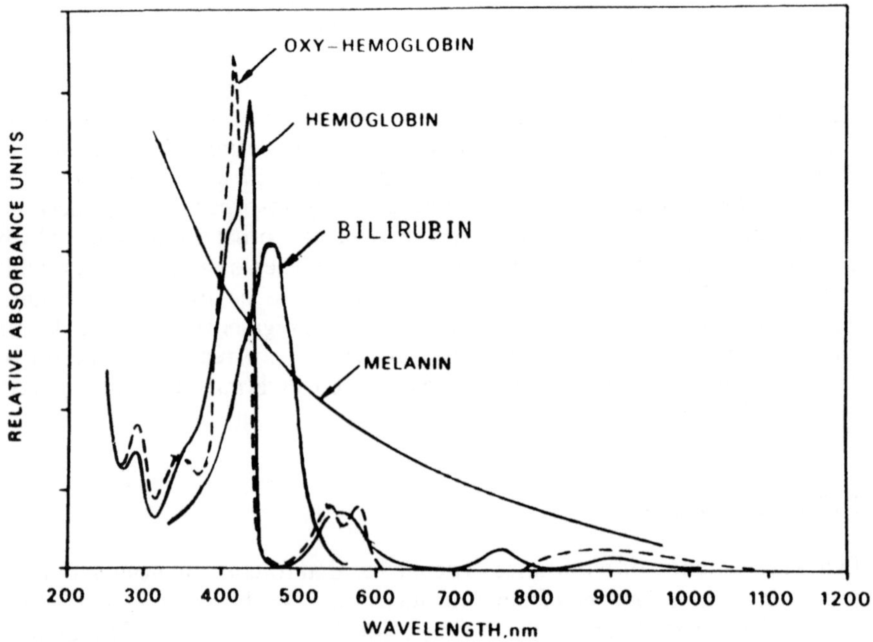

Figure 28.1 Absorption spectra of skin pigments

The more successful of these studies have expressed the diffuse reflectance in optical density units ($OD_R = -\log R$), chosen wavelengths corresponding to an absorption maximum and nearby absorption minimum for the pigment of interest, and established a relative scale for ΔOD_R ($OD_{R,\lambda max} - OD_{R,\lambda min}$) at these wavelengths, and looked for a correlation between ΔOD_R and some other measurement of the pigment.

The purposes of this study were to examine the optical properties of the various layers and pigments in human skin, to propose simple but quantitative models describing cutaneous diffuse reflectance spectra, and to offer explanations for some apparent anomalies in these spectra.

MATERIALS AND METHODS

All diffuse reflectance spectra were measured with a Beckman 5270 double-monochromator, double-beam spectrophotometer, fitted with a 15 cm diameter, barium sulphate-coated, integrating sphere (spectral range 250–3000 nm). Freshly prepared barium sulphate plates were used as reflectance references. This spectrophotometer was interfaced with a Hewlett–Packard 9825 A calculator system for data acquisition, subsequent computations, and plotting of spectra. For *in vivo* diffuse reflectance spectra, a 7.5 cm diameter integrating sphere was coupled to the spectrophotometer's sample and reference beams by means of matched optical quartz fibre bundles. This enabled a smaller, lightweight and movable sphere to be used, which was capable of measuring relatively inaccessible sites without discomfort to the subject.

Separation of the stratum corneum from fresh abdominal skin obtained at autopsy was achieved by incubation at 37 °C for 12 hours in 1 % staphylococcal scalded-skin syndrome epidermolytic toxin[17], in 20 mol/l Hepes buffer (pH 7.4) with 20 % fetal calf serum. When cleavage of the epidermis at a plane just above the basal cell layer was desired, samples were heated for 30 seconds in water at 60 °C and the epidermis subsequently teased off. 100–300 μm thickness sections of dermis were cut by dermotome after removal of the entire epidermis.

RESULTS AND DISCUSSION

Figure 28.2 is a diagrammatic cross-section of human skin showing the various components contributing to diffuse reflectance. On encountering the skin at near-normal incidence, about 5 % of the incident radiation is reflected owing to the refractive index of stratum corneum ($n_D \simeq 1.55$)[18]. In practice, this so-called *regular reflectance component* may be as high as 7 % for perpendicularly-incident radiation, presumably because of the non-planar surface of the skin which produces off-normal angles of incidence. At 280–300 nm and 1950 nm, the epidermis has high absorbance caused by protein and water, respectively, and the diffuse reflectance should consist chiefly of the regular reflectance

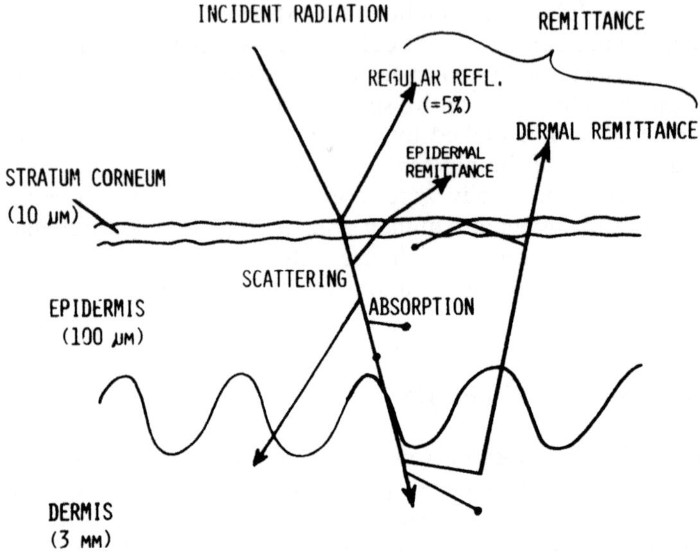

Figure 28.2 Schematic representation of major optical pathways in human skin

component. Indeed, it was found that the diffuse reflectance of *in vivo* skin, whether Caucasoid or Negroid, is 5–7% at these wavelengths.

The 95% of incident radiation entering the skin may be absorbed or scattered within any of the layers of skin. Absorption by the various chromophores mentioned above accounts for the spectral extinction of radiation within each layer. Scattering is the only process by which radiation may be returned to contribute to the diffuse reflectance of skin. To a first approximation, 'epidermal' and 'dermal' components may be assigned to this back-scattered radiation. However, multiple scattering, optical interfaces between layers, and the refractive interface between the skin and air lead to higher order combinations of these initial components which also contribute to the total diffuse reflectance. Despite this, a simple model for the contributions of the epidermis to diffuse reflectance of wavelengths over 350 nm can be proposed and verified.

The thickness of the epidermis and stratum corneum is only 50–150 μm over most of the body, as compared with the 1–4 mm thickness of the dermis. The dermis is comparatively acellular and 70% of its dry weight is collagen fibres, which are known to be highly birefringent. From these considerations scattering within the epidermis would, over at least part of the optical spectrum, be expected to contribute very little to the remittance of skin. Figure 28.3 compares the remittance of full-thickness Caucasian skin *in vitro* with the remittance of the heat separated epidermis alone from the same sample. Over almost the entire spectrum studied, the remittance of the isolated epidermis is accounted for by the regular reflectance component, suggesting minimal scattering within this tissue. The higher values of remittance for full-thickness skin must therefore be a result of radiation transmitted through the epidermis,

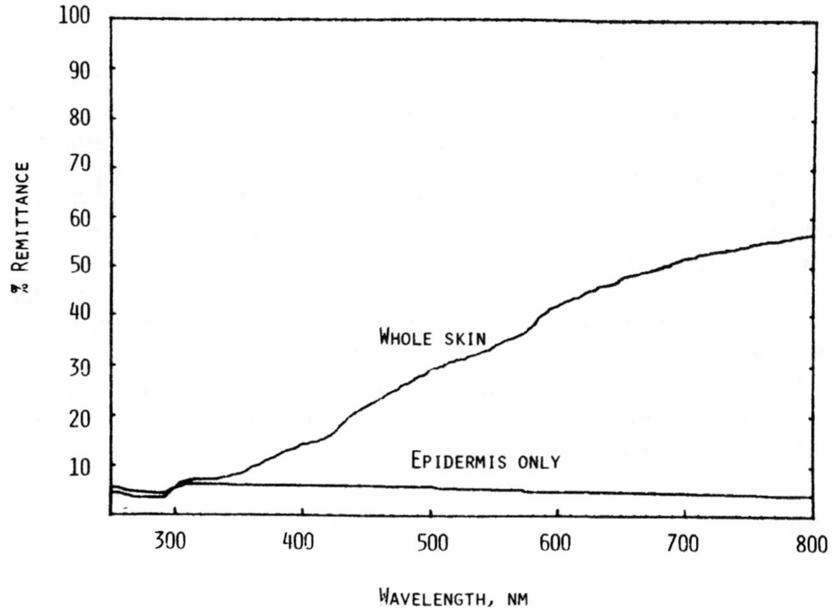

Figure 28.3 Spectral remittance of full-thickness Caucasian skin *in vitro*, and of heat-separated epidermis from the same site. The large difference between these two curves indicates an important dermal remittance component (Figure 28.2)

back-scattered within the dermis, and transmitted again through the epidermis to the skin's surface – i.e. the 'dermal remittance' in Figure 28.2.

This observation immediately suggests a simple model and experimental means for studying absorption within the epidermis by remittance spectroscopy. Neglecting the 'epidermal component', the remittance of full-thickness skin can be approximated as:

$$R_{FT} \simeq R_{REG} + (1 - R_{REG})\,[T_e^3 R_d - R_i T_e^3 R_d + (T_e^4 R_d)(T_e^3 R_d) R_i - (T_e^4 R_d)(T_e^3 R_d) R_i^2 + \ldots] \tag{1}$$

$$= R_{REG} + (1 - R_{REG})(T_e^3 R_d)(1 - R_i) \sum_{n=0}^{\infty} (R_i T_e^4 R_d)^n \tag{2}$$

where R_{REG} is the regular reflectance for normally incident radiation
T_e is the epidermal transmittance for collimated radiation, neglecting regular reflectance
R_d is the dermal remittance, neglecting regular reflectance
R_i is the internal reflectance for diffuse radiation at the skin/air interface.

This expression is derived by inspection, and is the sum of diffuse radiation fluxes contributing to the skin's remittance with each pass through the epidermis. The fluxes at the surface of the skin contributing to the total remittance, R_{FT}, appear as positive terms, while the effect of internal reflectance at the skin–air interface, R_i, gives rise to negative terms. R_{REG} is clearly a positive term. Thus a fraction of the incident radiation $(1 - R_{REG})$ enters the skin. The radiation is assumed to remain collimated on the first pass through the epidermis, but after back-scattering by the dermis (R_d), the returning flux is diffuse. Because the average pathlength for diffuse radiation is twice that for collimated radiation[19], the first positive term is $(1 - R_{REG})T_e^3R_d$. Internal reflection of this returning flux prevents a fraction $(1 - R_i)$ from contributing to R_{FT} and leads to the higher order terms. R_{REG} and R_i are calculated from Fresnel's equations and the refractive index of stratum corneum to be about 0.05 and 0.57 respectively. By convention, we have chosen lower case subscripts to denote model parameters and upper case subscripts to denote measurable or known quantities.

If the epidermis is removed entirely, T_e becomes 1.0 and disappears from the expression. Denoting the measured remittance of the dermis without epidermis as R_D, we have:

$$R_D = R_{REG} + (1 - R_{REG})(R_d)(1 - R_i) \sum_{n=1}^{\infty} (R_iR_d)^n \quad (3)$$

If the higher order terms are neglected, the ratio of $(R_{FT} - R_{REG})/(R_D - R_{REG})$ is simply T_e^3, from which, by analogy, remittance scans before and after removal of any epidermal layer can be used to calculate the transmittance of the layer removed. This makes it possible to determine the absorbance of epidermal layers which cannot be separated *per se*, such as the basal layer.

To test the validity of this approach, we measured remittance spectra of Caucasian skin samples at various stages of removal of the epidermal layers. After measuring R_{FT}, the stratum corneum was removed by staphyloccocal scalded-skin syndrome epidermolytic toxin, which is non-cytotoxic[17] and specifically cleaves at the stratum granulosum. After another measurement of remittance of the skin after removal of its stratum corneum, the remaining epidermal cells were removed by scraping with a scalpel and R_D measured. The optical density of the stratum corneum and of the viable epidermal layer were then calculated using the above model.

The results in a typical sample are presented in Figure 28.4. In all of the samples tested so far, the absorbance of the viable epidermal cell layers was higher than that of the stratum corneum. This presumably reflects the fact that the melanocytes and the resting basal cells of the basal layer contain stores of melanin. Indeed, as seen in Figure 28.4, the calculated absorbance of the viable epidermal cell layers closely matches that of the synthetic DOPA melanin. We chose to synthesize DOPA melanin because of the insolubility of intact human eumelanins and the apparent lack of quantitative absorption spectra for melanins in the literature.

It seems that the above model quantitatively describes the role of the epidermis in determining remittance of full-thickness skin, and is useful for mapping the deposition and amount of melanin in the epidermis and stratum corneum. Even in relatively fair-skinned Caucasians, melanin is the major pigment determining epidermal absorbance in the wavelengths 300–800 nm and occurs mainly in the basal cell layer.

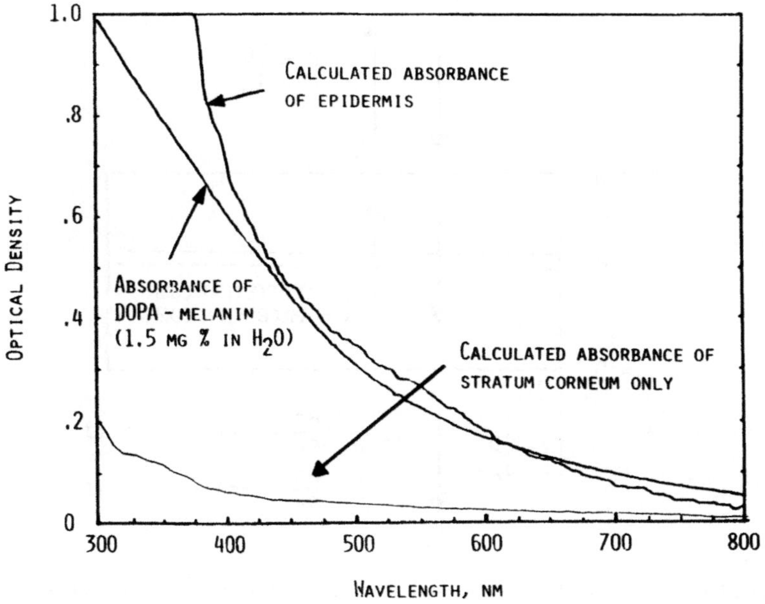

Figure 28.4 Calculated absorbance of entire epidermis, and of stratum corneum alone in Caucasian skin sample. See text for model used

Quantitative consideration of the dermal remittance term R_d in this model requires the study of both scattering and absorption within this connective tissue. To do this, we have used a modified form of the general radiation transfer theory originally proposed by Shuster[20] to describe transmission through interstellar dust. A specific case of Shuster's general theory was popularized by Kubelka and Munk[21] in the 1930s. This phenomenological model is presented below. Our model is a modification of the Kubelka–Munk model.

A *diffuse*, monochromatic flux, I_0, is incident on a sample of thickness *d*. The scattering and absorbing centers of this inhomogeneous sample are assumed to be much smaller than the sample thickness. For the moment, we shall neglect reflection at the interfaces between air and sample. As I penetrates the sample, back-scattering within the sample gives rise to a flux, J, in the opposite direction (the minus x direction of Figure 28.5). Similarly, back-scattering of J contributes to I. Absorption operates on both fluxes. S and K are defined as back-scattering and absorption coefficients, respectively, for the diffuse radiation fluxes I and J. Within any incremental layer of thickness $\mathrm{d}x$, we can write that $\mathrm{d}I$, the incremental change in flux I, is equal to the loss from I by back-scattering and absorption, plus the gain by back-scattering from J.

$$\mathrm{d}I = -(S + K)I\,\mathrm{d}x + SJ\,\mathrm{d}x \tag{4}$$

Similarly,

$$-\mathrm{d}J = -(S + K)J\,\mathrm{d}x + SI\,\mathrm{d}x \tag{5}$$

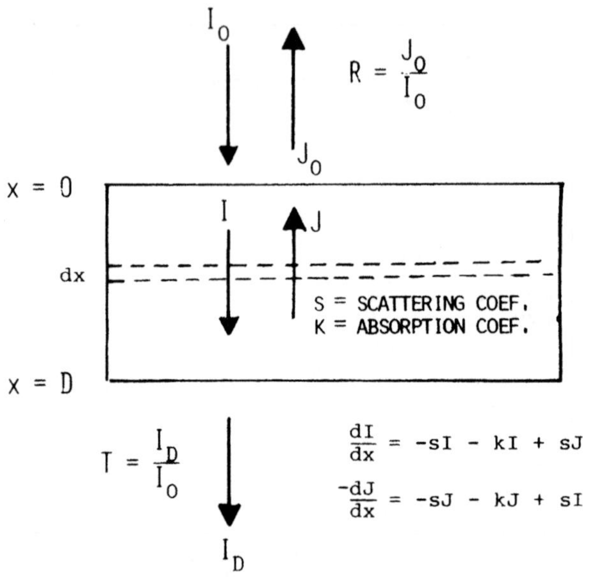

Figure 28.5 Model used by Kubelka and Munk to determine diffuse-flux scattering (s) and absorption (k) coefficients for turbid, absorbing media

The minus sign in equation (5) is necessary because J is defined as positive in the minus x direction. The differential equations (4) and (5) are solved to give a general solution of the form:

$$I = A(1 - \beta)e^{\alpha x} + B(1 + \beta)e^{-\alpha x} \tag{6}$$

$$J = A(1 + \beta)e^{\alpha x} + B(1 - \beta)e^{-\alpha x} \tag{7}$$

where $\alpha = [K(K + 2S)]^{1/2}$ and $\beta = [K/(K + 2S)]^{1/2}$.

The boundary conditions are that $R = J_0/I_0$, $J_d = 0$, and $T = I_d/I_0$, where R and T are remittance and transmittance of the sample respectively. Substitution of these boundary conditions gives a particular solution, which can then be rearranged to express S and K (our two unknowns) in terms of R and T (our measurable quantities):

$$K/S = [(1 + R^2 - T^2)/2R] - 1 \tag{8}$$

$$S = \frac{1}{d}[K/S(K/S + 2]^{-1/2} \coth\left[\frac{1 - R(K/S + 1)}{[K/S(K/S + 2]^{1/2}R}\right] \tag{9}$$

To use this model practically, one must also account for reflection at the sample boundaries, and use diffuse incident radiation. Ryde[22] and others[23,24] have considered models which apply to the use of collimated incident radiation and include reflection at the sample boundaries, but the number of experimental measurements necessary to determine S and K becomes, for our purposes, impractical. We have therefore developed an experimental technique which minimizes reflection at the sample boundaries and provides a diffuse incident flux, allowing use of the simple two-parameter model above.

This is accomplished by mounting a thin (100–300 μm) dermal sample between and in optical contact with two thin (1 × 25 × 25 mm) ground and etched quartz diffuser plates. The collimated incident beam of the spectrophotometer is made essentially diffuse by the first plate before striking the sample. The refractive indices of quartz and skin tissues are sufficiently close to allow us to neglect reflectances occurring at the skin–quartz interfaces. Incident beam and diffuse internal spectral reflectances at the quartz–air interfaces can be measured directly, and are used to calculate the transmittance T or remittance R of the sample in conditions applicable to the Kubelka–Munk model. The model used in these calculations is described below.

In a manner similar to that previously used to derive equation (1), the remittance of the quartz–skin–quartz 'sandwich' can be expressed as the summation of infinite series arising from multiple passes through the skin sample:

$$R_T = \left[\sum_{n=1}^{\infty} \sum_{m=1}^{n} \frac{n!}{m!\,(n-m)!} T^m R^{n-m} R_2^{n-1} \right] (1 - R_1)(1 - R_2) \tag{10}$$

where m is even numbers only. Similarly, the transmittance is given by:

$$T_T = \left[\sum_{n=1}^{\infty} \sum_{m=1}^{n} \frac{n!}{m!\,(n-m)!} T^m R^{n-m} R_2^{n-1} \right] (1 - R_1)(1 - R_2) \tag{11}$$

where m is odd numbers only, and

where R_T is the measured remittance of the quartz–skin–quartz sandwich
T_T is the measured transmittance of the quartz–skin–quartz sandwich
R is the remittance of the skin sample in conditions of diffuse incident radiation, without reflection or refraction at the skin sample boundaries
T is the transmittance of the skin sample in conditions of diffuse incident radiation, without reflection or refraction at the skin sample boundaries
R_1 is the reflectance of collimated incident radiation at the first encounter with the first quartz plate
R_2 is the internal reflectance of diffuse radiation at the quartz-to-air boundaries.

The summation in equation (10) appears only for even-numbered values of m because only even-numbered passes transmitted through the sample contribute to the measured remittance R_T. Similarly, only odd numbered passes (equation (11)) through the sample contribute to T_T, the measured transmittance. Summing equations (10) and (11) yields an exact solution to the series:

$$R_T + T_T = R_1 + \frac{a(1 - R_1)(1 - R_2)}{1 - aR_2} \tag{12}$$

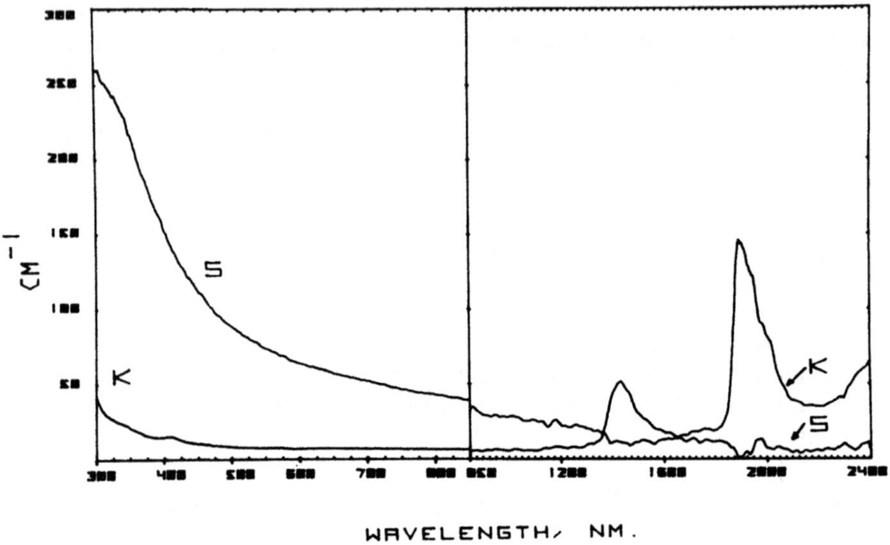

Figure 28.6 Diffuse-flux scattering (s) and absorption (k) coefficients for human dermis *in vitro*

where $a \equiv R + T$. Similarly, subtracting (11) from (10) yields:

$$R_T - T_T = R_1 + \frac{b(1 - R_1)(1 - R_2)}{1 - bR_2} \tag{13}$$

where $b \equiv R - T$.

R_1 and R_2 were measured experimentally over the 250–3000 nm range and used to calculate, by means of equations (12) and (13), the values of R and T for thin samples of human dermis. Because R and T are values fitting the conditions of the Kubelka–Munk model, S and K were then calculated directly from equations (8) and (9).

Results in a typical dermal sample are given in Figure 28.6. Note that the spectral character of S is about $S \propto \lambda^{-2.5}$ over the spectral region 300–800 nm, suggesting that the scattering may arise primarily from structures with sizes in the order of 100 nm. Most scattering centers in the dermis are probably collagen fibres, which have diameters of this order. The inverse relation between S and λ largely accounts for the fact that shorter wavelengths do not usually penetrate the dermis to the same extent as longer wavelengths[2].

K, the diffuse-radiation absorption coefficient for dermis, is remarkably spectrally flat and less than S over the visible and near infrared wavelengths, but shows bands in the infrared corresponding to absorption by water. The magnitude of K at these bands is in good agreement with the known extinction coefficients of water, lending a quantitative check on the validity of our model and techniques. The large increase of K in the ultraviolet spectral region may be due to many chromophores present in dermis. Because the data shown in Figure 28.6 are for a thin section of dermis *in vitro*, not much absorption by the blood-borne pigments haemoglobin, oxyhaemoglobin,, or bilirubin is seen. However, the determination of S, and of K for dermis without blood, allows a

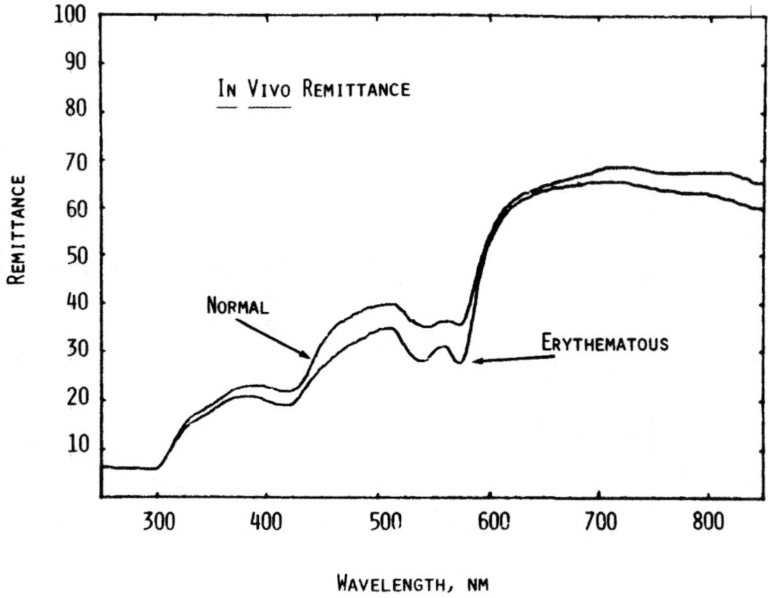

Figure 28.7 Remittance spectra of human skin *in vivo* before and after induction of erythema by UV-B radiation

more quantitative determination of blood-borne pigments from *in vivo* remittance spectra.

As discussed before, scattering within the dermis accounts for most of the 350–2500 nm, remittance of full-thickness Caucasian skin. Because of the spectral character of *S*, the average pathlength experienced by shorter wavelengths within the dermis is much less than that for longer wavelengths. We would therefore expect that the effective absorbance of dermal pigments such as oxyhaemoglobin, when determined by remittance spectroscopy on living skin, would not be proportional to ε, where ε is an extinction coefficient for the pigment, measured in the standard manner. Deviations from ε dependence could, of course, also be caused by non-uniform distribution of such pigments.

The concept that effective absorbance of dermal pigments is inversely dependent upon *S* explains much of the otherwise anomalous spectral character of *in vivo* remittance spectra of erythematous (vasodilated) skin. As an example, we have measured the remittance of Caucasian skin before and 24 hours after UV-B (290–320 nm) exposures to the forearm, which resulted in delayed erythema. These *in vivo* remittance spectra are shown in Figure 28.7. When the regular reflectance component was subtracted and the difference in 'optical density' between the erythematous and control skin calculated, the data shown in Figure 28.8 was obtained. The absorption bands noted can be assigned to oxyhaemoglobin and bilirubin shown in the figure. For comparison, the absorbance of oxyhaemoglobin is also shown in the same figure (relative units). Note that the longer wavelength (542, 576 nm) absorption bands of oxyhaemoglobin are greatly accentuated relative to the Soret band at

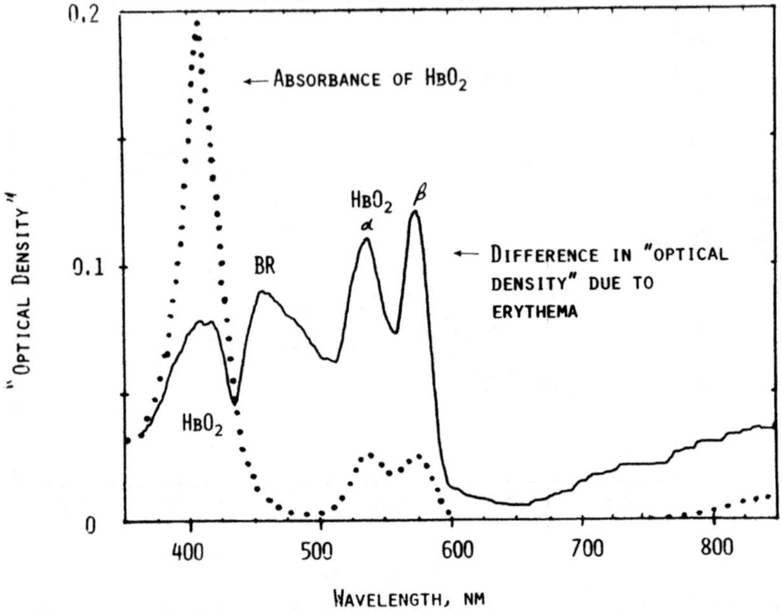

Figure 28.8 Comparison of the change in 'optical density' (text) of skin resulting from erythema with the absorption spectrum of oxyhaemoglobin (HbO_2). BR = Bilirubin

416 nm. This behavior is largely due to the effects of dermal scattering (S) in defining the average pathlength at the various wavelengths.

The fact that shorter wavelengths penetrate the dermis superficially compared with longer wavelengths raises the possibility of monitoring changes in blood volume or flow as a function of depth. That is, it is theoretically possible that an inverse transform could be obtained to provide a means of discriminating between vasodilatations occurring in different vessel plexi uses. Our current research in this area is focused on such practical, clinical and investigatory uses of *in vivo* remittance spectroscopy, and on further refining optical transfer models for data analysis.

We have reported some exploratory and largely preliminary observations leading to the formulation of reasonable conceptual models to describe radiation transfer and remittance spectra of human skin. This work is still in progress, and we hope to develop better non-invasive means for quantifying cutaneous vascular responses and pigmentation, and to understand the distribution and fate of optical radiations within the skin. Although the models presented above are approximate, we hope they may lead toward these goals.

Acknowledgements

The authors thank Dr Peter Elias for kindly providing us with the epidermolytic toxin, and Ben Ocko for his programming and mathematical assistance. This work was supported by National Institutes of Health Grant No. R01 AM25395-01.

References

1. Edwards, E. A. and Duntley, S. Q. (1939). The pigments and color of human skin. *Am. J. Anat.*, **65**, 1
2. Hardy, J. D., Hammel, H. T. and Murgatroyd, D. (1956). Spectral transmittance and reflectance of excised human skin. *J. Appl. Physiol.*, **9**, 257
3. Jacquez, J. A., Kuppenheim, H. F., Dimitroff, J. M., McKeehan, W. and Huss, J. (1956). Spectral reflectance of human skin, the region 235–700 nm. *J. Appl. Physiol.*, **8**, 212
4. Jacquez, J. A., Huss, J., McKeehan, W., Dimitroff, J. M. and Kuppenheim, H. F. (1956). Spectral reflectance of human skin in the region 0.7–2.6 μm. *J. Appl. Physiol.*, **8**, 297
5. Edwards, E. A., Finkelstein, N. A. and Duntley, S. Q. (1951). Spectrophotometry of living human skin in the ultraviolet range. *J. Invest. Dermatol.*, **16**, 311
6. Ballowitz, L. and Avery, M. E. (1970). Spectral reflectance of the skin. *Biol. Neonate*, **15**, 348
7. Hannemann, R. E., DeWitt, D. P. and Wiechel,, J. F. (1978). Neonatal serum bilirubin from skin reflectance. *Pediatr. Res.*, **12**, 207
8. Lucas, N. S. (1930). The permeability of human epidermis to ultraviolet radiation. *Biochem. J.*, **25**, 57
9. Everett, M. A., Yeargers, E., Sayre, R. M. and Olson, R. L. (1966). Penetration of epidermis by ultraviolet rays. *Photochem. Photobiol.*, **5**, 533
10. Jansen, M. T. (1953). A reflection spectrophotometric study of ultraviolet erythema and pigmentation. *J. Clin. Invest.*, **32**, 1053
11. Daniels, F. and Imbrie, J. D. (1958). Comparison between visual grading and reflectance measurements of erythema produced by sunlight. *J. Invest. Dermatol.*, **30**, 295
12. Breit, R. and Kligman, A. M. (1969). Measurement of erythemal and pigmentary responses to ultraviolet radiation of different spectral qualities. In Urbach, F. (ed.) *The Biologic Effects of Ultraviolet Radiation (with Emphasis on the Skin)*, pp. 267–275. (Oxford: Pergamon Press)
13. Frank, L., Rapp, Y. and Bergman, L. V. (1962). An instrument for the objective measurement of erythema. *J. Invest. Dermatol.*, **38**, 21
14. Brinkman, R., Zijlstra, W. G. and Koopmans, R. K. (1949). A method for continuous observation of percentage oxygen saturation in patients. *Arch. Chir. Ned.*, **1**, 333
15. Brinkman, R., Cost, W. S., Koopmans, R. K. and Zijlstra, W. G. (1949). Continuous observation of percentage oxygen saturation of capillary blood in patients. *Arch. Chir. Ned.*, **1**, 184.
16. Goldzieher, J. A., Roberts, I. S., Rawls, W. B. and Goldzieher, M. A. (1951). 'Chemical' analysis of the intact skin by reflectance spectrophotometry. *Arch. Dermatol. Syphilol.*, **64**, 533
17. Elias, P. M., Fritsch, P., Dahl, M. V. and Wolff, K. (1975). Staphylococcal toxic epidermal necrolysis: pathogenesis and studies on the subcellular site of action of exfoliation. *J. Invest. Dermatol.*, **65**, 501
18. Scheuplein, R. J. (1964). A survey of some fundamental aspects of the absorption and reflection of light by tissue. *J. Soc. Cosmetic Chem.*, **15**, 111
19. Kortüm, G. (1969). *Reflectance Spectroscopy*, pp. 107—108. (New York: Springer-Verlag)
20. Shuster, A. (1905). Radiation through a foggy atmosphere. *Astrophys. J.*, **21**, 1.
21. Kubelka, P. and Munk, F. (1931). Ein Bietrag zur Optik der Farbanstriche. *Z. Technichse Physik*, **12**, 593
22. Ryde, J. W. (1931). The scattering of light by turbid media. Part 1. *Proc. Soc. London*, **A131**, 451
23. Duntley, S. Q. (1942). The optical properties of diffusing materials. *J. Opt. Soc. Am.*, **32**, 61
24. Atkins, J. T. (1969). Optical properties of turbid materials. In Urbach, F. (ed.) *The Biologic Effects of Ultraviolet Radiation (with Emphasis on the Skin)*, pp. 141–150. (Oxford: Pergamon Press)

29
Alternating current electrical properties of human skin measured *in vivo*

D. C. SALTER

INTRODUCTION

To help meet the considerable medical need to quantify and thus characterize skin *in vivo*, we have sought to determine its electrical properties, since electrical changes with skin structure and function are much larger than other physical changes when expressed as a percentage of the mean population value. Moreover, physical characteristics can usually be more rapidly measured than chemical or biological ones, and with non-invasive techniques too, which is a great clinical advantage.

We have sought to avoid the various pitfalls in methodology or interpretation which have tended to confuse and spoil similar work, probably because of its extremely interdisciplinary nature. This has led us in the past 5 years to develop, at a cost of £25 000, novel electronic equipment and technique which we think powerful and rigorous enough to make possible a thorough re-exploration of the *in vivo* electrical characteristics of skin. The approach provides enough sufficiently fundamental physical indices to permit the quantification and comparison *in vivo* of electrical properties of skin. Applications include the monitoring of disease and healing processes in a single skin site, and objectively investigating the effects of different treatments by making inter-site comparisons on one or more patients.

The important though often neglected question of what constitutes an adequate basis with which to characterize the electrical properties of skin has been discussed elsewhere[1]; this paper briefly considers some preliminary clinical results; their observed degree of variability, and, more especially, the biophysical interpretation of the properties particular form. This last, equally important consideration has also been neglected or only superficially dealt with in the past. Extant interpretations are almost entirely untested, and we find that they do not stand up to testing by experiment or by theory. A more promising, although less simple, interpretation is offered.

METHODS

These methods have been more fully reported elsewhere[1–3]. Six electrodes are used non-invasively in a way which makes possible, apparently for the first time, the independent derivation of the properties of two areas of epidermis, the dermis and deeper tissue between them, and of the properties of the measuring electrodes themselves. Other non-invasive multielectrode methods of attempting to do this seem to have many more flaws[1].

Since the electrical properties of skin are not necessarily linear, a discrete Fourier transform procedure is used to analyse the shape as well as the amplitude and phase of the measured voltage waveforms resulting from the selected sinusoidal stimulating current. The first harmonic terms of the transform can be used to construct, on the complex impedance plane[1], the locus of the electrical impedance between 0.1 Hz and 100 kHz for epidermis, dermis and electrodes separately. After adding the contribution of any important higher harmonics, the voltage–current characteristic curve of epidermis measured as if it were isolated may be constructed, even though it is still *in vivo*. This has never been done non-invasively before, and our results should be correspondingly less affected by experimental procedure. Finally, the bias potential with respect to electrical earth is measured at the six electrodes by means of the zeroth harmonic, so that we can derive an estimate of the trans-epidermal potential difference at the two epidermal sites of interest.

RESULTS

Skin is not ohmic

Examination of all the voltage–current characteristic curves we have constructed for particular epidermal sites *in vivo* has always shown that considering them to be linear would be only an approximation to the truth, even at low current densities (below 1 $\mu A/cm^2$). Since all agree that skin is non-linear at high current densities, perhaps those investigators[4] who have claimed that there is a truly linear region at low current densities were simply using techniques which for such current densities were unable to show departures from linearity.

The non-linearity is almost entirely due to the fixed charge on the proteins of the stratum corneum[5,6], and indeed the resistance and impedance contribution of skin layers below the corneum is negligible[5,7,8], except that they affect the corneum's condition. However, the potential difference across the region of the stratum granulosum, visible as a colour change in sections stained with haematoxylin and eosin, probably outweighs the effects[9] of the fixed charge of the corneum in determining the epidermal contribution to the skin potential level[4,10].

The AC distribution of relaxation times is of Cole–Cole form

For a particular amplitude of sinusoidal stimulation, the alternating current complex electrical impedance of skin has usually been assumed to obey Cole's equation[1,11]:

$$Z = R_s + jX_s = R_\infty + \frac{R_0 - R_\infty}{1 + (j\omega\tau_p)^{1-m}} \qquad (0 < m < 1) \tag{1}$$

in which ω is 2π times the stimulating frequency, m is related to the width of the distribution of relaxation times present, τ_p is a mean relaxation time, R_∞ is the value of the equivalent series resistance R_s found at very high frequencies ($\omega\tau_p \gg 1$), R_0 is the value of R_s at very low frequencies and for direct current, and X_s is the equivalent series reactance. Equation (1) does *not* describe a Cole–Cole dielectric[12], since though the algebra is analogous, the physics is not.

Previous testing of the particular form of equation (1) against experimental results has been scarce and probably insufficiently discriminatory. For example, the method of Yamamoto and Yamamoto[7] is not very sensitive to the particular frequency dependence of the results. Consequently, values of R_0 and R_∞ were determined by least squares for each set of our data and the Cole–Cole 'log v/u against log ω' graph was plotted[12]. The best-fit log–normal distribution was also determined from the tables given by Yager[13] and the corresponding theoretical log v/u against log ω locus was superimposed. It was always found that the experimental locus was more linear than the log–normal prediction, and often considerably so. We may therefore conclude that the (apparent) distribution of relaxation times is Cole–Cole rather than log–normal or log–uniform, and that equation (1) is closely obeyed. This is important because no extant theory predicts a complex impedance sufficiently close to the form of equation (1). The theory outlined later, however, predicts equation (1) exactly, and also interprets it reasonably.

Variability of the parameters determining Z

Working with a specially developed minimally-hydrating electrode jelly[3] and Ag/AgCl electrodes of area 1 cm^2, we have observed an enormous variation of parameter values with body site, and a lesser one between similar sites on different subjects, all for the same conditions of ambient temperature and relative humidity. On dorsal forearms, for example, the maximum and minimum values of R_0 measured so far for a single epidermal site have been 1.4×10^6 and 1.85×10^4 ohms, with τ_p values of 2.27×10^{-2} and 1.33×10^{-3} seconds respectively. On this scale, R_∞ was indistinguishable from zero. Also, in any one subject, it seems that the same site measured again under the same conditions after a period of 24 hours can show a five- to ten-fold change in R_0 and τ_p.

There is much discussion of the 'phase angle' θ and its value at a single frequency as a function of body site, pathology, age, sex and so on[14–16]. I have pointed out elsewhere (reference 1, equation 13) that if equation (1) is obeyed such measurements are extremely difficult to interpret, because θ at a single frequency can vary with so many factors. A plot of θ over many decades of frequency is useful, however, and complementary to the Z locus plot from which the distribution width parameter m can be more easily obtained. This parameter m is much more fundamental than the phase angle.

The parameter m varies considerably from site to site in the same subject, but its variation with time is much less than that of R_0 and τ_p. It seems to be systematically increased in psoriatic plaques, as can be deduced for example by using equation (1) to interpret Figure 4 of Clar *et al.*[17], or the high-frequency phase angle findings of Gougerot[15].

Another indication that the value of m is controlled by the state of the cell layers of the epidermis can be obtained by measuring a site under sweating and non-sweating conditions. When the sweat glands are brought into action, R_0 and τ_p are greatly reduced, but m is almost unchanged. In fact, it seems that the value of m is determined by the corneum, since by stripping off the corneum layer by layer as far as possible, the impedance of skin without corneum can be shown to be negligible. By subtraction the electrical properties of successive layers of the corneum may also be obtained[7], and the result is that *each layer* seems to obey equation (1), with a non-zero value of m which may change a little or not at all with depth. This finding is very difficult to interpret with traditional models of skin impedance, even allowing for the errors inherent in stripping experiments, which alter the skin via the inflammatory response and in other ways during the process of measuring it. The difficulty arises because each cell layer of the corneum seems to obey equation (1) with a non-zero value of m. This however is a confirmation of, rather than a difficulty for, the new theory proposed below.

Models of Z

In 1938 Barnett noted[14] that a distribution of relaxation times would arise in skin if the intercellular resistivity varied exponentially with depth, and that this would reproduce some of the features of skin impedance which he had observed. However, he does not seem to have completed the analysis of the theory behind this, and Tregear when following up the idea resorted to constructing model networks and measuring their properties (reference 18, Figure 28). The Barnett–Tregear model consists of a series chain of relaxing elements. Each element is composed of a resistor and a capacitor in parallel, which are meant to represent the local resistivity and permittivity respectively of a layer of cells in the stratum corneum. To simulate a gradient of resistivity resulting from the gradient of hydration which is known to exist, the resistors change in value depending on the 'cell layer' concerned.

If an exponential dependence of resistance upon cell layer number is chosen, and as for the Tregear model only phase angle is plotted against frequency, then this model does indeed seem to reproduce the impedance behaviour of skin reasonably accurately. However, if the Z locus plot is constructed instead, it is skewed, and the discrepancies in the fit of the model are clearly shown. If the resistors (representing the resistivity) vary exponentially with distance and the capacitors (representing the permittivity) stay constant, then:

$$\frac{(R_s - R_\infty)}{(R_0 - R_\infty)} = \frac{\arctan(\omega\tau_2) - \arctan(\omega\tau_1)}{\omega(\tau_2 - \tau_1)} \tag{2}$$

and

$$\frac{-X_s}{(R_0 - R_\infty)} = \frac{1}{2\omega(\tau_2 - \tau_1)} \ln\left[\frac{1 + (\omega\tau_2)^2}{1 + (\omega\tau_1)^2}\right] \tag{3}$$

where $\tau_1 = CR_{min}$ and $\tau_2 = CR_{max}$. This Z locus of the Barnett–Tregear model is quite unlike that of skin. If instead a model is used in which the capacitors vary exponentially with distance and the resistors are constant, a circumstance not at all likely in skin, we find that the fit is rather better, the locus becoming symmetrical but flattened – in fact it is a semi-ellipse, as for all log–uniform distributions.

Nevertheless, it is still a locus evidently unlike that given by equation (1), which is a circular arc. Yamamoto and Yamamoto[19] analytically investigated a Barnett–Tregear type of model in which the resistivity (arbitrarily) varied inversely with distance and the permittivity was constant. They deduced from this model the same results that we have obtained from the (unlikely) model in which the permittivity varies exponentially with distance and the resistivity is constant. However, checking the behaviour of computer-simulated networks confirms the present analysis, and to prove this, computer simulations of discrete versions of these models together with algebraic details of the continuum forms will be given elsewhere[3] as part of a more general review which seeks to show that no presently available theory is capable of generating equation (1) without either making biologically unreasonable assumptions or appealing to unexplained physics.

An alternative theory is considered below.

THE SHUNTED LOCALIZED-STATE MODEL

Proteins in pure form have a large energy gap between their valence and conduction bands, but by means of charge–transfer interactions[20] or surface effects, localized states can be produced in the bandgap, increasing the conductivity by many orders of magnitude and also making dielectric relaxation possible if charges 'hop' between these states by multiphonon activation, or tunnelling, or both[20,21].

Dielectric relaxation within a neoexponential density of (localized) states has been shown by Lewis[21] to lead to a (real) conductivity which is a power function of frequency. Such behaviour is often shown by proteins[20], so let us rework and amplify Lewis's theory. By truncating his assumed density of states, combining it with some work of Macdonald[22], and finally bringing everything back into the frequency domain by Stieltjes transformation, we find that for a density of localized states in the bandgap,

$$g\left(\frac{\Delta E}{kT}\right) = C\exp\left[-\frac{\alpha\,\Delta E}{2\,kT}\right], \qquad \Delta E_1 \leqslant \Delta E \leqslant \Delta E_2$$
$$= 0 \qquad \text{otherwise} \tag{4}$$

not near the Fermi energy, where ΔE is the multiphonon activation energy for conduction, C is a normalization constant and α is also a constant, the complex permittivity (ε^*) of the material is given by:

$$\frac{\varepsilon^* - \varepsilon_\infty}{\varepsilon_s - \varepsilon_\infty} = \frac{(1-\alpha)\pi}{\sin(\alpha\pi)}(j\omega\tau_2)^{\alpha-1}, \qquad \omega_{\min} \leqslant \omega \lesssim \omega_{\max}, \quad 0 < \alpha < 1 \tag{5}$$

assuming that $\Delta E_1 \to 0$, so $\omega_{\max} \simeq \infty$, and τ_2 is an exponential function of $(\Delta E_2/kT)$. The value of $\omega_{\min}$ depends on τ_2 and α, and can be found numerically[3]. Incidentally, equation (5) well describes the complex permittivity of those biological membranes containing biologically active protein modules, and I doubt that this agreement is accidental. The complex impedance equivalent of equation (5) was used by Cole to describe a cell membrane as long ago as 1940[11], though without interpretation.

Assuming that a DC shunt conductivity (σ_0) is also present, and that the effect of ε_∞ may be neglected in the frequency range of interest, the total (complex) conductivity of the system (σ^*) reduces to the simple form:

$$\sigma^* = j\omega\varepsilon^* = \sigma_0 + \frac{\varepsilon_s(1-\alpha)\pi}{\tau_2\sin(\alpha\pi)}(j\omega\tau_2)^\alpha, \qquad \omega_{\min} \leqslant \omega \lesssim \infty, \; 0 < \alpha < 1 \tag{6}$$

which is a complex analogue of Lewis's result. Using $R_0 = 1/\sigma_0$, the corresponding complex impedance is:

$$Z = \frac{R_0}{1 + (j\omega\tau_p)^{1-m}}, \qquad 0 < m = (1-\alpha) < 1 \tag{7}$$

so long as $(\omega\tau_p) > (\omega\tau_p)_{\min}$ where

$$(\omega\tau_p)_{\min} = \tau_2\omega_{\min}\left[\frac{R_0\varepsilon_s(1-\alpha)\pi}{\tau_2\sin(\alpha\pi)}\right]^{1/\alpha} \tag{8}$$

With reasonable parameter values, the square bracket in equation (8) represents a very small number and equation (7) holds down to very small values of $(\omega\tau_p)$.

Some evidence in favour of this theory is as follows. Firstly, it enables us to derive equation (1), which we know to be exactly what we want, from a plausible model based upon the known behaviour of those materials (proteins) which largely compose the cell membranes and cell interiors of the stratum corneum. Secondly, when we have subtracted the effect of least-squares-fitted values of R_0 from the sets of data we have obtained so far, and transformed the results onto the complex permittivity plane, we see what can be called an (inclined) straight line to within experimental error, as is predicted by equation (5). Thirdly, considering the effect of water, Yamamoto and Yamamoto[23] showed that applying physiological saline to the skin not only reduced R_0 as would be expected, but also increased *m*, the *reverse* of what would be expected from the hydration gradient model of Barnett and Tregear and also apparently from all other available theories. Such effects can, however, be interpreted with the protein localized state theory.

As long ago as 1945, Rosendal[24] found that a dry 'membrane' of pulverized corneum proteins behaved as described by equation (1) with $m \simeq 0$ and, as the membrane hydration increased, so did *m*. Indeed, the alpha power dependence of conductivity usually decreases, and $m = (1 - \alpha)$ increases, as the hydration of proteins increases. The localized state theory would interpret this as resulting from a 'filling out' of the neoexponential density of states towards a flatter, more box-like shape. Rosendal may also have been the first to obtain evidence of the existence of these states. He showed[24] that adding water to the structureless protein 'membrane' increased the capacitative conductivity by an unaccountably large amount, an effect which may well result from the formation of localized states, as postulated here, by charge–transfer interactions between the water and the proteins[20]. Fourthly, the model can readily interpret the obedience of each cell layer to equation (1) with a non-zero value of *m* which changes little or not at all with depth simply by the reasonable hypothesis that the proteins of one cell layer in the corneum are much the same as those of other cell layers in the corneum. Fifthly, the model allows *m* to be independent of R_0, which agrees with the sweating/non-sweating results in the section on variability of parameters determining *Z*, and with the commonsense interpretation of R_0 as being due to the combined effects of unbounded transcellular and intercellular conduction and conduction through shunts such as sweat glands. Sixthly, the theory implies that the value of *m* depends upon the ionization of the side groups and thus upon the molecular environment of the proteins of the cell membranes and the cell interiors in the stratum corneum.

The theory therefore predicts that *m* can be expected to change with local pH and with cell protein composition. This agrees with our observation above that *m* is usually increased in psoriatic plaques, since it is well known that the proteins of the corneocyte membrane and interior are altered in this case. It will certainly be useful if *m* can serve as a non-invasive index of some cell kinetic parameter.

Acknowledgements

I would thank Professor K. S. Cole, For. Mem. R.S., for his kindness and hospitality to me in San Diego, making possible many discussions which paved the way towards the new theory outlined here. Discussions with Professor T. J. Lewis and Dr Ron Pethig were just as helpful, and I thank all three for their encouragement and enthusiasm.

References

1. Salter, D. C. (1979). Quantifying skin disease and healing *in vivo* using electrical impedance measurements. In Rolfe, P. (ed.) *Non-Invasive Physiological Measurements*, pp. 21–64. (London: Academic Press)

2. Salter, D. C. (1979). A means of rapidly quantifying and comparing skin without biopsy, and a fundamental way of non-invasively deriving the charge gradient across it. Presented at the Meeting of the Royal Society of Medicine Section of Measurement in Medicine, April 23, London
3. Salter, D. C. (1981). Studies in the measurement, form and interpretation of some electrical properties of normal and pathological human skin *in vivo.* (In three volumes) *DPhil Thesis.* University of Oxford
4. For a review, see for example Edelberg, R. (1971). Electrical properties of skin. In Elden, H. R. (ed.) *Biophysical Properties of the Skin,* pp. 513–550 (New York: Wiley-Interscience)
5. Rosendal, T. (1943). Studies on the conducting properties of the human skin to direct current. *Acta Physiol. Scand.,* **5**, 130
6. Rosendal, T. (1944). Further studies on the conducting properties of human skin to direct and alternating current. *Acta Physiol. Scand.,* **8**, 183
7. Yamamoto, T. and Yamamoto, Y. (1976). Electrical properties of the epidermal stratum corneum. *Med. Biol. Eng.,* **14**, 151
8. Scheuplein, R. J. (1976). Percutaneous absorption after twenty-five years: or 'old wine in new wineskins'. *J. Invest. Dermatol.,* **67**, 31
9. Teorell, T. (1953). Transport processes and electrical phenomena in ionic membranes. *Prog. Biophys. Biophys. Chem.,* **3**, 305
10. Edelberg, R. (1977). Relation of electrical properties of skin to structure and physiologic state. *J. Invest. Dermatol.,* **69**, 324
11. Cole, K. S. (1940). Permeability and impermeability of cell membranes for ions. *Cold Spring Harbor Symp. Quantitative Biol.,* **4**, 110
12. Cole, K. S. and Cole, R. H. (1941). Dispersion and absorption in dielectrics. I. Alternating current characteristics. *J. Chem. Phys.,* **9**, 341
13. Yager, W. A. (1936). The distribution of relaxation times in typical dielectrics. *Physics,* **7**, 434
14. Barnett, A. (1938). The phase angle of normal human skin. *J. Physiol.,* **93**, 349
15. Gougerot, L. (1947). Recherches sur l'impédance cutanée en courant alternatif de basse fréquence au cours de différentes dermatoses. *Ann. Derm. Syph.,* **7**, 101
16. Rothman, S. (1954). *Physiology and Biochemistry of the Skin,* pp. 9–25. (Chicago: University of Chicago Press)
17. Clar, E. J., Cambrai, M., Heid, E. and Grosshans, E. (1976). Les paramètres électrophysiologiques de l'épiderme dans le psoriasis. (Etude preliminaire). *Ann. Derm. Syph.,* **103**, 291
18. Tregear, R. T. (1966). *Physical Functions of Skin,* pp. 55–72. (London: Academic Press)
19. Yamamoto, T. and Yamamoto, Y. (1976). Dielectric constant and resistivity of epidermal stratum corneum. *Med. Biol. Eng.,* **14**, 494
20. Pethig, R. (1979). *Dielectric and Electronic Properties of Biological Materials.* (Chichester: John Wiley & Sons)
21. Lewis, T. J. (1977). The dielectric behaviour of non-crystalline solids. In *Dielectric and Related Molecular Processes,* vol. 3, pp. 187–218. The Chemical Society Specialist Periodical Reports. (London: The Chemical Society)
22. Macdonald, J. R. (1963). Transient and temperature response of a distributed, thermally activated system. *J. Appl. Phys.,* **34**, 538
23. Yamamoto, T. and Yamamoto, Y. (1977). Analysis for the change in skin impedance. *Med. Biol. Eng. Comput.,* **15**, 219
24. Rosendal, T. (1945). Concluding studies on the conducting properties of human skin to alternating current. *Acta Physiol. Scand.,* **9**, 39

30
A theoretical and experimental study of the optical properties of *in vivo* skin

J. W. FEATHER, J. B. DAWSON, D. J. BARKER and
J. A. COTTERILL

INTRODUCTION

Many studies have been made of skin colour[1–3] but to date few of these techniques have been applied clinically; when some index of colour change is required an arbitrary subjective grading system is usually used.

We propose that by presenting the spectrum of light reflected from the surface in an appropriate form, the pigments and structures within skin which give rise to its colour may be identified and quantified. This approach is based on a simplified model of skin as a layered, light transmitting, absorbing and scattering structure and on a parameter which is the logarithm to the base 10 of the inverse of the reflectance (LIR) of the skin. From the wavelength dependence of the parameter (i.e. the LIR spectrum) indices may be derived which correlate closely with subjective impressions of erythema, blanching and pigmentation. In the model it is assumed that three layers containing fibrous protein, melanin and haemoglobin respectively, lie on a base of collagen and fat (Figure 30.1). These layers are assumed to transmit, absorb and scatter light in accordance with formulae originally developed to describe the optical properties of diffusely reflecting powders[4]. From these formulae it may be shown that when scattering within a layer is small, and where T is transmittance, R is reflectance, k is 2 (fraction of light absorbed/unit path length in layer) and d is the thickness of layer, then $T \rightarrow e^{-kd}$ and $R \rightarrow 0$.

The intensity (I) of light reflected from a layered structure is the sum of the contributions from each layer:

$$I = I_0(R_1 + T_1^2R_2 + T_1^2T_2^2R_3 + T_1^2T_2^2T_3^2R_4)$$

where I_0 is the incident intensity.

In the simplest case it is assumed that the reflectance of the upper three layers approaches zero, hence:

$$I = I_0T_1^2T_2^2T_3^2R_4$$

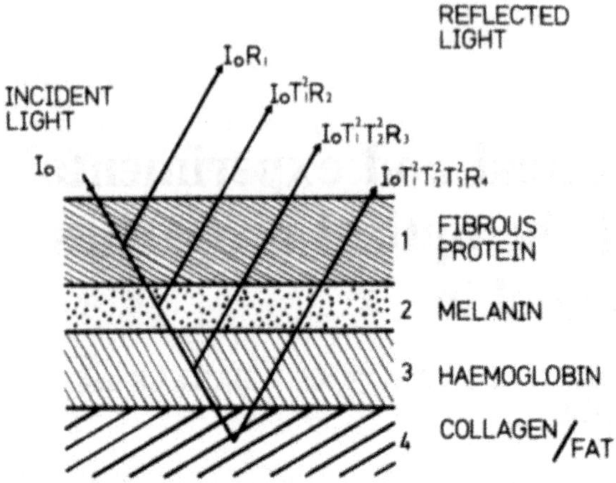

Figure 30.1 Simplified model of skin structure. For key see text

and thus the reflectance of skin is: $R_s = \dfrac{I}{I_0} = T_1^2 T_2^2 T_3^2 R_4$

It follows that the value of LIR for skin *in vivo* is approximately $\log(T_1^2 T_2^2 T_3^2 R_4)^{-1}$. By substituting $T = e^{-kd} = 10^{-A/2}$, where A is a parameter corresponding to absorbance in transmission spectrophotometry, a simplified expression for the LIR of skin (L) is obtained:

$$L = A_1 + A_2 + A_3 - \log R_4$$

where A_1, A_2 and A_3 are the absorbance equivalent parameters for the layers of fibrous protein, melanin, haemoglobin respectively, and R_4 ($\geqslant 0.9$) is the reflectance of the collagen and fat layer. According to this expression we have a system in which the logarithm of the inverse of the reflectance is simply the sum of the absorbances of the skin's constituent layers. To examine the valdity of this system it was necessary to construct an instrument for the measurement of the LIR spectra of skin *in vivo*.

INSTRUMENTATION

A schematic diagram of the instrument is shown in Figure 30.2. Fibre optics transmit light from a tungsten filament lamp to the skin and then to a monochromator for spectral analysis. The fibre optics are mounted in a blackened metal viewing head supported on a counter-balanced arm. The head rests gently on the skin with the fibre optics at an angle of 45° to the surface to avoid direct specular reflection. The transmitting and receiving fibres were

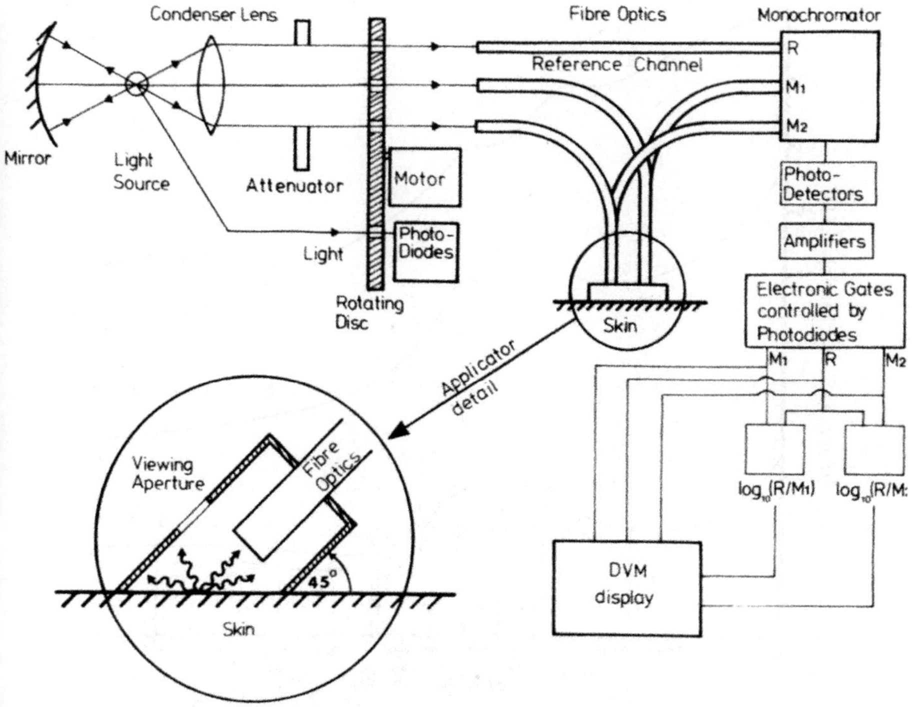

Figure 30.2 Diagram of reflectance spectrometer. R = Reference signal. m_1, m_2 = Signals received from skin

randomly mixed at the end proximal to the skin thereby ensuring that the same area of surface is both illuminated and viewed. Electronic signal processing gives a digital display and printout directly in LIR units.

The instrument has an operating range of about 300 nm (450–750 nm) and is trolley-mounted for use in clinics. The reproducibility of the instrument has been assessed *in vitro* with coloured tiles and *in vivo* by repeating measurements, at 1-minute intervals, on the same patch of skin. In the former case, LIR values were repeatable to within 0.005 units, while in the latter the repeatability was 0.01 units.

OBSERVATIONS AND DISCUSSION

The LIR spctra of the skin of subjects with differing degrees of melanization are shown in Figure 30.3. There are two notable features of these spectra: firstly, the peak lying in the 530–590 nm region of the spectrum, which may be attributed to haemoglobin; and secondly, the change in the underlying slope of the curve with change in the melanin content of the skin. By measuring the LIR spectra of suitable areas of skin containing large amounts of other

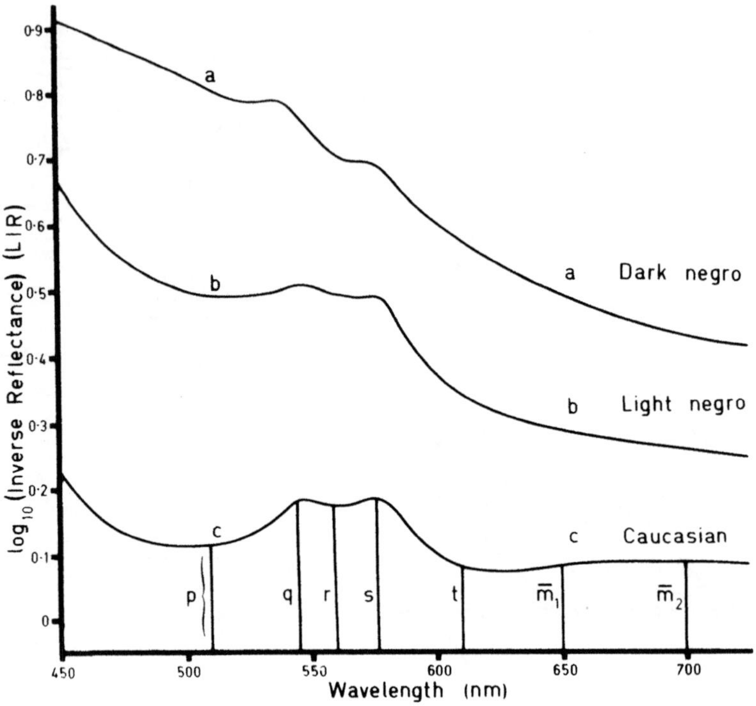

Figure 30.3 LIR spectra of *in vivo* human skin

pigments and by means of a mathematical spectrum-stripping procedure, the *in vivo* spectrum of those pigments, such as bilirubin and carotene, may be deduced. The results of such calculations are compared with the absorbance spectra measured *in vivo* in Figure 30.4; in the case of melanin no satisfactory *in vitro* spectrum was available. The general agreement between these *in vivo* and *in vitro* spectra justified further investigation of the possible uses of LIR spectra as a means of quantifying the pigment content of skin. We have used the haemoglobin absorption peak as a means of quantifying erythema. The proposed erythema index (E) is proportional to the area under this peak and above a baseline created by joining the points on the spectrum at 510 and 610 nm; it is calculated from the LIR values at 510, 543, 560, 576 and 610 nm (p, q, r, s and t respectively) as follows:

$$E = 100\,(r + 1.5(q + s) - 2.0(p + t))$$

To quantify the melanin content of skin an index is proposed which uses the gradient of the LIR spectrum in a spectral range where melanin is the dominant contributor to the LIR value. This index reduces the effect of measurement conditions and instrument response. The melanin index (M) is calculated as the difference between the mean of the LIR values at 645, 650 and

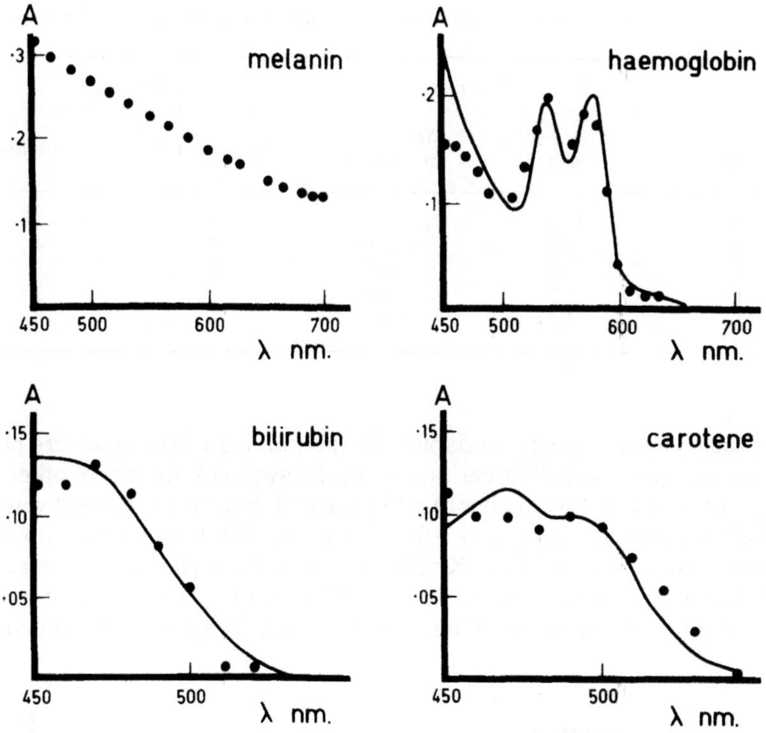

Figure 30.4 *In vitro* absorbance (———) and *in vivo* LIR (● ● ● ● ●) spectra of skin pigments

655 nm ($\overline{M}_1$) and that of values at 695, 700 and 705 nm, ($\overline{M}_2$):

$$M = 100(\overline{M}_1 - \overline{M}_2 + \alpha)$$

where α is an instrumental correction term (0.015). The information provided by the erythema and melanin indices was compared with subjective assessment in trials where subjects were treated to produce erythema, blanching and melanization.

Blanching was induced using topically applied corticosteroids and the erythema index was used to quantify the haemoglobin present in the skin. Subjective assessment of the same areas was made by seven observers on a scale of 0 to ± 6; positive corresponding to reddening and negative to blanching. The average results are presented in Table 30.1. As the erythema index and the subjective scales have differing numerical values, to facilitate comparison, the test areas were placed in rank order according to the degree of blanching as measured on each scale (1 = most blanched; 5 = least blanched). There is good agreement between the average subjective ranking and the instrumentally measured erythema index; this shows that the assessment by eye and the erythema index respond similarly to changes in the blood content of the skin. However, when the results for individual observers were

Table 30.1 Comparison of mean subjective and instrumental assessments of skin blanching

	Grading		*Ranking*	
Test area	*Erythema index*	*Subjective assessment*	*Instrumental*	*Subjective*
A	13.7	−3.6	3	3
B	14.2	−2.7	4	4
C	9.2	−4.7	1	1
D	9.2	−4.4	1	2
E	21.7	+0.9	5	5

examined there was greater variability in the ranking (three observers placed two or more areas in a different rank than the average). Results of other tests using coloured tiles (five shades of pink) showed that the instrument was able to distinguish small changes in colour more reliably than a single observer. The results were assessed using Kendall's rank correlation test; the correlation coefficient using the instrument was never less than 0.93 and in over half the trials absolutely correct ranking was achieved. Subjective assessments of

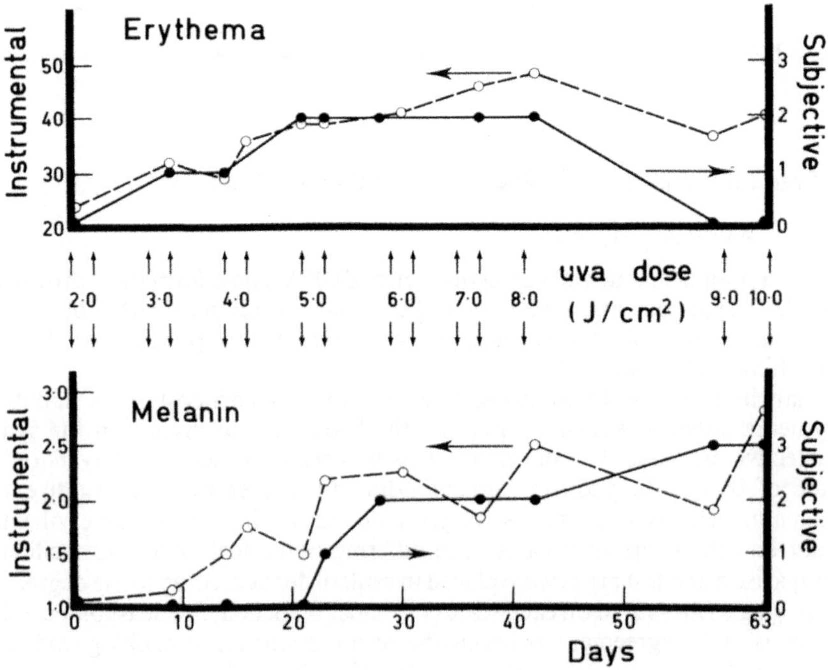

Figure 30.5 Subjective (—●—) and instrumental (--○--) assessment of erythema and melanization during 63 days' treatment with PUVA

ranking were made by eight observers on two occasions; the average ranking correlation coefficient was 0.48 with individual values ranging from 0.32 to 0.67.

The value of the instrumental approach in the quantifying skin colour was further demonstrated in a study of the changes occurring during PUVA treatment. In this experiment both erythema and melanin indices were measured and 'redness' and 'brownness' assessed subjectively. In Figure 30.5 the changes over 63 days of treatment are plotted. There are two aspects of the response curves which should be noted: (1) the general shape of the instrumental and subjective curves are closely similar; (2) the subjective assessments detected neither the early development of melanin pigmentation nor the residual erythema at the end of the treatment. The more sensitive assessment achieved by the instrument may be used for evaluating the dose response and effectiveness of the treatment in the individual patient.

CONCLUSION

Based on a simple theory of the optical properties of skin we have shown that objective clinical measurements of skin colour may be made and used to monitor treatment of patients. In particular, two indices have been proposed which may be used to describe the haemoglobin and melanin content of skin. The value of these indices has been demonstrated in trials of topical steroid blanching and PUVA therapy. The excellent reproducibility of measurements made by the instrument suggest that it is well suited to monitor the progress of treatment taking place over many weeks and months. Other applications of the instrument, such as observing the LIR spectra of diseased skin, are being investigated.

References

1. Edwards, E. A. and Duntley, S. Q. (1939). The pigments and colour of living human skin. *Am. J. Anat.*, **65**, 1
2. Monash, S. (1961). Skin colour. *Arch. Dermatol.*, **84**, 170
3. Wasserman, H. P. (1971). The colour of human skin. *Dermatologica*, **143**, 166
4. Kubelka, P. (1948). New contributions to the optics of intensely light scattering materials. Part I. *J. Opt. Soc. Am.*, **38**, 448

31
Dermatological photoacoustic spectroscopy

E. PINES and T. CUNNINGHAM

SUMMARY

Photoacoustic spectroscopy, a new and developing technique, has generated considerable interest as a spectroscopic tool for the studying of solid, semi-solid and biological samples. This technique allows the depth profiles of both the optical and thermal properties of skin to be determined in cases where other techniques can be applied only with difficulty.

Photoacoustic spectroscopy was used to measure the effects of hydration on normal and delipidized skin. The strength of the photoacoustic signal as a function of chopping frequency and, therefore, skin depth was shown to be dependent on type ('bound' or 'free') and amount of water present. The static and dynamic photoacoustic hydration data showed that: (1) as a function of skin depth the moisture content is non-linear; (2) skin regions underwent thermal transitions which were dependent on type and amount of water present; (3) a change in the distribution gradient of water in skin occurred with increasing moisture content; (4) normal and delipidized skin possessed vastly different hydration properties; and (5) the uppermost skin layers were more sensitive to a change in water content and hence suggested the importance of this region to dry skin. An *in vivo* photoacoustic spectrometer was described, its potential uses in dermatology discussed, and *in vivo* skin spectra presented.

INTRODUCTION

Skin can be thought of as an inhomogeneous, multilayered and somewhat opaque membrane comprising the stratum corneum, viable epidermis and dermis. Studying the thermal and optical properties of skin as a function of depth would be most informative. We show that this is precisely one of the capabilities of photoacoustic spectroscopy.

One of the most effective means for studying the properties of matter non-destructively is to observe how matter interacts with photons – that is, perform conventional optical spectroscopy. At present the two most common spectroscopic techniques are absorption and reflection spectroscopy.

In biology material must often be dealt with which in its intact unmodified state cannot be readily studied by these conventional techniques because of the sample's opacity, light-scattering properties, and surface characteristics. Recently, a new spectroscopic technique – photoacoustic spectroscopy (PAS) – which overcomes the drawbacks associated with opaque and highly effective light-scattering systems and allows spectroscopic investigation to be made *in situ* has been developed[1,2].

In PAS, the sample is placed in a sealed chamber, a PAS cell, which also contains a highly sensitive microphone and is filled with air at ambient temperature and pressure. The sample is irradiated with periodically chopped monochromatic light. If the sample absorbs any of the incident radiation some energy level in the sample is excited, and this energy level – because of the periodic nature of the incident radiation – must subsequently de-excite, usually by a non-radiative or heating mode.

The periodic input of light thus results in a periodic heating of the sample and subsequent periodic heat flow from the sample to the surrounding gas. The gas at the sample–gas interface responds to this periodic heat flow with an oscillating motion that produces a periodic pressure change in the sealed PAS cell. The microphone in turn detects this pressure change as an acoustical signal which is then processed electronically and recorded.

Unlike conventional spectroscopy where the strength of the signal is governed only by the optical properties of the sample, in PAS the strength of the signal is also governed by heat flow parameters – that is, the thermal properties of the sample. In general, the PAS signal is a complex function of the thermal and optical parameters of the sample and of their magnitude and phase relative to each other[1].

To illustrate the capabilities of PAS, we will arbitrarily divide matter into two categories. The first category consists of material which is optically transparent (Figure 31.1). There is not much to be gained from PAS when considering the full thickness of optically transparent material. However, as indicated previously, the strength of the PAS signal is a function of both the optical absorption coefficient (β) – which governs how much heat is generated

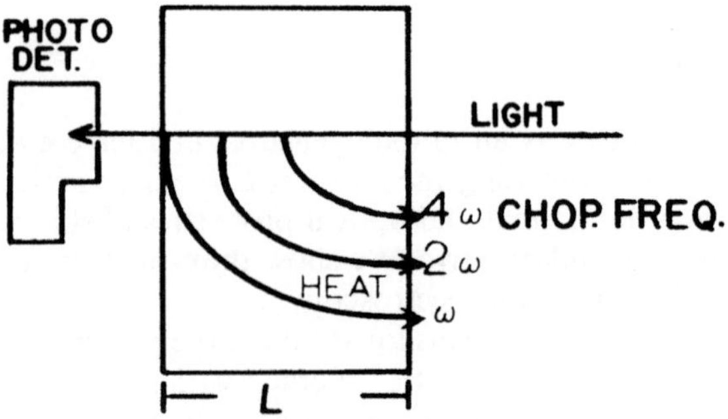

Figure 31.1 Schematic of optically transparent materials and depth profiling. For key see text

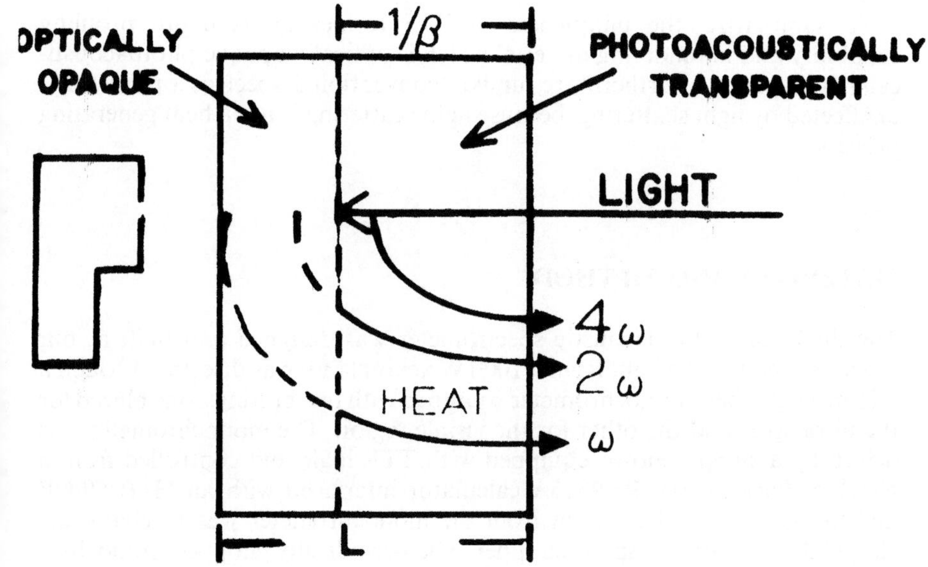

Figure 31.2 Schematic of optically opaque material made PAS transparent. For key see text

within the sample – and the thermal properties of the material which govern what portion of the heat generated within the sample will reach, within the allotted time, the sample–gas interface. These give rise to the photoacoustic effect. Thermal diffusion length[1] (μ_s), the important thermal parameter, which determines how far a periodic heat wave can travel in a solid before dissipating, is given by $\mu_s = (2\alpha/\omega)^{1/2}$; where α is the thermal diffusivity, and ω is the frequency of the heat wave, and in our case simply the frequency at which the light beam is being chopped. Thus by varying the chopping frequency, one can vary how far the heat wave can travel within the sample before dissipating and hence vary the thickness of the sample responsible for the photoacoustic effect. By chopping at a frequency ω information identical with that from conventional spectroscopy is obtained (information on full-thickness material; (Figure 31.1); and by further increasing the chopping frequency (2ω, 4ω etc.) thermal as well as optical information associated with thinner and thinner sections of the intact sample can be obtained. The depth profile of both the optical and thermal properties of a material can therefore be determined.

The second category is that of an optically opaque material (Figure 31.2). The entire light beam is absorbed within the material and hence conventional spectroscopy is inapplicable. However, by using PAS the chopping frequency can be increased to the point where the thermal diffusion length is less than the inverse of the optical absorption coefficient ($\mu_s < 1/\beta$); in that region the material, although optically opaque, has been rendered photoacoustically transparent. Therefore, as before the depth profile of both the thermal and optical properties of the material within the photoacoustically transparent region, can be obtained.

To summarize, the uniqueness of PAS resides in its depth profiling capability and in rendering material which is optically opaque photoacoustically transparent. Furthermore, unlike 'conventional spectroscopy' PAS is unaffected by light scattering, because light scattering is not a heat generating process.

MATERIALS AND METHOD

The dual-beam photoacoustic spectrometer was designed and built at our research facility. The light from a 1000 W xenon lamp was directed through a 0.25 m $f/3.5$ Ebert monochrometer equipped with two gratings, one blazed for the ultraviolet and the other for the visible region. The monochrometer was driven by a stepper motor equipped with TTL logic and controlled from a Hewlett–Packard (H–P) 9825A calculator interfaced with an H–P 6940B multiprogrammer. The output from the monochrometer was mechanically chopped by a variable-speed chopper. The periodically chopped monochromatic light beam was split into two beams by a reflective beam splitter. The beams were directed into two photoacoustic cells, one containing the sample and the other containing carbon-black. The acoustic signal generated in each cell was detected with a high-sensitivity microphone whose output was processed by preamplifiers and magnitude- and phase-sensitive lock-in amplifiers (one for each cell). The outputs from the lock-in amplifiers were directed to a ratio meter operating in the A/B mode–i.e. sample signal/carbon-black signal. Hence the sample's signal was normalized against the lamp power spectrum at all wavelengths to provide a normalized spectrum. The ratioed output was processed by the multiprogrammer and calculator to yield the photoacoustic spectrum which was plotted on a digital plotter.

For the hydration study, the skin (stratum corneum and/or full-thickness skin) was placed in a dry box for 48 hours and subsequently cut into 12 samples, with pairs of samples equilibrated in closed constant humidity chambers of 11, 33, 52, 75, 86 and 93 % relative humidity over saturated salt solutions for 24 hours. From each pair one sample was used for a gravimetric determination of water content (mg H_2O/mg dry skin), while the other was placed in the photoacoustic cell and its photoacoustic signal measured at 300 nm.

RESULTS AND DISCUSSION

To fully characterize the physicochemical properties of skin, it is essential to know the amount, type and distribution of water within it. We believed that PAS and its dependence on the thermal properties of the material, which are in turn affected by the degree of hydration, was uniquely suited for determining the state of water in skin.

Static water measurement

Samples equilibrated to known constant water content.

Figure 31.3 shows the strength of the PAS signal as a function of the gravimetrically determined water content in stratum corneum. The shape of the curve can be readily understood in terms of the photoacoustic theory[1,2]. Conditioning the skin to less than about 0.15 mg H_2O/mg dry skin results in only small changes in the cohesive forces maintaining the stratum corneum's rigidity[3]. Results from differential scanning calorimetry[4] as well as nuclear magnetic resonance studies[5] show that as water is initially added to the stratum corneum it goes into 'bound' sites and, in particular, into the matrix micropores. Since this 'bound' water is tightly bound to specific sites and minimally affects the matrix's rigidity, its effects on the thermal properties are small. With further hydration (>0.15 mg H_2O/mg dry skin), the matrix is plasticized, the cohesive nature of the stratum corneum is compromized, the additional sorbed water behaves more like bulk or 'free' water[2,4,5] and hence the thermal properties change very rapidly.

By using the theory and some preliminary calibration the water content in biological systems, such as skin, can thus be determined not only *in vitro* but also *in vivo*[6,7].

The depth-profiling capability can also be used to determine the amount, type and distribution of water as a function of skin depth merely by varying the chopping frequency. Using this method, curves as shown in Figure 31.4 can be generated, which show the change in the strength of the PAS signal as a

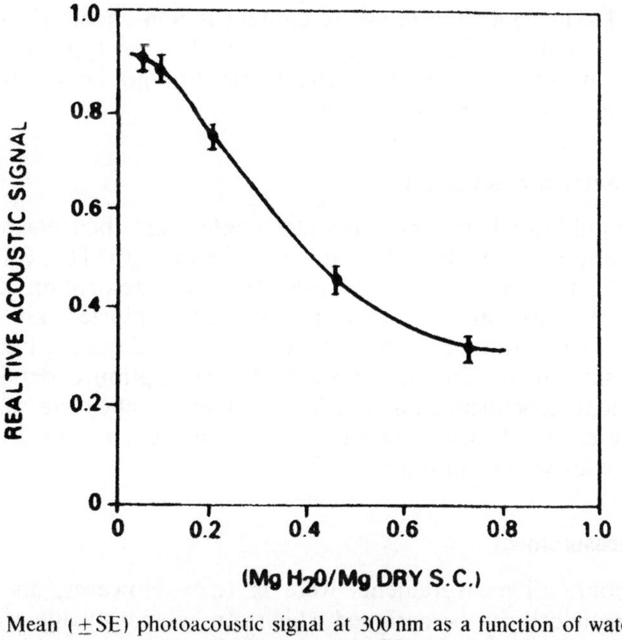

Figure 31.3 Mean ($\pm$SE) photoacoustic signal at 300 nm as a function of water content in stratum corneum (SC)

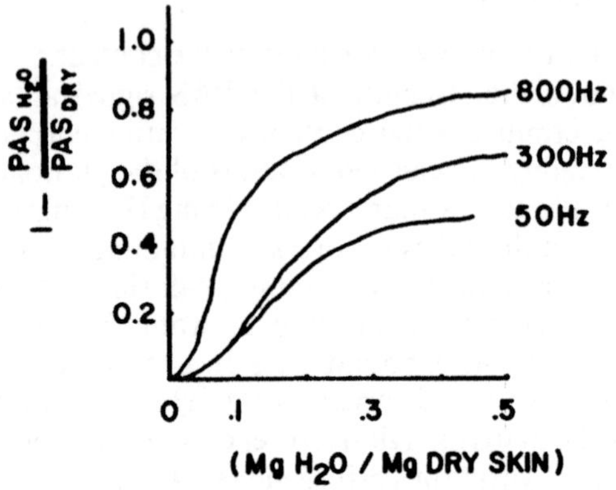

Figure 31.4 Depth profile of photoacoustic signal as a function of water content and chopping frequency of full-thickness skin

function of water content and chopping frequency. Interestingly, the uppermost skin layers, associated with 800 Hz, are the most sensitive to a change in water content (Figure 31.4). These findings are important because they show that the uppermost layers, the ones associated with the dry-skin syndrome, have disproportionately less 'bound' water than deeper layers and hence their thermal properties change more rapidly with hydration.

To summarize, the data given in Figures 31.3 and 31.4 show that: (1) as a function of skin depth the moisture content is non-linear; (2) skin regions undergo thermal transitions which are dependent on type and amount of water present; and (3) a change in the distribution gradient of water in skin occurs with increasing moisture content.

Dynamic water measurement

Samples equilibrated to a known water content are then placed in a PAS chamber containing Dri-Rite. The change in the strength of the PAS signal as a function of elapsed time is then recorded to yield a desorption curve. Figure 31.5 shows the desorption curves for normal and delipidized skin (a model for dry skin) as a function of initial water content and chopping frequency and, therefore, skin depth. The figures show that desorption is depth as well as water-content dependent and that lipids either enhance the water binding and/or are essential in maintaining a macromolecular organization which accommodates water more intimately.

In vivo measurement

To this point all measurements were *in vitro*. However, an open-ended photoacoustic cell can be designed which can be acoustically sealed to skin and thereby yield *in vivo* spectra on optically opaque human skin. Figure 31.6

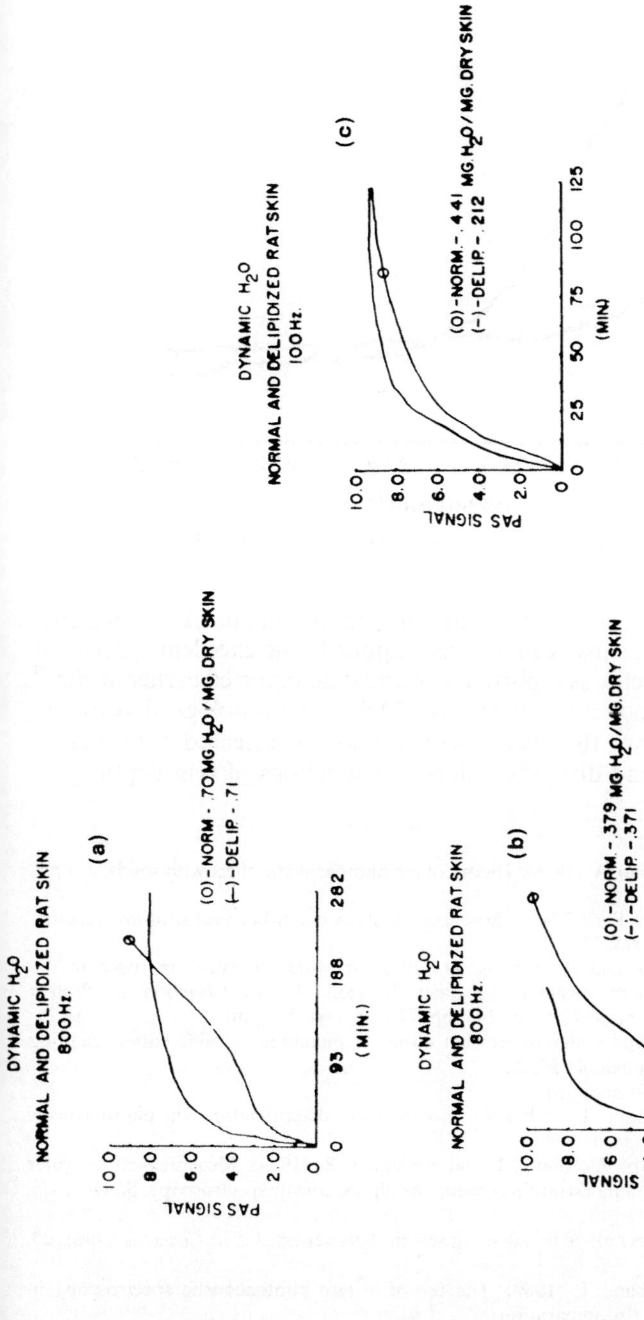

Figure 31.5 Photoacoustic desorption spectra (PAS signal against elapsed time) for normal (----) and delipidized (——) rat skin at different water contents and chopping frequencies. (a) At 800 Hz; normal skin 0.70 mg H_2O/mg dry skin, delipidized skin 0.71 mg H_2O. (b) At 800 Hz; normal skin 0.38 mg H_2O, delipidized skin 0.38 mg H_2O. (c) At 100 Hz; normal skin 0.44 mg H_2O, delipidized skin 0.21 mg H_2O

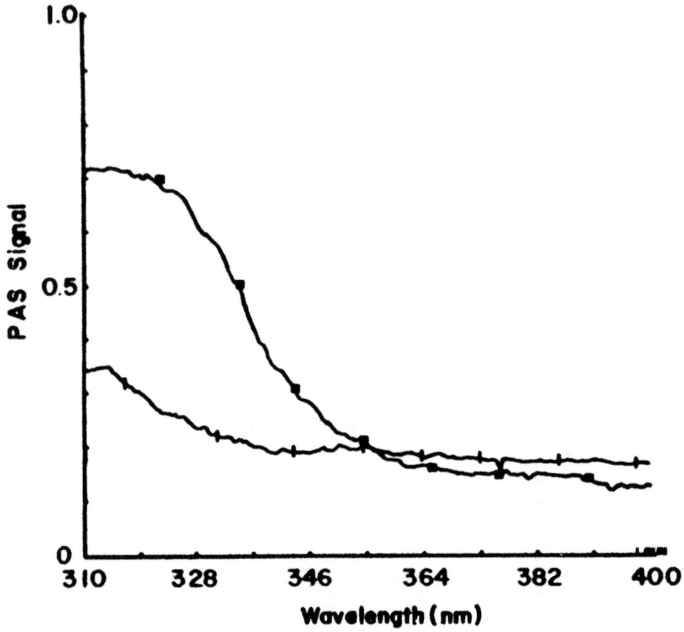

Figure 31.6 *In vivo* photoacoustic spectra for normal (|) and sunscreen treated human skin (› (as on graph))

shows examples of an *in vivo* PAS spectra of human skin and skin to which a commercially available sunscreen has been applied. The excellent agreement between the *in vivo* spectra as reported here and data reported earlier *in vitro*[8] clearly illustrate the potential of *in vivo* PAS in dermatological research. Specifically, PAS allows the spectroscopic study of diseased skin, percutaneous absorption[9] and drug disposition[9] as functions of skin depth.

References

1. Rosencwaig, A. and Gersho, A. (1976). Theory of the photoacoustic effect with solids. *J. Appl. Phys.*, **47**, 65
2. Rosencwaig, A. and Pines, E. (1977). A photoacoustic study of newborn rat stratum corneum. *Biochim. Biophys. Acta*, **493**, 10
3. Wildnauer, R., Miller, D. and Humphries, W. (1975). A physicochemical approach to the characterization of stratum corneum. In Baier, R. (ed.). *Applied Chemistry at Protein Interfaces. Advances in Chemistry series 145*, pp. 74–125. (Washington:)
4. Walkley, K. (1972). Bound water in stratum corneum measured by differential scanning calorimetry. *J. Invest. Dermatol.*, **59**, 225
5. Pines, E. Unpublished observations
6. Pines, E. and Cunningham, T. (1980). *In vivo* moisture determination via photoacoustic spectroscopy. (In preparation)
7. Simon, I., Emslie, A., Apt, C., Blank, I. and Anderson, R. (1979). Measurement of water concentration profile in human stratum corneum by photoacoustic spectroscopy. *Bioeng. Skin*, **2**, 50
8. Pines, E. (1978). A new technique to assess sunscreen effectiveness. *J. Soc. Cosmetic Chem.*, **29**, 559
9. Pines, E. and Cunningham, T. (1980). The use of *in vivo* photoacoustic spectroscopy in pharmaceutical science. (In preparation)

32
Ellipsometric measurement of skin refractive index *in vivo*

P. T. PUGLIESE and A. J. MILLIGAN

SUMMARY

This technique measures two angles (ψ and Δ) of a light beam, and the changes in these angles because of the surface reflection. ψ is a trigometric ratio function of the parallel and perpendicular beam amplitudes (relative to the plane of incidence). Δ is the related phase shift between these two orthogonal beam components. After the beam strikes a surface these two values change according to the electrical properties of the material. Measurement of these two parameters can be accomplished by various techniques and in this investigation measurement was performed using phase-modulated ellipsometry. These two values can be used to determine the real and imaginary components of the refractive index which in turn describes both the light-bending and light-absorption properties of the medium. With continued experimentation this instrument will enable non-invasive detailed exploration to be made of skin thickness, changes due to sun damage, and effectiveness of various cosmetic preparations.

INTRODUCTION

The use of refractive index measurements to characterise organic solutions is well established in biochemistry. The measurement of serum protein and urinary protein concentrations by refractive index methods is a routine procedure in medicine. The measurement of protein and water concentration in living cells by the refractive index method was developed by Barer and Joseph[1], their technique being essentially a modification of the Becke line-immersion method[2]. The purpose of this study was to extend the use of refractive index measurements to the skin surface to determine moisture content of the epidermis *in vivo*. Since we were unable to use the immersion technique, a non-invasive optical method was adopted. No attempt will be made to explain the methodology in great detail as there are many references available for this purpose. We shall provide the rationale for our method and present our data as a starting point for other investigators interested in this method.

RATIONALE OF METHODOLOGY

The fundamental equation relating refractive index to protein concentration with the immersion method is:

$$n - n_{\mathrm{m}} = \alpha C$$

where n is the refractive index of the measured protein, n_{m} is the refractive index of the immersion medium, α is the specific refractive increment – which is defined as the increase in refractive index per 1% increase in concentration (0.00185) – and C is the concentration of the protein in 100 ml solution. This concentration can be converted to 100 g protoplasm (c) by:

$$c = \frac{C - C^2}{400}$$

With this formula it is possible to extend these meaurements to refractive index determined by other methods if we assume that the body protein is dissolved in water and that the effect of dissolved salts and other organic compounds is negligible compared with the protein concentration.

The reported refractive index of the stratum corneum is 1.55, which represents the refractive index of the dried, flaky horny layers[3]. If we take the refractive index of water as 1.33, then:

$$1.55 - 1.33 = 0.00185C;$$

$$C = 118.92\ \mathrm{g/100\ ml\ solution}$$

Obviously there is no moisture present and this refractive index represents keratin protein. If, however, there is some moisture on the skin surface, or the cells contain some moisture, the refractive index should lie between 1.55 and 1.33. To determine the refractive index of skin, we chose to use reflected light with ellipsometric techniques.

METHODOLOGY

The skin is classified as a dielectric, by most biologists, and as such is subject to the laws of reflection and refraction. One of the major problems encountered in reflectance studies of the skin is the scattering phenomenon. Light interaction with a non-absorbing molecule induces an oscillation in the outer electron shells of the same frequency as the incident light but a different phase. These vibrating electrons emit secondary light waves which interfere with each other in a lateral direction but not in the forward direction. This produces the effect known as scattering. In the forward direction, these secondary waves combine with the primary wave producing a phase change which results in decreased velocity. This decrease in velocity compared to the velocity of light in a vacuum is defined as the refractive index (n):

$$n = c = \frac{\text{speed in vacuum}}{\text{speed in medium}}$$

While polarized light does not eliminate scatter, it does reduce this phenomenon. For this reason, we chose ellipsometric techniques to measure the refractive index. Ellipsometry is the branch of optical spectroscopy that deals with the measurement and interpretation of the change in polarization state in terms of the physical properties of the element[4].

Ellipsometry measures the change in the state of polarization as a result of reaction with a surface and interprets it in terms of the reflecting surface. The two parameters measured are the change in relative amplitude, ψ, and the relative change, Δ, of the two orthogonal components of the reflected light. Monochromatic linearly polarized light is directed onto the skin with the direction of vibration at 45° to the plane of incidence. The elliptically polarized reflected light passes through a quarter wave plate which is adjusted until the light energy from it becomes linearly polarized, as evidenced by its being extinguished by a second polarizer. Measuring the angular setting of both the quarter wave plate and the polarizer at the extinguished condition allows the orientation and ellipticity of the reflected light to be determined. This is classic ellipsometry and allows properties of the reflecting surface to be deduced[5–7]. This method, however, is slow, tedious and requires long and complicated equations for final values. New methods have been developed that take advantage of solid-state electronics[8,9].

Our method was essentially a modification of the ellipsometric technique developed by Jasperson and Schatterly[10] and reported elsewhere[11]. Monochromatic light passes through a plane polarizer oriented at 45° to the plane of incidence (Figure 32.1). The output of the polarizer is directed into a Morvue photoelastic modulator, consisting of a piezoelastic crystal oscillating at 50 kHz. This modulator constantly changes the phase angle between the orthogonal components of the polarized light beam. The modulation level is set so that this angle varies from −138° to +138° thus greatly easing the otherwise voluminous calculations. The output of the modulator passes through the collimator side of the ellipsometer onto the skin surface of interest. The reflected light enters the telescope side of the spectrometer and is directed through a second polarizer (analyser) onto an EMI Gencom number 6129 photomultiplier tube (PMT). The output of the PMT enters the signal side of a PARC number 5101 lock-in amplifier while a reference signal of 50 kHz from the modulator enters the reference side. All signals are normalized to DC light levels to eliminate changes caused by line voltage fluctuation.

The values for the ellipsometric parameters ψ and $\cos \Delta$ are determined by intensity readings from the lock-in amplifiers and are entered into a programmed TI-59 computer to yield the real and imaginary components ($n + \kappa$ respectively) of the refractive index.

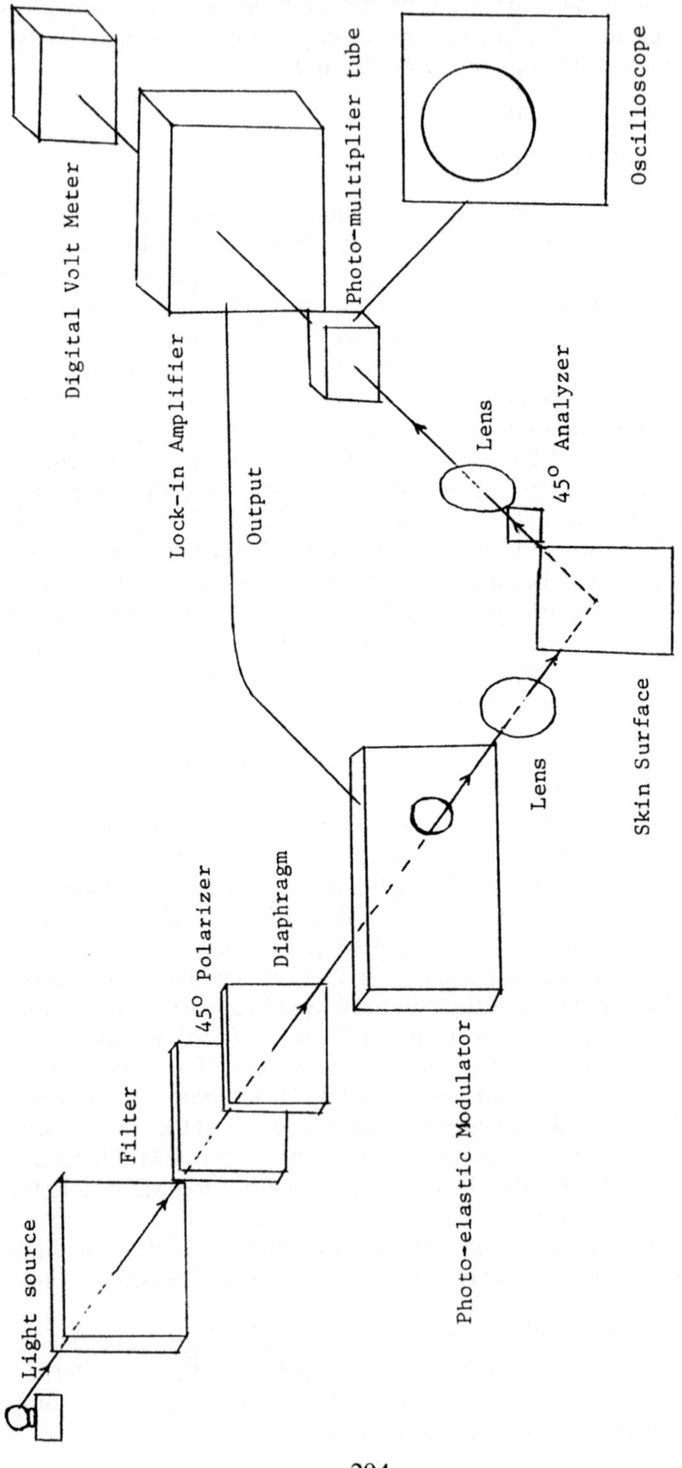

Figure 32.1 Schematic diagram of instrument components

Subjects were men and women aged between 20 and 80 years with no obvious disease. Measurements were made on the volar surface of the forearm. All topically applied materials were either over-the-counter preparations or commercial raw materials.

RESULTS

Figure 32.2 shows the refractive index *n* value as a function of age for an incident angle of 45° and a wavelength of 540 nm. The expression obeys the relationship:

$$n = 0.0054A + 0.633$$

where n is the real component of the refractive index and A is the subject's age in years. This equation shows that the normal refractive index increases as the subject's age increases.

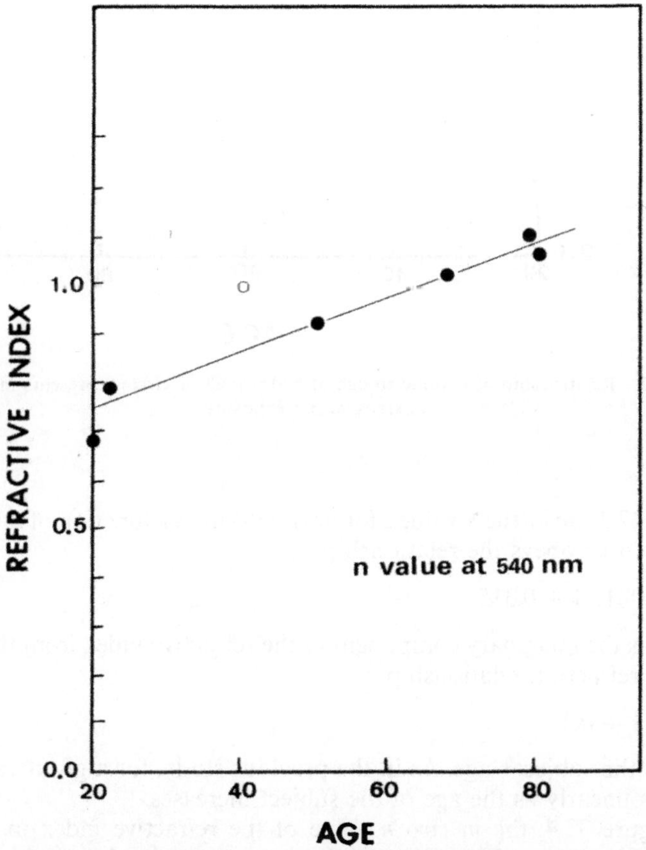

Figure 32.2 Relationship of refractive index to age at 540 nm (7 subjects). ○ = value in individual with extensive sun exposure

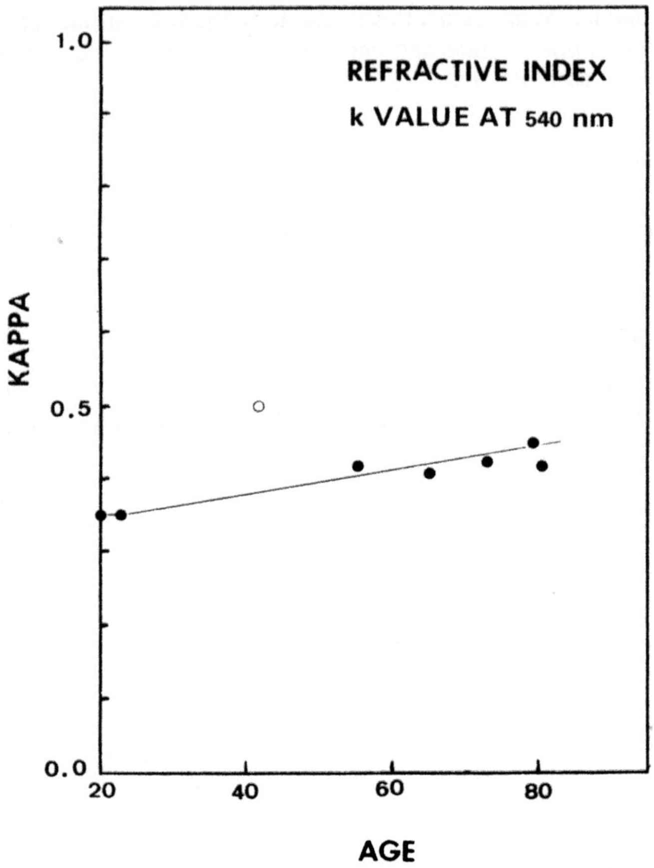

Figure 32.3 Relationship of κ value to age at 540 nm. Clear circle represents individual with extensive sun exposure

Figure 32.3 shows the κ values, for normal skin as a function of the subject's age. The curve obeys the relationship:

$$\kappa = 0.0019A + 0.315$$

where κ is the imaginary component of the refractive index from the complex index of refraction relationship[12].

$$n = n(1 + i\kappa)$$

and A is the subject's age. As in the previous study, for a given value of n, κ increases linearly as the age of the subject increases.

In Figure 32.4, the *in vivo* n value of the refractive index in three skin conditions is shown. Normal conditions represent the skin of the forearm at ambient temperatures and humidity of ten healthy subjects. No creams,

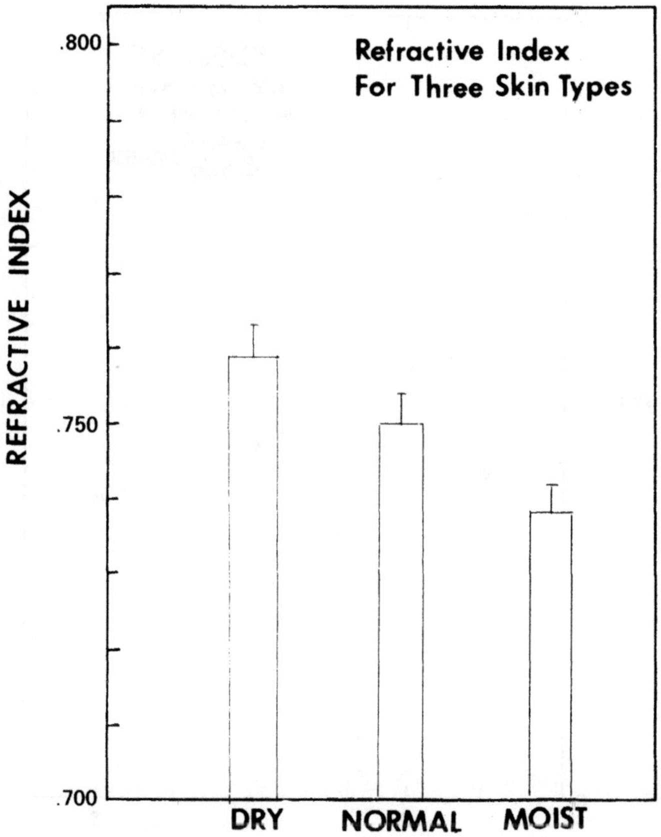

Figure 32.4 Mean ($\pm$SE) refractive index measurements in three hydration states of the stratum corneum ($n = 10$)

lotions, or other cosmetic preparations were used for at least 12 hours before the study.

'Dry' conditions were created by placing the arm in a heated dessication chamber for 10 minutes with a temperature of 100 °F and a relative humidity of 20 %. 'Moist' conditions were created by placing the arm in a 98.6 °F water bath for 10 minutes and then lightly towel-dryed.

The n value of the refractive index decreased under moist conditions and increased under dry conditions as compared with the control. These values represent a change in water–protein ratio and thus a change in the real component of the refractive index.

Measurements made with applied cosmetic material were slightly more difficult because of changes in reflectivity associated with application of the topical agent. For this reason values of n were normalized by the κ value, since κ is a measurement of the absorption characteristics of the material. The normalization eliminates the error introduced by applying highly reflective substances.

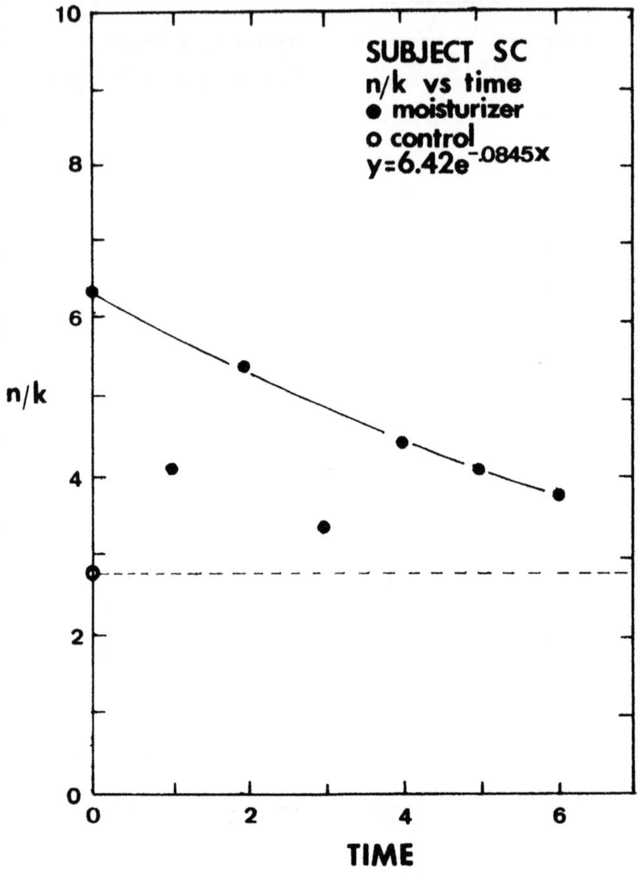

Figure 32.5 n/κ values of the stratum corneum in the presence of a commercial moisturizer in one subject. ○ = Control value

Figures 32.5, 32.6 and 32.7 show results generated from studies on biopsy samples in one subject. This subject had an over-the-counter moisturizer labelled with fluoroscein applied to his forearm. Ellipsometer measurements at 560 nm were made at this site at various times after application. Immediately after the measurement was performed a 5 mm biopsy sample was taken from the measured site and examined under a microscope. Figure 32.5 shows the n/κ value versus time after applying this moisturizer. This indicates an increase in the level of moisture caused by the product in this time and a subsequent decrease in the n/κ value. This high control n/κ value was due to the addition of the fluoroscein label, but the trend of the decrease in n/κ shows changes in the moisturizer status.

Figure 32.6 shows the depth of penetration of the moisturizer as a function of time after application measured by microscopic examination of the biopsy

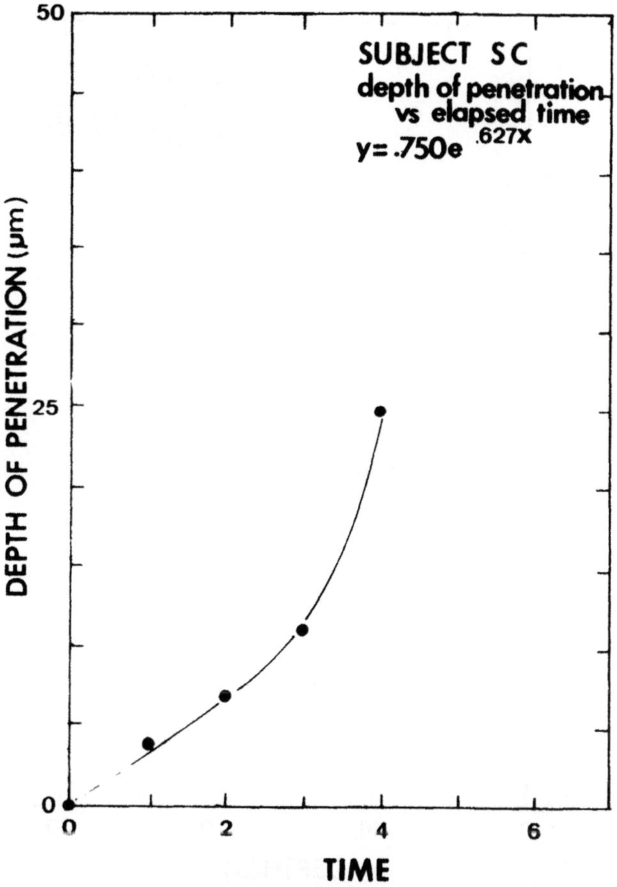

Figure 32.6 Penetration of fluoroscein-labelled moisturizer into epidermis in one subject. Serial biopsies were performed on the forearm and the depth of penetration determined microscopically

samples. After four hours no line of demarcation existed, hence it was no longer possible to determine the depth of penetration of the moisturizer. When the n/κ value was plotted as a function of depth of penetration (Figure 32.7), the resulting curve was biphasic with a break occurring at 10 μm. This implies that the maximum effective depth of penetration of the light beam used in this experiment (560 nm) was 10 μm and that once the material reached this depth additional changes in the refractive index were not detectable.

DISCUSSION

Changes in the refractive index suggest a change in the composition of the measured material[13]. In this case, we believe this change is in the water–

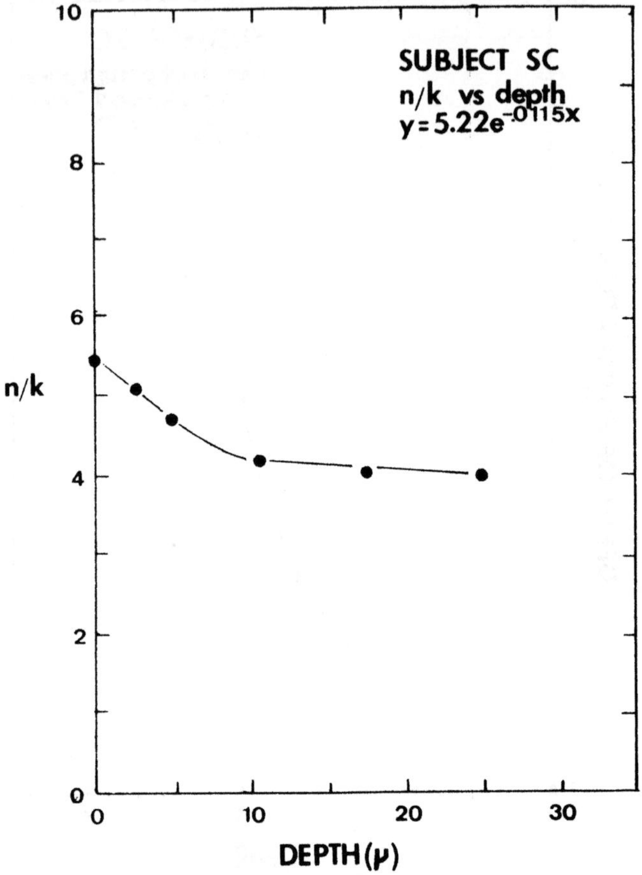

Figure 32.7 Relationship of n/κ values to depth of penetration of moisturizer in one subject

protein ratio. The n/κ ratio for the equation which defines both the n and κ value is:

$$n/\kappa = \frac{0.0054A + 0.633}{0.0019A + 0.335}$$

and this yields the expected normal n/κ range as:

$$2.01 \leqslant n/\kappa \leqslant 2.85$$

for A between 0 and infinity respectively. As a first approximation these values seem reasonable for a normal range of n/κ and values of n/κ greater than 2.85 would suggest 'dry skin' while values of n/κ less than 2.01 would indicate overly hydrated skin.

There are several questions raised by this study which remain to be answered. Firstly, assuming that the reflected light measured is a surface phenomenon and that penetration did not exceed 10 μm, why should there be a difference with age if only the refractive index of stratum corneum is measured? Secondly, it has been shown by Kligman's group that aged skin contains more moisture per unit than young adult skin[14]; why then should the refractive index increase with age? If more moisture is present, it should go down. Thirdly, one 40-year-old subject had a refractive index (Figure 32.2) equal to that of a 68-year-old subject. The first subject was an avid sunbather, having spent his youth tanning his skin for long hours for many sunny days on the beaches of Southern California. There are no reported occult biochemical

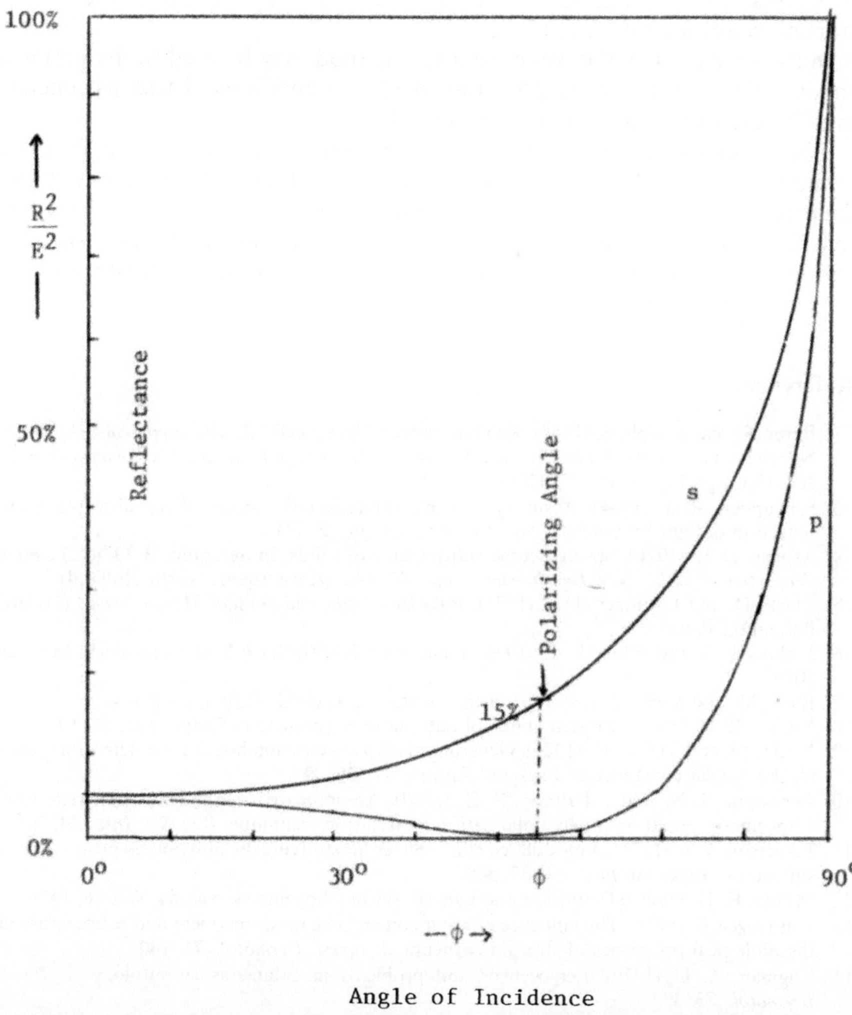

Figure 32.8 Change of reflectance with change of angle of incidence and type of incident polarized light. s represents perpendicular polarized light and p parallel polarized light

changes in the epidermis because of sunlight that are demonstrable 10 years after exposure, while such changes in connective tissue in the dermis have been reported[15].

It appears that we are measuring reflected polarized light from the 'body of the skin'; light that has been transmitted and refracted through many layers of cells and partially reflected back to the surface. This is similar perhaps to the 'pile of plates' type of polarization. Measurements made on the skin by reflectance techniques must take into account the angle of incidence and state of polarization of both the incident or reflected light (Figure 32.8). While our study is preliminary and raises more questions than it has answered, we believe that accurate, meaningful measurements of the complex refractive index of skin can be made and related to the physiological status of the body. Optical constants of the skin remain to be completely determined but will serve as a starting point for future studies.

In its present state of development the method may be used to determine n and κ values with reliable precision in given conditions. These parameters may be used to study many aspects of skin.

The finding of a refractive index below unity suggests that the skin is a semiconductor rather than a true dielectric, though there are obvious dielectric properties present. This technique further suggests an increased sensitivity to the presence of free electrons in the skin. The possible connections between these parameters and various states of health warrant further investigation.

References

1. Barer, R. and Joseph, S. (1954). Refractometry of living cells. *J. Microscopical Sci.*, **95**, 399
2. Needham, G. H. (1977). *The Practical Use of the Microscope Including Photomicrography*, p. 204. (Springfield: Charles C. Thomas)
3. Scheuplein, R. J. (1964). A survey of some fundamental aspects of the absorption and reflection of light by tissue. *J. Soc. Cosmetic Chem.*, **15**, 111
4. Aspnes, D. E. (1976). Spectroscopic ellipsometry of solids. In Seraphin, B. O. (ed.) *Optical Properties of Solid: New Developments*, pp. 801–842. (Amsterdam: North-Holland)
5. Clark, D. and Grainger, J. E. (1971). *Polarized Light and Optical Measurement.* (Oxford: Pergamon Press)
6. Jenkins, F. A. and White, H. E. (1976). *Fundamentals of Optics*, 4th ed. (New York: McGraw Hill)
7. Born, M. and Wolf,, E. (1975). *Principle of Optics.* (Oxford: Pergamon Press)
8. Muller, R. H. (1976). Present status of automatic ellipsometers. *Surface Sci.*, **56**, 19
9. Malin, M. and Vefam, K. (1976). Generalized ellipsometric method for the determination of all the optical constants of a system. *Surface Sci.*, **56**, 49
10. Jasperson, S. N. and Schatterly, S. E. (1969). An improved method for high reflectivity ellipsometry based on a new polarization modulation technique. *Rev. Sci. Inst.*, **40**, 761
11. Jasperson, S. N. (1973). A modulated ellipsometer for studying the film optical properties and surface dynamics. *Surface Sci.*, **37**, 548
12. Muller, R. H. (1969). Definition and convention in ellipsometry. *Surface Sci.*, **16**, 14
13. Van Duzee, B. (1978). The influence of water content, chemical treatment and temperature on the biological properties of stratum corneum. *J. Invest. Dermatol.*, **71**, 140
14. Kligman, A. L. (1979). Perspectives and problems in cutaneous gerontology. *J. Invest. Dermatol.*, **73**, 39
15. Lariker, R. M. (1979). Structural alterations in exposed and unexposed aged skin. *J. Invest. Dermatol.*, **73**, 59

33

Influence of thermal properties and epidermal thickness on pain threshold in conductive heating

A. M. STOLL, M. A. CHIANTA and J. R. PIERGALLINI

SUMMARY

Measurements of time to pain threshold during contact of bare skin with materials at various increased temperatures were made to provide maximum permissible temperatures for safe contact in given circumstances, determinable solely from a knowledge of the thermal properties of the material. The profound influence of epidermal thickness on the heat flow during contact led to a method for determining this thickness, given the initial temperature of the material productive of threshold pain within the first five seconds of contact.

We concluded that the system used in deriving the data is valid, the epidermal thickness may be determined within $\pm 10\%$, and a simple method for performing this measurement may be developed from the present data.

INTRODUCTION

Earlier work[1] described the prediction of material temperature which would produce a blister after momentary (0.3 s) contact with skin of minimal epidermal thickness and verified this prediction by using a good conductor (aluminium). The present report shows similar results with Masonite, a good insulator, and with materials of intermediate properties. These verifications again confirm the validity of the method used in relating pain and burn thresholds to heat transfer through skin of various epidermal thicknesses and provide a means of broadening our knowledge of the dimensions and physical properties of skin.

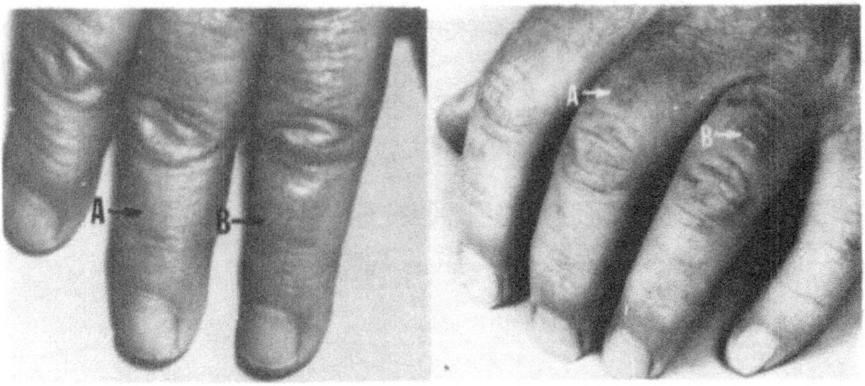

I II

Figure 33.1 Verification of blister effect in contact with aluminium (I), and with Masonite (II) at predicted material temperatures (132 °C and 195 °C (respectively). I: (A) At 132 °C for 0.2 s, transient erythema; (B) At 132 °C for 0.45 s, full blister, II: (A) At 194.4 °C for 0.29 s, discoloration without blister; (B) at 194.6°C for 0.35 s, blister slightly larger than threshold

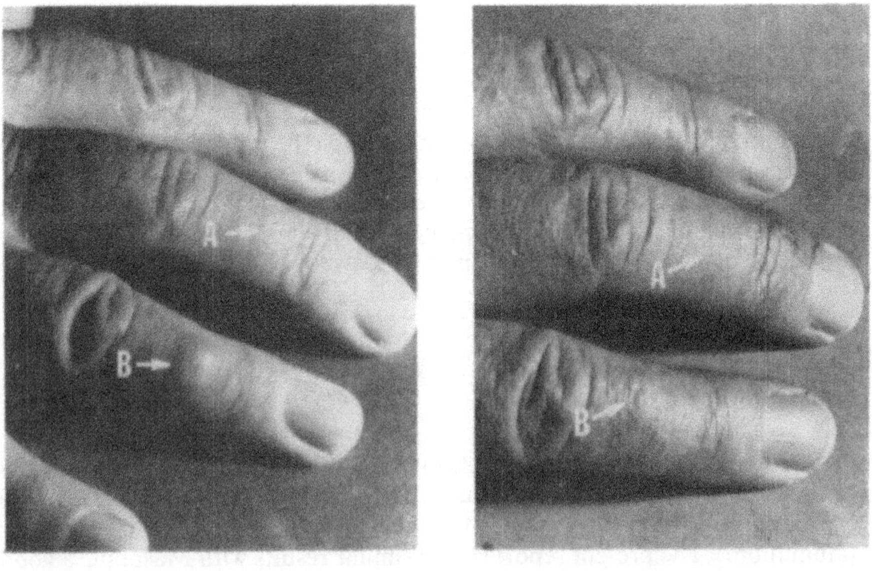

I II

Figure 33.2 Verification of blister effect in contact with Hercuvit (I), and with glass (II) at predicted material temperatures (149 °C and 159.6 °C respectively). I: (A) At 149.3 °C for 0.2 s, transient erythema; (B) at 149.6 °C for 0.29 s, full blister; II: (A) At 157 °C for 0.26 s, transient erythema; (B) at 158 °C for 0.3 s, threshold blister

EXPERIMENTAL DATA

The experimental apparatus and method have been described[1]. Briefly, six materials representing the entire range of thermal properties – from good conductors to good insulators – were heated to various temperatures; measurements were then made of the time to pain threshold during contact of the skin with each specimen. Pain threshold time was converted to time-to-blister and the resultant values were used to predict material temperatures which would produce blistering after 0.3 s of contact[2].

Figures 33.1 and 33.2 show the experimentally produced blisters, verifying the predicted temperatures for the four materials used. The predicted temperatures were usually correct within 1 °C – considerably better than might be expected, since the basic data may vary by as much as $\pm 6.6\%$ of the mean in areas of minimal epidermal thickness in any one individual[3].

After validation of the procedure, all the experimental data were analyzed and arranged to correlate temperatures of the materials with the thermal properties of the materials as functions of the pain threshold. As seen in Figure 33.3, the co-ordinates used were those described by Wu[4]: the temperature of

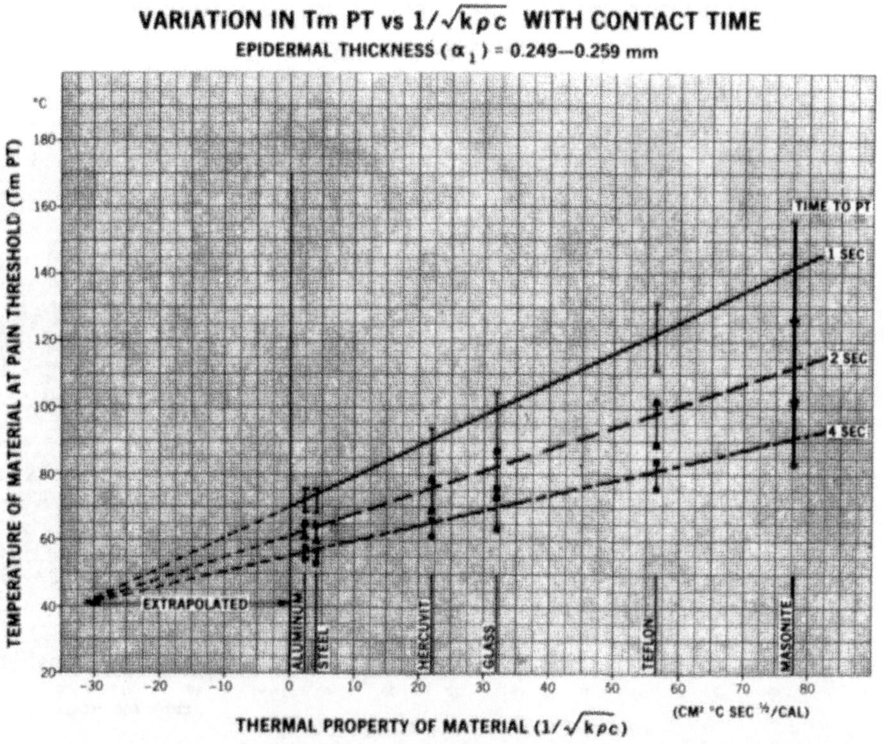

Figure 33.3 Temperature and thermal inertia ($k\rho c$) as functions of pain threshold (PT) for materials in contact with skin of given epidermal thickness (0.249–0.259 mm) at three contact times (1, 2 and 4 s)

the material at pain threshold (TmPT) against the properties expressed as the reciprocal of the square root of thermal inertia (which is a product of conductivity, density and specific heat, $k\rho c$). The data shown refer to a single thickness of epidermis (α_1) and three different contact times, the range markers indicating that all averaged data fall within $\pm 10\%$ of the mean value for TmPT throughout the range of material properties. As the properties approach those of the best insulator, the TmPT rises and four-fold increase in contact time lowers the TmPT of the best insulator (Masonite) by 50 °C. Extrapolation of these lines to a point of convergence yields coordinates of $-31.5\ \mathrm{cm^2\ °C\ s^{1/2}\ cal^{-1}}$ on the abscissa and 41 °C on the ordinate (y intercept on the translated x axis), which will be used in subsequent calculations.

Similarly, Figure 33.4 presents the data, expressed in the same co-ordinates, showing variation in TmPT with variation in epidermal thickness. A five-fold increase in thickness results in an increase of 40 °C in the TmPT for the same material, Masonite, at a single exposure time of 3 s. This effect occurs because of the increase in heat pathway from the surface of the skin to the level of the pain receptors in the dermis as the epidermal thickness increases.

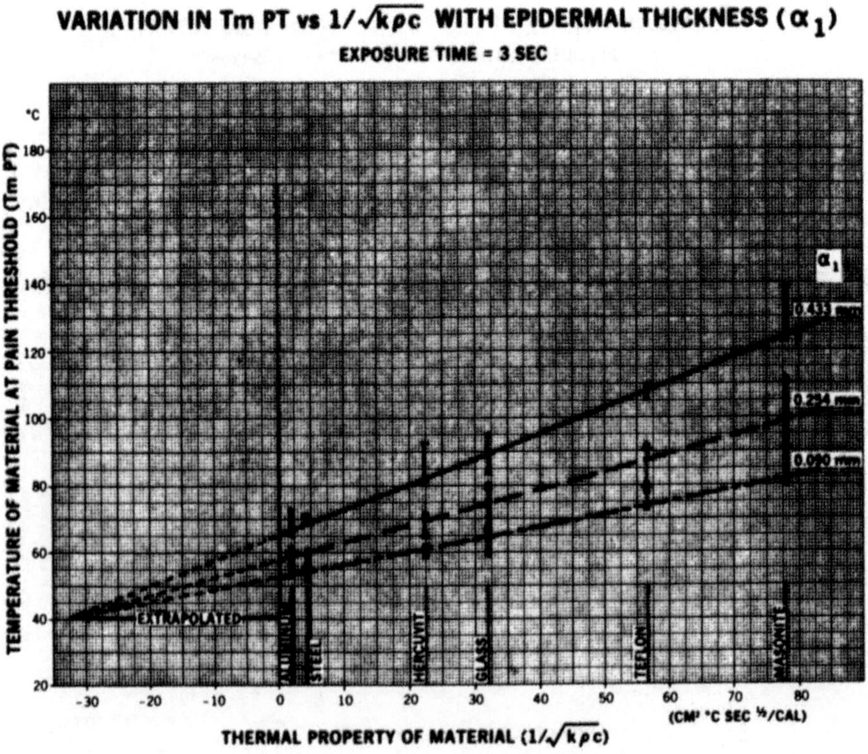

Figure 33.4 Temperature and thermal inertia ($k\rho c$) as functions of pain threshold for materials in contact with skin at a given contact time (3 s) at three epidermal thicknesses (α) (0.090, 0.254 and 0.433 mm)

Extrapolation of these lines to the point of convergence yields the same coordinates as in Figure 33.3. Again, the range markers indicate that all data lie within $\pm 10\%$ of the mean. More detailed treatment of these data appears elsewhere[3].

RESULTS AND DISCUSSION

The experimental verification of the predicted material temperatures for blistering validates the extrapolation of the radiation and convection data on which the predictions were made. These data are shown in Figure 33.5 as tolerance time to the endpoints of pain and blister at energy input levels determined experimentally between 0.1 and 0.4 cal cm^{-2} s^{-1} and extrapolated to 1.2 cal cm^{-2} s^{-1}. All of the conduction blister data and much of the pain threshold data fall within the extrapolated area of this chart so that confirmation of time-to-blister derived from measured time-to-pain also confirms the level of energy absorbed in each instance. Thus the rate of conductive heat absorption during contact with each material is made available by these comparisons; this information, together with the measured

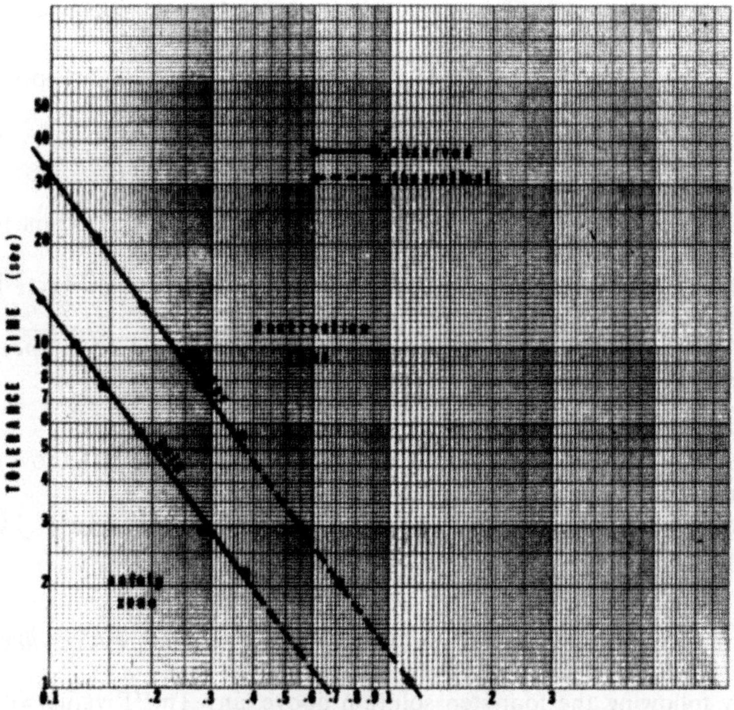

Figure 33.5 Threshold pain and blister parameters, shown as tolerance time against energy absorption rate

interface temperatures, provides the means for calculating the coefficient of heat transfer in each skin contact.

The present data relating material properties and temperature at pain threshold may be described by straight lines in rectilinear co-ordinates, which greatly simplifies their use in the practical application on this knowledge to the definition of thermally safe surfaces for contact with bare skin[3]. Another consequence of this straight-line relationship is that, within the experimental limits in which the data hold, the thermal inertia ($k\rho c$) of the skin must be constant if the $k\rho c$ of the material is constant. These data may then be used to derive the thermal inertia for the skin from surface to pain receptor level within the first 5 seconds, when neither dessication nor vasodilation interfere. Preliminary results indicate a value in the vicinity of 1.0×10^{-3} cal^2 cm^4 s °C^2, a reasonable figure to be explored more thoroughly[3].

Applying these data to the non-destructive estimate of epidermal thickness is of more interest. As described[3], the slopes of the lines in Figures 33.3 and 33.4 yielded families of straight lines for different contact times when plotted in semi-log co-ordinates against epidermal thickness, and the slopes and intercepts of these semi-log lines yielded straight lines when plotted against contact time in log–log co-ordinates. Thus all the data may be expressed mathematically by using three forms of the equation of a straight line as seen below, where they are presented in the order of solution for calculation of TmPT from given data on contact time (t), epidermal thickness (α_1) and thermal properties of the material contact ($1/\sqrt{k\rho c}$).

The equation form for the log–log plot of intercepts and slopes against contact time is:

$$y = ax^b$$

where y is either the intercept or slope of the semi-log curves, x is time (s) and a and b are constants found by least squares fit.

Hence, where yIII is the intercept (1), $a = 0.490$ and $b = -0.412$; where yII is the slope (2), $a = 1.094$ and $b = -0.184$.

The equation form for the semi-log plot of slope (TmPT v $1/\sqrt{k\rho c}$) against α_1 is:

$$\log y = mx + \log b \qquad (3)$$

where y is the slope TmPT against $1/\sqrt{k\rho c}$, m is the slope (yII above), x is α_1 (mm) and b is the intercept (yIII above).

The equation form for the rectilinear plot of TmPT against $1/\sqrt{k\rho c}$ is:

$$y = mx + b \qquad (4)$$

where y is TmPT (°C), m is slope (anti-log of $\log y$ above), x is $1/\sqrt{k\rho c} + 31.5$ (cm^2 °C s$^{1/2}$ cal^{-1}), and b is 41 (°C).

By following the four-step solution above, any TmPT value within the experimental limits of $t = 1$–5 s, $\alpha_1 = 0.09$–0.50 mm and $1/\sqrt{k\rho c} = 2$–80 cm^2 °C s$^{1/2}$ cal^{-1} may be regenerated and checked against observed data as plotted in Figures 33.3 and 33.4.

To apply the above system to skin, the order of solution may be rearranged to yield a solution for epidermal thickness given a TmPT of a material of known properties and a contact time within 1–5 s.

To calculate α_1 where PT is reached in 3 s in contact with Masonite at a temperature of 82 °C:

Substituting in equation (1),
$y\text{III} = 0.490\ (3)^{-0.412} = 0.3116$

Substituting in equation (2), $x = t = 3$;
$y\text{II} = 1.094\ (3)^{-0.184} = 0.894$

Substituting in equation (4), $x = 1/\sqrt{kpc} + 31.5) = (77.8 + 31.5)$ and $y = \text{TmPT} = 82$, $b = 41$, and solving for m:
$82 = m(109.3) + 41;\ m = 0.375.$

Finally, substituting in equation (3), where y is the slope (0.375), m is yII above (0.894) and b is yIII above (0.3116) and solving for $x = \alpha_1$:

$$\log 0.375 = 0.894x + \log 0.3116$$
$$-0.4260 = 0.894x + (-0.5064);$$
$$x = 0.090\,\text{mm} \qquad \text{(compare with Figure 33.4).}$$

The accuracy of this system does not exceed ±10%, since this is the overall experimental error for the observed data[3]. However, as suggested earlier[1], in a clinic where skin grafting procedures are carried out routinely it would be a relatively simple matter to accumulate direct measurements of epidermal thickness from skin specimens in which measurements of pain threshold had been made before removal from the donor sites. Many such observations might improve the present accuracy, although it is also possible that the accuracy of such measurements might prove to be no better. In any event, within the limitations noted, α_1 could be readily standardized with respect to specific areas of interest by applying the pain threshold measuring procedure as described to a population of both sexes over a wide range of age, occupation, disease condition, body surface location, etc.

For the specific purpose of measuring α_1 a standard simplified technique may be developed using an apparatus to provide a surface of a material of known thermal properties, maintained at a given constant temperature and applied to the skin site at a controlled, reproducible pressure. It would still be necessary to assure that the initial temperature of the skin falls within ±0.5 °C of 32.5 °C so that the present data would be applicable.

Alternatively, depending on the extent of concern and need for such information, a similar set of data could be generated at different initial skin temperatures (T_0) to provide more flexibility. Theoretically, it should be possible to adjust the present data for variations in T_0 with appropriate heat transfer equations but this course would require considerable effort to validate the calculated results by experimental observations – a procedure which experience has shown[5] is vital to the reliability of such data.

References

1. Stoll, A. M. (1977). Thermal properties of human skin related to nondestructive measurement of epidermal thickness. *J. Invest. Dermatol.*, **69**, 328
2. Stoll, A. M., Chianta, M. A. and Piergallini, J. R. (1979). *Thermal Conduction Effects in Human Skin: II. Experimental validation and application of data in selection of materials.* Phase Report No. NADC-79034-60
3. Stoll, A. M., Piergallini, J. R. and Chianta, M. A. (1979). *Thermal Conduction Effects in Human Skin: III. Influence of epidermal thickness and exposure time.* Phase Report No. NADC-79035-60
4. Wu, Y-C. (1972). Material properties criteria for thermal safety. *J. Materials*, JMLSA, **7**, 573
5. Stoll, A. M. and Greene, L. C. (1959). Relationship between pain and tissue damage due to thermal radiation. *J. Appl. Physiol.*, **14**, 373

34
Quantification of thermoregulatory mechanisms in relation to psoriasis: thermometry, thermography, and venous flow studies

T. G. WARSHAW, H. ALEXANDER and G. CURTIS

SUMMARY

Thermal measurement techniques were applied to the study of psoriasis. Evidence is presented for several thermal characteristics of psoriasis associated with thermoregulation.

(1) There was an increased thermal gradient and possibly an increased thermal emissivity in the psoriasis lesion.
(2) The response to a 0 °C cold challenge test was greatly diminished in cases of psoriasis of the hand.
(3) The response to heat challenge of 44–46 °C was the same in normal and in psoriatic subjects.
(4) Preliminary studies on vasomotor response in psoriasis were inconclusive.

Hence, psoriasis is associated with several abnormalities in thermoregulation.

INTRODUCTION

The cause of psoriasis is unknown. With multiple technologies now available, temperature regulation in psoriatic and normal subjects was investigated. The complexity of thermoregulatory physiology has been a formidable barrier to its study. Skin behaves variably in its several roles in temperature regulation, so simple explanations of skin disorders in thermoregulation are unlikely.

Previous work has shown that temperature findings in psoriatic subjects were different from observations in normal subjects[1]. Whether the aberrations are symptoms, or the underlying defect itself, will ultimately be resolved.

To understand the role of skin in normal and psoriatic subjects, surface temperature, ambient temperature, thermograms, and responses to thermal challenges such as 0 °C and 44 °C were recorded. Doppler techniques were used to study venous blood flow in hands of psoriatic and normal subjects.

An extensive review of thermoregulatory physiology would be encyclopaedic, and wholly outside the scope of this discussion; only a minimal outline will be presented.

METHOD AND RESULTS

Ten cases of psoriasis were studied from July 1978 to June 1979. A normal control subject was studied with each psoriatic patient.

Daylight photographs were taken and afforded ready comparisons with thermograms. A thermistor thermometer (Gulton Industries) measured skin temperature at selected sites on dorsae of hands and on dorsae of fingers. In nine of 10 cases, the thermistor measurements were lower in psoriatic lesions than in the unaffected adjacent skin by 1.33 ± 0.86 °C. The time required to make a thermistor reading ranged from 14 to 50 s.

Thermograms were taken in a room with an ambient temperature of 17–18 °C with no wind velocity, but at unmeasured humidity. Thermograms document the infrared radiation emitted. Since the skin behaves as a black body it is almost totally emissive of heat, and thermograms accurately document temperature at the body surface. The Spectrotherm 2000 (General Electric) equipment we used converts infrared signals with a cadmium telluride cell into a visual image. Additionally, a permanent temperature read-out of one horizontal scanning line is provided. The thermograms of fingertips in nine of 10 psoriatics were at a higher temperature than in controls. An example of a hand thermogram is given in Figure 34.1.

The hands were subjected to thermal stresses[2]. Hands were immersed in ice water at 0 °C for 60 seconds as a cold challenge. Immersion in well mixed water serves as a heat clamp, enveloping and holding skin at a known temperature, Additionally, cold thermogenesis begins below 5 °C, since this is below the critical temperature at the lower end of the thermoneutral zone. Below the thermoneutral range, the heat of metabolism is partly dependent on air temperature.

After cold challenge, thermistor readings were made and thermography immediately repeated. In four psoriatic subjects, they were repeated again after 2 minutes and after 4 minutes.

After 0 °C cold challenge, fingertip temperatures in seven psoriatic subjects dropped. Temperatures were still higher than in the fingertips of controls (Table 34.1).

After cold challenge there was a temperature drop, but insufficient time elapsed to show the expected cold thermogenesis.

Heat challenge consisted of immersing the hand in hot water, usually at 44–46 °C, but at 52 °C water in one case. Subsequent thermograms showed no difference in response between psoriatic and normal subjects, all of whom showed sharply demarcated skin temperature elevation.

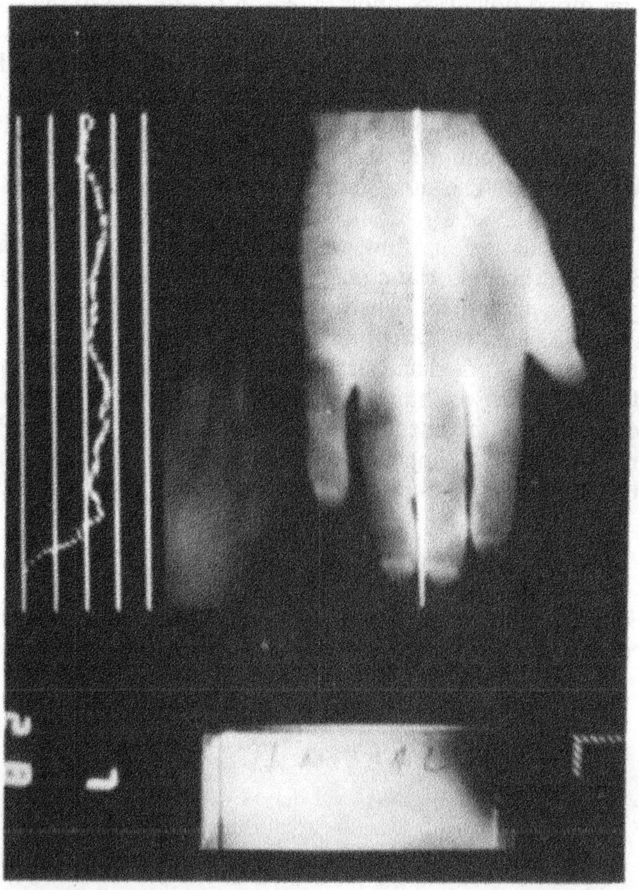

Figure 34.1 Thermogram of hand with psoriasis; result for hand of control subject shown in lower half of field. (White = hot; black = cold)

Table 34.1 Response to cold challenge test in seven psoriatic subjects and in normal subjects

Case No.	*Fingertip temperature* (°C) *At* 17–18 °C *ambient*	*After* 0 °C *ambient*	*Control*
1	28	19	<19
2	23	18	<18
3	29	25	<24
4	23	20	<19
5	32	18	<18
6	20	19	<19
7	25	22	<22

Studies of blood flow were undertaken on the dorsae of the hands of five psoriatic patients (two women and three men) and in three normal controls.

Ultrasonic Doppler flow detector equipment (Parks Electronic Laboratory Model B12, Beaverton, Oregon) was used for making recordings of hands of five psoriatic and three normal subjects. The equipment records flow velocity rather than the true volume flow rate, but the correlation is good enough for qualitative purposes. The area under the tracing on the record represents volume.

Large superficial venous flow was documented in two female patients (Figure 34.2). In three psoriatic males, the hand venous flow was below the sensitivity of the equipment. Their artery flow readings seemed to indicate vasoconstriction. Among the three normal controls, (two men and one woman) the woman studied had sizable superficial hand venous flow. Doppler studies of her hand showed hunting phenomenon after immersion in ice water. Lewis[3] described hunting phenomenon as rapid fluctuations from 1–8 °C after ice water immersion of fingers. It is an inefficient mechanism, probably mediated by a vasoactive peptide, of maintaining tissue temperature.

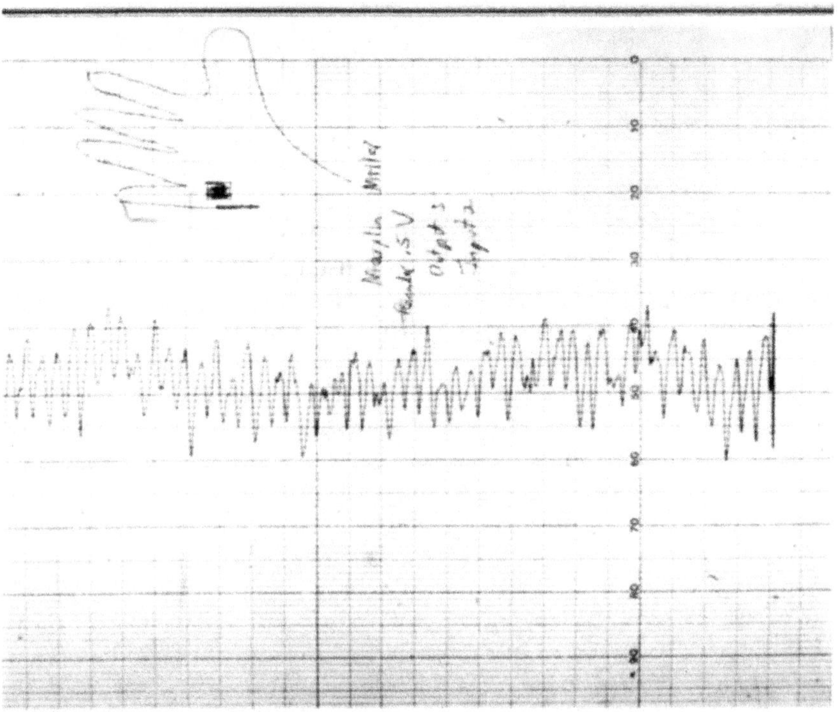

Figure 34.2 Doppler study of venous flow in female subject with psoriasis (case No. 5). Area under the tracing represents vein flow volume

AN OVERVIEW OF THERMOREGULATION

Any model of man as a homeotherm must take many facts into account. For one thing, skin is an interface between the environment and the body core. It incorporates thermal sensors of the environment, and probably also of body temperature. A skin temperature of 33 °C has been considered the neutral point from which displacements are perceived; these affect feelings of pleasure or displeasure[4,5].

Afferent thermal stimuli from skin, spinal column, viscera, and other anatomic sites converge and are integrated primarily at the preoptic nuclei in the hypothalamus[6]. This is considered the position of the set point, the thermostat for the body, and it begins appropriate responses for temperature regulation[7]; where hypothalamic autonomic response is suppressed, there is evidence of thermoregulatory centers situated further caudally in the brain[8].

The efferent stimuli, the next step in thermoregulation, go to muscles, which perform thermogenesis. At peak efficiency 30 % of muscle metabolism results in work and 70 % in production of heat[9]. Food, drink, and hormones, especially of thyroid, affect this process.

The skin then acts as a shell of almost total emissivity of heat[10]. As the outermost envelope of a closed-loop system, the skin participates in Newtonian heat transfer (by radiation, convection, and conduction) and in evaporative heat loss or gain by sweating and transepidermal water loss.

The energy budget for the body is derived from the first law of thermodynamics, which implies that the net rate of energy gain equals rate of energy storage.

The energy budget for the body[11] is:

$$M + W + K + R + C + E = S$$

where M is the metabolic rate or rate of chemical energy gain; W is the work rate or rate of energy gain from action of physical forces; K is the rate of conductive heat gain; R is the rate of radiant heat gain; C is the rate of convective heat gain; E is the rate of evaporative heat gain; and S is the rate of heat storage in the body: K, R and C constitute Newtonian heat transfer. At thermal equilibrium, $S = 0$.

The physical properties of the environment include temperature, relative humidity, and wind velocity; all enter the energy equation.

In addition to considering the components of the closed-loop system of human thermal regulation, and considering the energy budget equation, evolution of thermal physiology must be mentioned. Poikilotherms (animals that assume the temperature of the environment) maintain preferred body temperature by behavioral means. For example, the lizard basks on a sunny rock ledge to raise body temperature. Birds, with their high metabolic rate required for flying, have developed an autonomic nervous system for thermal regulation.

For an optimal temperature man uses behaviour such as environmental cooling or heating, sunbathing, and selection of clothing. Also, autonomic responses – metabolism, shivering, sweating and vasomotor responses – are constantly in operation.

Finally, the purpose of body temperature regulation comes into question. Is it brain temperature (or simply hypothalamus temperature), or mean body temperature that is being controlled? Could mean body temperature be regulated to achieve the desired control of brain temperature[12]?

DISCUSSION

Two sets of observations in psoriatic subjects are distinctive.

Firstly, psoriasis plaques are colder than surrounding unaffected skin as determined with contact thermometry. In the same subjects, thermograms of the same plaques are hotter than the surrounding skin; therefore, psoriatic skin must show an altered thermal gradient and must also be studied for possible altered thermal emissivity.

Thermography is a suitable technique for thermal emissivity studies. Materials of known emissivity can serve as markers for comparing skin emissivity over periods of time.

Secondly, psoriatic response to cold challenge is less than in normal people. In 7 of 10 subjects (at 17–18 °C ambient, and at 0 °C ambient), fingertips on thermograms were demonstrably hotter than in controls.

Where there is hand psoriasis, the subject does not adequately respond to exposure to cold.

In the extremities, peripheral vasomotor tone can be a response to local ambient temperature[13]. This is not a thermoregulatory response through the central nervous system. Open-loop local control can alter the diameter and volume of blood vessels to maintain a constant internal temperature. Thus there may be a fault in local vasomotor tone in psoriatic subjects, since their fingertips do not get as cold as do normal subject's when challenged with 0 °C. This possibility has led us to perform Doppler venous flow studies.

Two female psoriatic subjects had large venous flow, and three males had imperceptible venous flow. Obviously, it is too early to draw conclusions, but further investigation will be performed on blood flow.

The short outline of physiology of thermoregulation was intended as a framework for interpreting data.

Recent reports on increased transepidermal water loss is psoriasis[14,15] imply evaporative heat gain at the skin surface. In psoriasis, sweating is impaired; therefore, larger transepidermal water loss is a compensatory mechanism to maintain evaporative heat in the energy equation.

Out of our 10 patients, two were undergoing thyroid replacement therapy. Hypothyroidism was paralleled by poor thermoregulation, so coincidence of psoriasis and hypothyroidism in 2 of 10 of our subjects is of clinical interest.

Clearly, psoriasis is a relatively benign disorder because behavioral thermal regulation can compensate for many possible effects of the autonomic nervous system.

It remains to be seen whether psoriasis is caused by a variety of faults in thermoregulatory inputs and outputs, or whether these findings are only symptomatic of the disorder.

References

1. Warshaw, T. G. (1973). Thermal studies in psoriasis *J. Invest. Dermatol.*, **60**, 91
2. Warshaw, T. G.. and Lopez, F. (1979). Thermoregulation in Psoriasis. *Proc. Am. Therm. Soc.* (In press)
3. Lewis, T. (1930). Observations on the reactions of the vessels of human skin to cold. *Heart*, **15**, 177
4. Hardy, J. D. (1953–54). Control of heat loss and heat production in physiological temperature regulation. *Harvey Lectures*, **49**, 242
5. Cabanac, M. (1969). Plaisor ov displaisir della sensation, thermique et homeothermis. *Physiol. Behav.*, **4**, 359
6. Bligh, J. (1966). The thermosensitivity of the hypothalamus and thermoregulation in mammals. *Biol. Rev.* **41**, 317
7. Hammel, H. J. (1968). Regulation of internal body temperature. *Ann. Rev. Physiol.*, **30**, 64
8. Cabanac, M. (1974). In Shaw, D. R. (ed.) *MTP International Review of Science, Environmental Physiology*, p. 253. (London: Butterworth's)
9. Whipp, B. J. and Wasserman, K. (1969). Efficiency of muscle work, *J. Appl. Physiol.*, **26**, 644
10. Cohen, M. L. (1977). Measurement of thermal properties of human skin, a review. *J. Invest. Dermatol.*, **69**, 333
11. Mitchell, D. (1974). In Shaw, D. R. (ed.) *MTP International Review of Science, Environmental Physiology*, p. 3. (London: Butterworth's)
12. Mitchell, D., Atkins, A. R. and Wyndham, C. H. (1972). In Bligh, J. and Moore, R. E. (eds.) *Mathematical and Physical Models of Thermoregulation*, p. 37. (Amsterdam: Elsevier)
13. Bligh, J., Cottell, D. and Maskrey, M. (1971). Influence of ambient temperature on thermoregulatory response to 5HT, noradrenaline, and acetylcholine injected into lateral cerebral ventricles of sheep, goats, and rabbits. *J. Physiol.*, **212**, 337
14. Grice, K., Sattar, H., Baker, H. and Sharatt, M. (1975). The relationship of transepidermal water loss to skin temperature in psoriasis and eczema. *J. Invest. Dermatol.*, **64**, 313
15. Rajka, G. and Thune, P. (1976). The relationship between the course of psoriasis and transepidermal water loss, photoelectric plethysmography and reflex photometry. *Br. J. Dermatol.*, **94**, 253

35
The surface translucence meter: its use in studying human skin

C. W. HARGENS

SUMMARY

Surface translucence has applications in commerce and medicine. It influences the appearance of a substance, especially human skin, because it determines certain observable optical qualities of the material being viewed. *Surface* translucence implies optical effects occurring at a single surface, as distinct from light passing through from one side and being measured on the other. With this technique light is shone on the side that is of interest, and it must be internally diffused and reflected back to an adjacent area of the same surface. The quantity returned in this way is measured photometrically, and its magnitude depends on the ingredients of the material.

When the elements of the experiment just described are arranged so that both the input and the output – juxtapositional channels – are light-proof and each quantity measured reliably, a functional instrument is obtained. The ratio of output to input light as a percentage can be calculated and this quantity defined as surface translucence.

Certain surface effects are also measurable, because when the surface has opaque material its light is blocked partially or completely, depending on the optical density of the preparation.

A more elaborate instrument could include refinements in spectral illumination and measurement. That is, we know that the several physical mechanisms whereby light is transmitted from the point of surface entry through the material until it emerges as wavelength sensitive. Thus a complete expression of surface translucence must show how it varies with wavelength.

SURFACE TRANSLUCENCE METER

Basic concept

The outer layers of animal skins are translucent; in fact, primitive peoples used them for windows to admit light. Human skin shows this translucent property, although it is not thought of as such, and its color and fine features are perceived as inherent in the surface. In places where large vessels are evident because of their blue color, as in the palm or side of the wrist, it is clear that the image coming from inside, well beneath the surface, must be seen by virtue of this translucence. Skin could almost be said to be transparent, and indeed some people's skin does seem to have that quality to a remarkable extent.

Light falling on the skin's surface penetrates these translucent layers by varying amounts, is partially absorbed or scattered and some is reflected back to emerge again. This, of course, is what is seen, and it shows some of the details of what is underneath the skin. For example, the absorption in vein structures noticeably discloses their outlines. Less resolved images of capillaries impart color in the same way as redness.

When high intensity light is directed at the skin and the returned fraction observed in certain ideal viewing conditions, these effects become strikingly enhanced.

Scheme of measurement

It is not difficult to imagine a measurement that could show the differences in skin translucence, if they exist. It should be possible to show that a material is not translucent and blocks the passage of light. Thus there will be a scale of intermediate values resulting from degrees of translucence.

I have defined a 'surface translucence' as light entering and returning from the surface. This is distinguished from translucence from one side through to the other. In single-surface techniques the light must be returned by scattering, which is caused mainly by reflection of light by particles. In through translucence the particles impede the passage of light as well as reflecting it. There will also be particles which may impede, to some extent, the reflected light with single-surface techniques.

The apparatus used in measuring translucence is described below. A light projector concentrates light on the surface of the skin or other test material. This part of the system is contained in a light-proof tube as shown schematically in Figure 35.1. Thus none of this light can escape except by travelling under the barrier through the specimen. It becomes dispersed and travels in various directions in the vicinity of the area of entry. Some of the light is scattered or reflected back toward the surface. A portion emerges and enters a photometric volume where its intensity is measured. The exit surface is immediately adjacent to the place of entry, so the light does not have to travel very far within the test medium. Thus it is minimally attentuated, and one is able to obtain a reading in material that is fairly dense optically. The double aperture physically approximates a circle of 30 mm diameter.

There are two switchable sensitivity ranges in the photometric instrument. This permits readings to be made on substances varying in translucence from those plastics that are nearly transparent to Negro skin, which is the darkest material that it will be necessary to measure for prosthetic appliances.

The photometric instrument used in the translucence meter has a scale marked in EV (exposure value) units. This was done as a matter of convenience. However, with the data in Figure 35.2 EV values may be

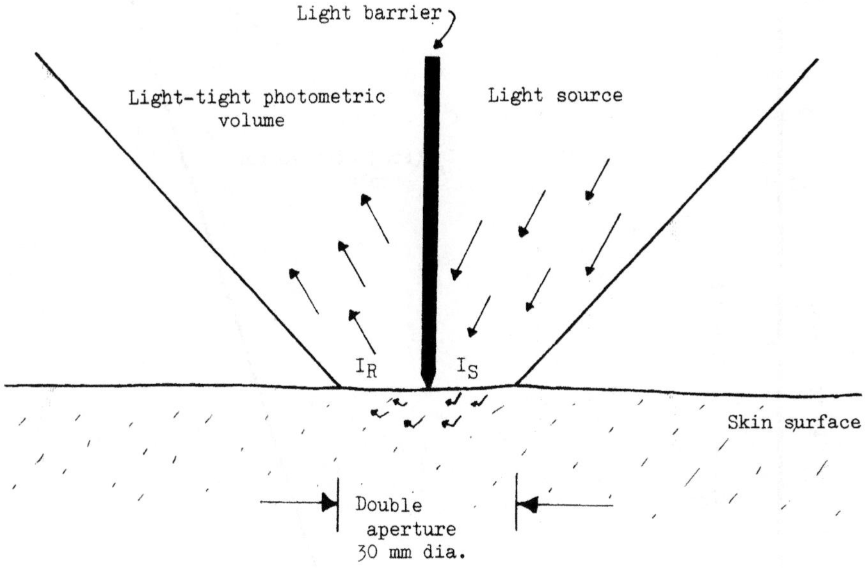

I_S = measured light intensity on sending side (foot candles)

I_R = measured light intensity on receiving side (foot candles)

Figure 35.1 Cross-section of light-proof tube. For key see text

converted to light intensity in foot candles, which are more relevant to this study. The foot candle magnitude (C) can also be computed from EV by the defining equation:

$$C = 2^{\mathrm{EV}-2}$$

The two instrument sensitivities mentioned above are 2–10 and 10–18 EV. These are indicated in Figure 35.2 as 'low range' and 'high range'; equivalent light intensities are also shown. The translucence of skin and skin substitutes lies largely in the mid-portion of the low range. However, substances whose translucence is higher may be easily made. In this figure I have also shown the 8000 foot candle intensity of the illumination projected on the skin by the instrument.

Operation of the instrument is simple. The test is done in a dimly lit room so that there is no other source of illumination, especially on the receiving side, apart from the internal source. The scale of the photometric instrument is viewed from the eyepiece (Figure 35.3) and is internally illuminated by the light received. It is readable down to values at the beginning of the useful scale – about EV = 2.

Figure 35.3 shows the two-sided measuring aperture at the bottom of the elbow structure. The barrier plate divides the opening into an illumination section on the right and the receiver section on the left. The instrument is placed squarely on the test spot (Figure 35.4) and the scale read while the sensitivity switch (the small button beside the eyepiece sleeve) is depressed by

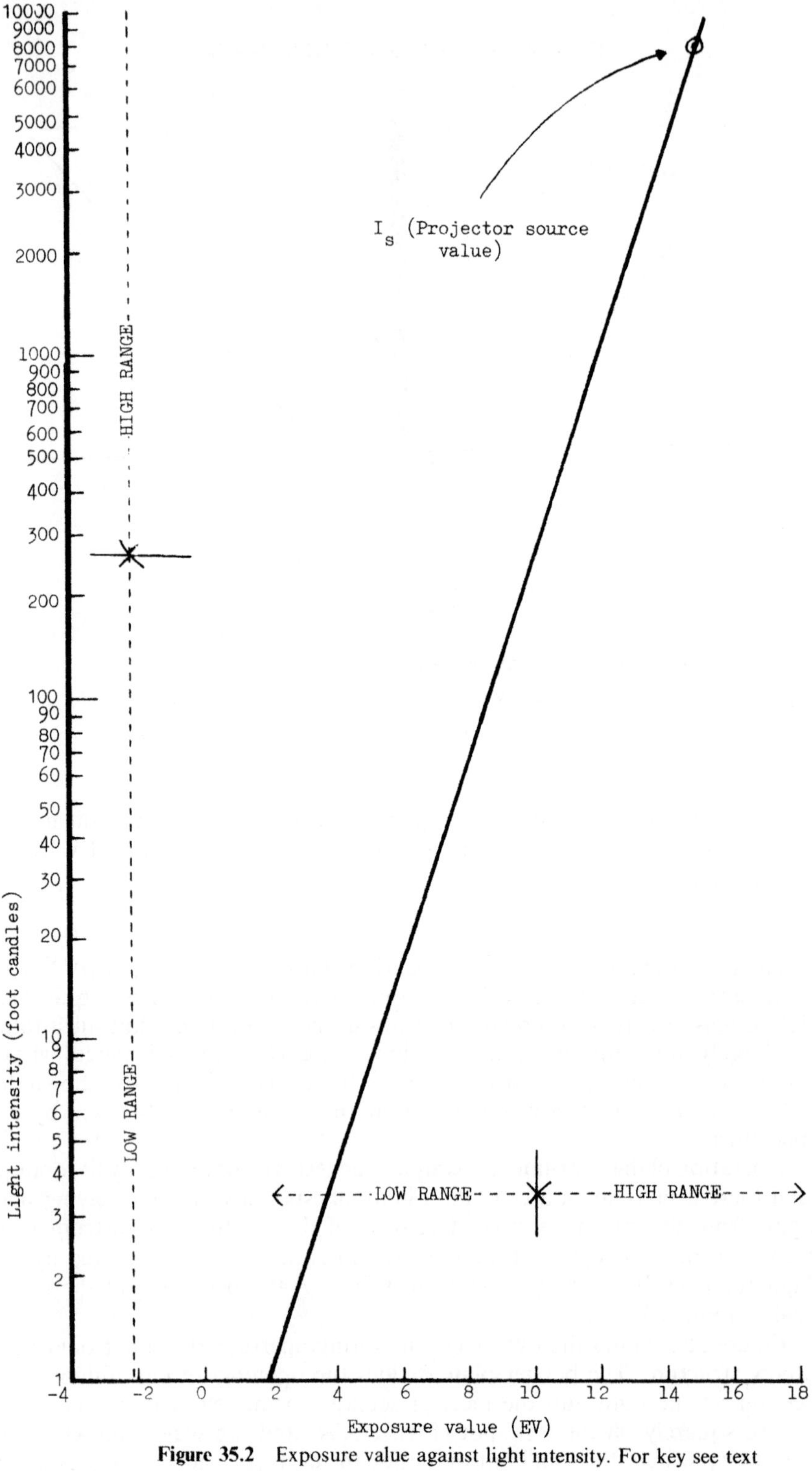

Figure 35.2 Exposure value against light intensity. For key see text

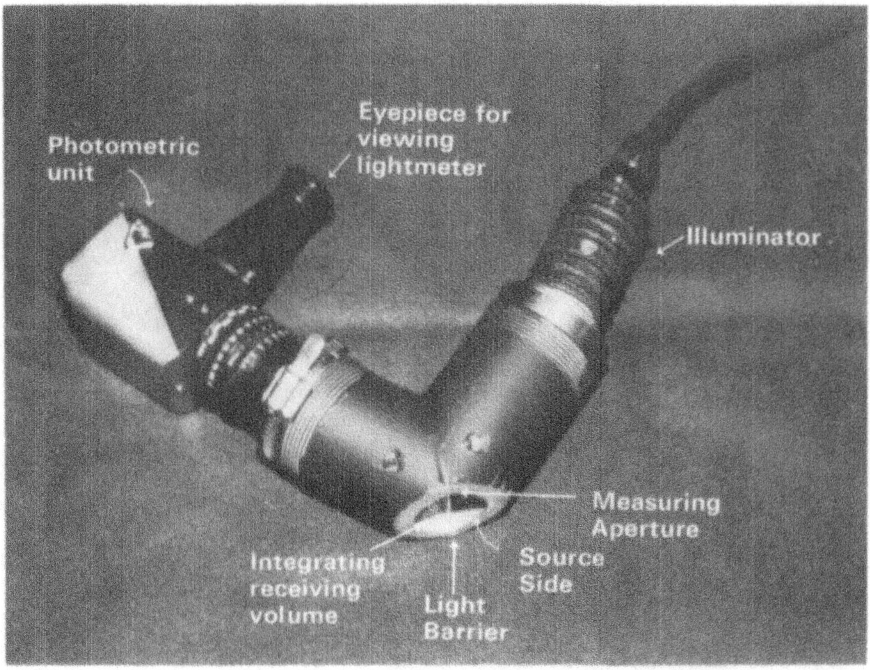

Figure 35.3 Photometric instrument, showing measuring aperture

the amount necessary to obtain a satisfactory reading. These are the EV readings, and they are then converted to foot candles using Figure 35.2. Surface translucence is then given by the equation:

$$T_S(\%) = \frac{I_R}{I_S} \times 100$$

where $T_S(\%)$ is translucence, I_R is the measured transmitted light intensity and I_R is the measured received light intensity.

Data taken using the surface translucence meter

The surface translucence meter (STM) has been applied to a sufficient number and variety of surfaces to give a range of readings for comparison. These indications may be summarised and specific values tabulated.

The range of values lies from 0 for a totally opaque medium, such as a black rubber stopper, to 4 or 5% for highly translucent, nearly transparent, material with some kind of reflective content or backing. White human skin has a value of about 0.3%, while darker Negroid types have values of about 0.07%. Skin simulants which have been formulated recently in these laboratories are able to give similar values to those in both skin types by this method.

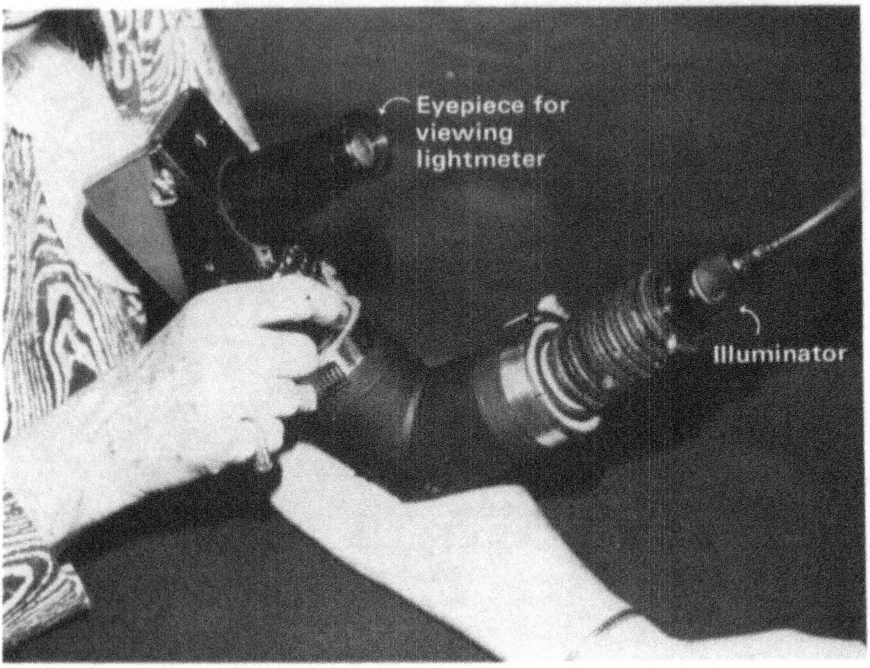

Figure 35.4 Photometric instrument in position

Standardization of surface translucence meter

To make the method universally applicable some type of standardization is needed. This means that the light flux should be known or measurable at several points in the optical system. The amount of light falling upon the receiving surface and the brightness of the emitting area must be known and the geometry of the illuminated and emitting surface areas and their relationship specified. The angles of incidence and observation should also be established. The spectral characteristics of the light source and the photometer will likewise influence the results.

In the present instrument an incandescent bulb is used as the light source (Medical Lamp, GE number 1493, 6.5 V, 2.75 A) in an AO projector. The angle of incidence with the surface is about 45° for both illuminator and photometer – i.e. a right angle is formed between them (Figures 35.3 and 35.4). The photometric system is based on a Honeywell $3\frac{1}{2}$° exposure meter, modified for close viewing.

For highly translucent materials the backing on which the specimen is placed during the tests can influence the result, particularly with thin samples. In such cases readings are taken on both flat black and grey backing. Most tests where the backing plays essentially no part (as for skin) use the flat black backing because only a small amount of light reaches it and returns to emerge at the front surface.

Index

GPSR Compliance

The European Union's (EU) General Product Safety Regulation (GPSR) is a set of rules that requires consumer products to be safe and our obligations to ensure this.

If you have any concerns about our products, you can contact us on ProductSafety@springernature.com

In case Publisher is established outside the EU, the EU authorized representative is:

Springer Nature Customer Service Center GmbH
Europaplatz 3
69115 Heidelberg, Germany

Batch number: 10755284

Printed by Printforce, the Netherlands